Technical Report Writing Today

NINTH EDITION

Technical Report Writing Today

Daniel G. Riordan

University of Wisconsin–Stout

Houghton Mifflin Company

Boston   New York

Publisher: Patricia A. Coryell
Executive Editor: Suzanne Phelps Weir
Sponsoring Editor: Michael Gillespie
Development Manager: Sarah Helyar Smith
Associate Editor: Bruce Cantley
Editorial Assistant: Lisa Littlewood
Senior Project Editor: Tracy Patruno
Manufacturing Manager: Karen Banks
Senior Marketing Manager: Cindy Graff Cohen
Marketing Associate: Wendy Thayer

Cover Image: Sea Platform wire, © PIXELPUZZLE LTD (John Harwood & Jos Hany)

Acknowledgments

Pages 130–131: J. C. R. Licklider, "The Computer as Communication Device." *Science and Technology,* April 1968. Reprinted as "The Internet Primeval."
Pages 210–211: "Corporate Express Goes Direct: The B2B Office Products Vendor Is Slashing Costs by Integrating Tightly with Customers' Procurement Systems" by Gary H. Anthes from *Computerworld,* September 1, 2003. Copyright © 2003. Reprinted by permission of Reprint Management Systems.
Pages 594–595: From *Web Style Guide: Basic Design Principles for Creating Web Sites,* 2/e by Patrick Lynch and Sarah Horton, pp. 154–156. Reprinted by permission of Yale University Press.

Printed in the U.S.A.

Library of Congress Control Number: 2003110185

ISBN: 0-618-43389-9

123456789-CW-08 07 06 05 04

Brief Contents

Contents

Chapter 3   **The Technical Communication Process**   **55**

Chapter 7   **Using Visual Aids**   **174**

Chapter 15   **Recommendation and Feasibility Reports**   **381**

Section 3   **Professional Communication**

Chapter 18   **Oral Presentations**   **474**

Chapter 18 in a Nutshell   **474**

To the Instructor

Since the first edition of *Technical Report Writing Today* appeared on instructors' desks some three decades ago, technical writing has changed dramatically. Yet with each successive edition, this book has consistently reflected an emphasis on the last word in its title—*today*. In every aspect of its up-to-date coverage, this book focuses on the state of the field right now and the tools technical communicators need in today's workplace. Upon mastering this text's principles for effective technical communication and learning the latest approaches and standards, students using *Technical Report Writing Today* will succeed not only in today's workplace but also in the workplace of tomorrow.

New to the Ninth Edition

The ninth edition of *Technical Report Writing Today* has been revised—as with past editions—to incorporate current issues important to teachers and students of technical communication in the new millennium. Sections on globalization, ethics, and electronic presentations keep the text abreast of new communication demands in the workplace. These new features, coupled with the text's accessible style and abundance of exercises, will help students prepare for the communication demands they will face in college and on the job.

New Coverage of Globalization.   The globalized world economy requires students to be able to assess how to interact with audiences across the world. No longer is the audience for memos and e-mails those people somewhere down the hall, on another floor, or at another company office within the United States. Now the audience is likely to be in India or South America. Students must learn to function effectively in this new situation. The ninth edition incorporates in Chapter 1 a lengthy section on globalization and then in many chapters provides sidebars that relate global issues to the contents of those chapters, for instance, global concerns in relation to Web design.

Expanded Coverage of Ethics.   The ethical role of the communicator also becomes more important in this new communication situation. Thus the book also includes in Chapter 1 a section on ethics and provides sidebars addressing ethical concerns throughout the chapters. These sidebars discuss important concerns such as ethical handling of résumés and e-mail.

Changes in Overall Structure.   The structure of the book has been reduced from four sections to three: Technical Communication Basics, Technical Communication Applications, and Professional Communication. The book continues to include appendixes on style and research citations.

The "repertoire" section of the book, Technical Communication Basics, has been changed to include ten chapters. These chapters are grouped more effectively than in previous editions. The new order is Definition of Technical Communication, Profiling Audiences, The Technical Communication Process, Technical Communication Style, Researching, Designing Pages, Using Visual Aids, Summarizing, Defining, and Describing. (Summarizing has been moved to a more appropriate position with other kinds of basic writing.) Section 2, Technical Communication Applications, now includes Instructions. Section 3, Professional Communication, now includes Oral Presentations.

Updated Exercises.   The exercises have all been categorized to indicate different types of goals for students—you create, you analyze, you revise, and group exercises. Many of the examples at the end of the chapters have been replaced. Many exercises have been revised.

Chapter-by-Chapter Changes.   Chapters 7 (Using Visual Aids), 14 (Formal Reports), and 19 (Letters) have remained the same. All other chapters have been revised, as outlined below:

▶ **Chapter 1, Definition of Technical Communication.** Chapter 1 has been expanded. It now includes complete sections on ethics and globalization. Theoretical sections have been updated to include current thinking about the role of communication. New exercises and examples are included.

▶ **Chapter 2, Profiling Audiences.** This chapter has been revised to include current thinking about defining audiences, including an emphasis on the tasks that audiences must perform after reading and a section on creating audience profiles. Worksheets have been revised and a section on meeting "quality benchmarks" has been added.

▶ **Chapter 3, The Technical Communication Process.** The description of the process of creating documents has been updated to include recent thinking on information design. The key graphic in this chapter now includes questions based on information design principles.

▶ **Chapter 4, Technical Communication Style.** The examples throughout this chapter have been largely replaced. A brief treatment of "garden path" sentences has been included. New sample papers are included.

▶ **Chapter 5, Researching.** This chapter now includes a focus section on using keywords and on using Google to research items on the Web.

▶ **Chapter 6, Designing Pages.** Chapter 6 has been simplified to make the rather difficult process of designing pages easier for students to grasp. In particular, the number of graphics has been reduced, thus giving more prominence to effective design principles.

▶ **Chapter 8, Summarizing.** A new article, explaining computerized ordering, is the basis for summarizing activities. New summaries and abstracts are included in the chapter and in the examples.

▶ **Chapter 9, Defining.** This chapter includes many new examples of definitions.

▶ **Chapter 10, Describing.** This chapter includes new examples.

▶ **Chapter 11, Sets of Instructions.** New examples have been added to this chapter.

▶ **Chapter 12, Memorandums and Informal Reports.** Chapter 12 now includes more emphasis on and new examples of the IMRD report ("Lab Report") genre. The chapter also includes a focus section that treats e-mail in depth.

▶ **Chapter 13, Developing Websites.** A new form for evaluating websites has been added.

▶ **Chapter 15, Recommendation and Feasibility Reports.** This chapter has two new feasibility reports; one of them was created by a small business to determine whether to market an item.

▶ **Chapter 16, Proposals.** This chapter includes new examples of effective proposals.

▶ **Chapter 17, User Manuals.** Chapter 17 includes a new student model as well as a complete usability report.

▶ **Chapter 18, Oral Presentations.** This chapter now focuses on PowerPoint presentations. The theory of oral reports is expanded to include recent criticisms of PowerPoint presentations and to give advice on the effective creation of PowerPoint presentations.

▶ **Chapter 20, Job Application Materials.** New letters of application and new résumés provide even more examples for students.

▶ **Appendix A, Brief Handbook for Technical Writers.** Most of the examples in this appendix are new to this edition.

▶ **Appendix B, Documenting Sources.** Most of the examples in this appendix are new to this edition. All sections have been updated in accord with the latest MLA and APA guidelines.

Overall Organization

The structure of the book remains a sequence from theory and skills to applications. Early chapters present current information on how to handle the repertoire of technical communication skills from audience analysis through research and design. However, the chapters are organized as modules so that you may assign chapters in the sequence that best meets the needs of your course. For instance, you could easily begin your course with a discussion of audience and design or you could start with applications such as descriptions or letters. Moreover, all of the introductory chapters have extensive exercise sections so that students can either practice the concepts covered or go right into writing memos and short reports.

The approach of this edition is the same. The book's emphasis continues to be on such skills as definition and description and on such common writing forms as memos, informal reports, proposals, and letters of application. Each chapter is self-contained, asking students to follow a process of creation that emphasizes analyzing audience, analyzing information design, and addressing problems related to creating the type of document under consideration. Each chapter contains exercises, assignments, models, planning worksheets, and evaluation worksheets designed to guide students through all phases of document creation. Exercises provide a variety of strategies to help students learn. For instance, in Chapter 20, Job Application Materials, students are encouraged to analyze or revise a letter, to create a letter, or to follow an extended process of group interaction to create and test the letter.

The basic question for a technical communication textbook is, "Will it help teachers help students understand the communication demands they will meet on the job in the near future?" The ninth edition blends instruction on traditional tools of the trade with new strategies and information to help your students develop not just their skills but also their "savvy." This book will position students as effective communicators in the early twenty-first century and will position you and teachers like you as effective mentors for those students.

Features

Technical Report Writing Today retains the features that have made it the useful and popular text it is—and adds new ones.

The following list highlights some of the features that have proven effective in previous editions of *Technical Report Writing Today* and continue to be highly praised by users of the book.

▶ **Clear and Concise Presentation.** The chapters in this book are designed as "read to learn to do" material. The book assumes the reader is a student with a goal. By providing short paragraphs and clear presentation, the text helps the student achieve the goal of becoming an effective technical communicator.

▶ **Pragmatic Organization.** The text proceeds from theory to skills to applications, but teachers may assign chapters in any sequence that fills their needs. For instance, teachers could easily begin their course with an application such as descriptions or letters. Because of the situational approach used in many chapters, students can start writing without having to read many theory chapters.

▶ **Helpful Chapter-Opening Features.** Each chapter opens with two features to help orient students to the material that follows. "Chapter Contents" provides an outline of the chapter's main sections. "In a Nutshell" briefly summarizes the chapter's most important concepts.

▶ **Focus Boxes.** The text contains numerous "focus" boxes (appearing in selected chapters), which discuss concepts that build on issues introduced in the chapter. These boxes discuss important topics such as credibility, research using Google, e-mail, and bias in language, all of which students must master to become effective professionals.

▶ **Worksheets.** Every major project has a worksheet that helps students organize their thoughts and prepare for the assignment. Each genre chapter also has an evaluation worksheet so that students working in groups have a basis for making helpful critical remarks.

▶ **Annotated Student Examples.** This edition contains more than 100 sample student documents, illustrating different writing styles and approaches to problems.

▶ **Numerous Professional Examples.** Professional examples in the book illustrate contemporary ways to handle writing situations. Many students and teachers have commented on the helpfulness of the examples, which appear both within chapters and at the ends of chapters. Numbered examples at the ends of chapters provide a greater level of detail than the necessarily brief examples within chapters.

▶ **Exercises (including Writing Assignments and Web Exercises).** Appearing sequentially at the ends of all chapters, Exercises, Writing Assignments, and Web Exercises balance individual and group work, as well as traditional and Web work, exposing students to different kinds of technical communicating problems and solutions. Exercises appear in all chapters, even theory chapters, making it easy to get students writing. In many chapters, the exercises are actually steps in the planning and drafting process required by the writing assignments for the chapter. As students complete the exercises, they also will be developing the project required for that unit.

▶ **Situational Approach.** Each of the genre chapters (e.g., proposals, instructions, job application letters) is built on situational principles. The student finds in the chapters all the necessary information, ranging from the audience to the rhetoric of the situation to the organization, format, and type of visual aids that work best in the situation. For instance, Chapter 15, Recommendation and Feasibility Reports, includes a discussion of generating criteria. Chapter 16, Proposals, includes a brief discussion of Gantt charts. Chapter 17, User Manuals, includes a discussion of storyboarding and of usability reports.

▶ **Appendixes.** The book's two appendixes provide easily accessible material on grammar and mechanics (Appendix A, Brief Handbook for Technical Writers) and MLA and APA documentation (Appendix B, Documenting Sources).

Ancillary Materials

Instructor's Resource Manual. The *Technical Report Writing Today* Instructor's Resource Manual retains its chapter-by-chapter organization, but it offers more features to help teachers teach. Each chapter provides an abstract of a chapter in the book, teaching suggestions (including suggested schedules for sequencing an assignment), and comments on the exercises and writing assignments. Over

50 student examples responding to assignments in the text will show your students how others have solved the problems posed in this book. The goal is to provide your students with material that they can sink their teeth into. Use these examples as models or as the basis for discussions and workshops on effective or ineffective handling of the paper in the situation. You may photocopy these examples and use them as class handouts or create transparencies from them.

Technical Report Writing Today **Website.**   The website has been expanded to include PowerPoint slides for each chapter and extensive lists of links to sites dealing with global and ethical materials. The PowerPoint slides provide topical coverage of each chapter and, in many places, include examples. For instance, the slides for Chapter 7, Using Visual Aids, annotate the parts of a table, a line graph, and a bar graph. In Chapter 20, Job Application Materials, the slides contain a brief annotated sample résumé. This extensive website is divided into two sections: Student Resources and Instructor Resources. The Student Resources section features chapter overviews, additional exercises, additional sample documents, links to professional technical writing organizations, and other materials that expand on the student text. The Instructor Resources section features an overview of the book, chapter outlines and abstracts, a transition guide outlining changes in the book since the previous edition, and other materials of use to instructors.

Other Houghton Mifflin Products for This Course

▌ *Creating Websites That Work,* **by Kathryn Summers and Michael Summers.** In this unique and inexpensive primer, Kathryn Summers, veteran technical writing instructor at the University of Baltimore, and her brother, Michael Summers, user experience specialist with Nielsen Norman Group, teach you to build a website that meets the needs of both the site owner and the site user. This easy-to-follow text provides a wealth of illustrations and examples to take you from creating a site map and considering credibility issues through designing content elements and site testing. An excellent companion website includes examples of effective work at each stage of the Web development process, summary checklists, links, exercises, and evaluation sheets.

▌ *A Guide to MLA Documentation,* **Sixth Edition, by Joseph Trimmer.** This concise guide to the documentation system of the Modern Language Association of America is briefer, cheaper, and easier to use than the MLA's own handbook. *A Guide to MLA Documentation* includes numerous examples, a sample research paper, an updated appendix on American Psychological Association (APA) style, and helpful hints on such topics as taking notes and avoiding plagiarism. The booklet is thin enough to slip into a notebook and inexpensive enough to serve as a supplement for a main text. A complete sample research paper on Internet chat rooms is annotated with explanations of proper MLA format.

▶ *Pocket Guide to APA Style,* **by Robert Perrin.** This quick reference to the APA documentation style provides an inexpensive, portable, and easy-to-use alternative to *The Publication Manual of the American Psychological Association.* In addition to a thorough overview of APA conventions, this guide features an overview of the research process, four chapters of sample APA-style citations, two sample APA-style papers, and appendixes on poster presentations, APA abbreviations for states and territories, and APA-style shortened forms of publishers' names.

▶ *The American Heritage Dictionary,* **Fourth Edition.** This standard reference is available in a hardcover, thumb-indexed College Edition or a briefer, less expensive, but still durable hardcover Concise Edition. Both dictionaries can be purchased at a deep discount when ordered in a shrinkwrapped package with *Technical Report Writing Today.*

Acknowledgments

I would like to thank the following technical writing teachers who offered valuable and insightful comments about the manuscript:

Harryette Brown, Eastfield College
Timothy R. Lindsley, Nicholls State University
Susan Malmo, St. Louis Community College at Meramec
Debbie Reynolds, Florida Community College at Jacksonville
Judith Szerdahelyi, Western Kentucky University
Carol Yee, New Mexico Institute of Mining and Technology

In addition, my appreciation goes to the many students who over the years have demanded clear answers and clear presentations and who have responded with quality writing. For allowing their material to be reprinted in this book, my thanks to these students:

Jill Adkins
Dan Alexander
Mark Anderson
David Ayers
Sandra Baker
Karin Broekner
Scott Buerkley
Josh Buhr
Fran Butek
Tony Bynum
Kevin Charpentier
Nikki Currier
Susan Dalton
Melissa Dieckman
Rodney Dukes
John Dykstra

Mark Eisner
Louise Esaian
Craig Ethier
Curt Evenson
Carol Frank
Greg Fritsch
John Furlano
Bret Gehring
John Gretzinger
Cheryl Hanson
Rob Hoffman
Tadd Hohlfelder
Jodi Hubbard
Kevin Jack
Rachel A. Jacobson
Jim Janisse

Holly Jeske
Kris Jilk
Mike Jueneman
Pat Jouppi
Kim KainzPoplawski
Cindy Koller
Mary Beth LaFond
Chris Lindblad
Chris Lindner
Jerry Mackenzie
Eric Madson
Matt Maertens
Todd Magolan
Richard Manor
Julie McNallen
Shannon Michaelis

(continued)

Heather Miller
Jim Miller
Tracy Miller
Keith Munson
Mark Olson
Hilary Peterson
Nicole Peterson
Steve Prickett
Steve Rachac
Cara Robida
Dana Runge

Ed Salmon
Aaron Santama
Jocelyn Scheppers
Jim Schiltgen
Stacy Schmansky
Brian Schmitz
Julie Seeger
Chad Seichter
Dave Seppala
Michelle Stewart
Brad Tanck

Sharon Tobin
M. R. Vanderwegen
Steve Vandewalle
Mike Vivoda
Steve Vinz
Rosemarie Weber
Michelle Welsh
Marya Wilson
John Wise

Special thanks to Jane and Mary for their patient work with a difficult manuscript. Jane has grown up with this book and has become a clear writer in her own right. Her contributions to this edition, both textual and editorial, are significant. Her knowledge, gained from travels throughout the world, added immensely to the globalization text of this edition. I also thank Nathan for his clear contributions to sidebars on ethics; my colleague, Michael Martin, for important contributions on ethics and citations; my colleague, Scott Zimmerman, for gracious permission to use his PowerPoint material; my former student Fran Butek for in-depth research on definitions and descriptions; and my former students Amy Reid and Michelle Stewart for permission to publish the kind of résumé that makes former teachers proud.

I am especially grateful for the insightful, even-handed editing of Maggie Barbieri. For their insight and patience I would also like to thank Michael Gillespie, Bruce Cantley, and Tracy Patruno at Houghton Mifflin and Nancy Benjamin at Books By Design. Although Dean Johnson has passed away since the last edition, I especially want to acknowledge the fine guiding hand he provided to this book through its first eight editions. As is so often true, the contributions of others has enabled me to produce a revision more extensive and insightful than I had imagined.

Finally, thanks once again to an understanding family who offered encouragement and support throughout this project—Mary, Jane, Simon, Kelly, Clare, Nathan, Shana, and Mike and Tim Riordan.

D.G.R.

Technical Report Writing Today

Technical Communication Basics

Definition of Technical Communication

Chapter 1
In a Nutshell

Here are the basics for getting started in technical communication:

Focus on your audience. Your audience needs to get work done. You help them. To help them, you must stay aware that your goal is to enable them to act.

Think of audiences as members of your community who expect that whatever happens will happen in a certain way and will include certain factors—your proposal is expected to include certain sections covering specific topics. When you act as members of the community expect other members to act, your message will be accepted more easily.

Use design strategies. Presenting your message effectively helps your audience grasp your message.

▶ Use the top-down strategy (tell them what you will say, then say it).

▶ Use headings (like headlines in newspapers).

▶ Use chunks (short paragraphs).

▶ Establish a consistent visual logic by making similar elements in your document look the same.

▶ Use a plain, unambiguous style that lets readers easily grasp details and relationships.

These strategies are your repertoire. Master them.

Assume responsibility. Because readers act after they read your document, you must present a trustworthy message. In other words, readers are not just receptacles for you to pour knowledge into by a clever and consistent presentation. They are stakeholders who themselves must act responsibly, based on your writing. Responsible treatment of stakeholders means that, among other things, you will use language and visuals with precision and hold yourself responsible for how well your audience understands your message.

Think globally. Much technical communication is distributed to audiences around the world. To communicate effectively, you must learn to localize. Radical localization allows you to consider the audience's broad-based cultural beliefs, taking into consideration thinking patterns that they use. General localization allows you to present the details of your document in accord with locally expected methods of description, for instance, designating the date as Day/Month/Year, or weights in kilograms.

Welcome! Technical communication is a large and important field of study and professional activity. Universities worldwide offer courses and programs in technical communication. Professionals are either technical communicators or produce technical communication documents as part of their jobs. The goal of this book is to make you an effective, confident technical communicator. This chapter introduces you to the basic concepts you need to know in order to communicate effectively. All the rest of the ideas in the book are based on three concepts: technical communication is audience centered, presentational, and responsible.

This chapter introduces the field with two major sections, A General Definition of Technical Communication and Major Traits of Technical Communication.

A General Definition of Technical Communication

What Is Technical Communication?

Technical communication is "writing that aims to get work done, to change people by changing the way they do things" (Killingsworth and Gilbertson, *Signs* 232). Authors use this kind of writing "to empower readers by preparing them for and moving them toward effective action" (Killingsworth and Gilbertson, *Signs* 222). This is a brief definition; later in this chapter, you will learn more about the implications of empowering readers.

What Counts as Technical Communication?

Technical communication is an extremely broad field. It encompasses a wide range of writing types. For instance, each year the Society for Technical Communication, a 25,000-member professional organization, sponsors a contest to recognize excellence in technical communication. You can see the breadth of the field in this list of 18 categories, all of which count as technical communication:

▶ Annual Reports
▶ Books
▶ Computer Hardware Guides
▶ Documentation Sets
▶ Hardware/Software Combination Guides
▶ Informational Materials
▶ Magazines
▶ Newsletters
▶ Noncomputer Equipment Guides
▶ Organizational Manuals

- Promotional Materials
- Quick Reference Guides
- Scholarly/Professional Articles
- Scholarly/Professional Journals
- Software Guides
- Technical Reports
- Trade/News Articles
- Training Materials (Currie 21)

Saul Carliner, a former president of the Society for Technical Communication, points out that technical communicators create such diverse documents as manuals, technical reports, articles, books, proposals, catalogs, brochures, videotapes, audiotapes, online cue cards and online coaches, newsletters, magazines, e-zines, websites, and multimedia CDs (Carliner).

With such a variety, technical communication is a part of almost everyone's lives on a regular basis.

Who Creates Technical Communication?

Two different types of writers create technical communication—technical communication professionals and those professionals who write as part of their jobs.

Professional technical communicators are hired to write the documents that companies need to explain their products or services, usually to help customers interact efficiently with the product or service. For instance, computer manuals are a major component of this aspect of technical communication; technical communicators work with software engineers to understand the program and then write the manual that the users need. But as the lists above indicate, technical communicators produce all kinds of materials. Whatever is needed to make information available to help people with their work, technical communicators produce.

Technical communicators are also those professionals who write about problems in their specific field or workplace. Sometimes these experts write for other experts. For instance, an engineer might write a report explaining the feasibility of a new airplane design; a dietitian could write a proposal to fund a new low-fat breakfast program at a hospital; a packaging engineer may offer a solution for an inefficient method of filling and boxing jars of perfume. Sometimes these experts write to help nonexperts with technical material. Dieticians, for instance, often write brochures explaining the components of a healthy diet to hospital patients. Engineers write manuals for nontechnical users and reports for nontechnical managers.

Both groups and their activities center on the basic definition of technical writing given by Killingsworth and Gilbertson. The goal is to empower readers who depend on the information for success.

How Important Is Technical Communication?

Communication duties are a critical part of most jobs. Survey after survey has revealed that every week people spend the equivalent of one or more days communicating. In one survey, professionals in the aerospace industry revealed that they spend 68 percent of their time—three and one-half days—communicating (Pinelli et al. 9). Bob Collins, a corporate manager, put it this way: "The most critical skill required in today's business world is the ability to communicate, both verbally and in writing. Effective communication has a direct impact on one's potential within an organization." Holly Jeske, an assistant technical designer for a department store chain, says "communication is my job." Her comments demonstrate the importance and complexity of everyday, on-the-job writing:

> I have to say that I depend a lot on my computer and e-mail for communicating with our overseas offices. I send and receive a lot of e-mails daily. A huge part of my job depends on writing and communicating in that way. I don't get the chance to hop on a plane every time there is a fit issue so that I can verbally communicate with them or even call them on the phone. . . . If I were never able to communicate through writing what I want the factory to change about a garment, I probably never would be moving from my current position. Communication is my job and pretty much anyone's job, especially now that e-mail is a huge part of the corporate world.

Major Traits of Technical Communication

Technical Communication Is Audience Centered

Let's return now to the implications of our brief definition of technical communication—"writing that aims to get work done" and writing "to empower readers." What does that imply? It means that technical communicators create documents that aim to help readers act effectively in the situations in which they find themselves. Janice Redish, an expert in communication design, explains that "a document . . . works for its users" in order to help them

> Find what they need
> Understand what they find
> Use what they understand appropriately (163).

In order to create a document in which readers can find, understand and use content appropriately, writers need to understand how writing affects readers and the interesting ways in which readers approach writing. *Audience centered,* in this larger explanation, means that technical communication

▶ Enhances relationships
▶ Enables readers to act
▶ Occurs within a community
▶ Is interactive
▶ Has definite purposes
▶ Is appropriate

Technical Communication Enhances Relationships

The starting point for creators of documents is the realization that their documents enhance relationships (Schriver, "Foreword"). Audiences don't exist in a vacuum. They exist in situations. Those situations mean that they have relationships with many people. Writing, and all communication, enhances those relationships. Audiences read documents because they need to relate to someone else.

This is not a commonly held concept about writing. Many beginners tend to see the goals of writing as "being clear" or "having correct spelling and grammar," both of which are fine and necessary goals. But the modern conception of writing asks you to consider the issues related to those goals later. First, you need to understand the relationship issue. Let's take a personal example. Suppose a father has to assemble a tricycle for a birthday present. To assemble it, he first opens the box it came in, reads the instructions included, collects the correct tools, and then puts the parts together. He is able to assemble the trike because you wrote clear instructions, identifying the parts and presenting the steps so that at the end the father has completed a functional toy ready for a child to ride.

If you think about the example for a moment, you can see that the father is using your instructions to enhance his relationship with his child. His goal in this situation is not just to turn a pile of parts into a working machine. It is to give a present to another person, someone with whom he has an ongoing relationship. This present will enhance that relationship and your instructions are a helpful factor to that end.

Now let's take a business example. Your department is in the process of upgrading its computer network. Your job is to investigate various vendors and models in order to suggest which brand to buy. When you finish your investigation and produce a report, the computers are purchased and the network upgraded. Here, too, if you think about it, the report is about enhancing relationships. The goal is not just to get the cheapest, best computer, but to facilitate the effectiveness of the work flow between people. If the system is effective, the people can interact more easily with one another, thus enhancing their relationships. Your report is not just about a brand of computer; ultimately, it is about the relationships people have with one another in the department.

In both examples, you can see the same dynamic at work. Documents enhance relationships. Documents function to make the interaction of people better, more effective, more comfortable. Documents then empower people in a rather unexpected way—not only is the tricycle assembled, the child rides it, and the gift is exciting. Not only are the computers installed efficiently, the office workers can cooperate in effective, satisfactory ways as they exchange and analyze their data.

Technical Communication Enables Readers to Act

According to Killingsworth and Gilbertson, it is helpful to view technical writing as "writing that authors use to empower readers by preparing them for and moving them toward effective action" (*Signs* 221–222). "Effective action" means that readers act in a way that satisfies their needs. Their needs include anything that they must know or do to carry out a practical activity. This key aspect of technical writing underlies all the advice in this book.

Figure 1.1 (p. 8) illustrates this concept in a common situation. The reader has a need to fulfill and a task that she must do. She must assemble a workstation. A writer, as part of his job, wrote the instructions for assembling the workstation. The reader uses the instructions to achieve effective action—she successfully assembles the workstation. This situation is a model, or paradigm, for all technical writing. In all kinds of situations—from announcing a college computer lab's open hours to detailing the environmental impact of a proposed shopping mall—technical writers produce documents that enable effective action. The writing enables the reader to act, to satisfy a need in a situation.

Technical Communication Occurs Within a Community

Action, however, occurs within a *community*, a loosely or closely connected group of people with a common interest. The key point for a writer to remember is that belonging to a community affects the way a person acts and expects other members to act (Allen; Selzer). This concept means that readers expect writing—all communication, actually—to flow in a certain way, taking into account various factors that range from how a document should look to what tone it projects. Effective writers use these factors, or *community values*, to produce effective documents.

Figure 1.2 (p. 9) illustrates the community basis for writing. If you and I are employees of a company, we belong to the community of the company. We depend on each other to get our work done. We each have roles. In one of my roles, I visit job sites to investigate items our company has installed. In one of your roles, you oversee installation, interact with clients, and make decisions about the effectiveness of our product line. When I visit a particular site, I perform research to carry out some of my responsibilities. I examine all the appropriate items, speak to the appropriate people, and take appropriate notes. However, my responsibilities also include enabling you to carry out your responsibilities. So, when I return from the site visit, I write a memo that will enable you to act after you read it.

Figure 1.1

Writing Makes
Action Possible

Need: To assemble object.

Writing makes possible…

…effective action (assembly).

As I write that memo, community values affect the way I write. I know that you expect memos to appear in a certain format because the company has a policy about format. I know, too, that you need the information I have found. Therefore, I will write the memo in the tradition that a person in this company expects, briefly but succinctly explaining what I found. You will read the

Figure 1.2

Writing Occurs Within a Community

Writer must research site information.

Reader needs site information.

Members have roles.

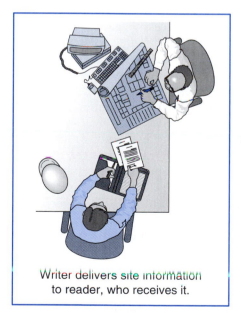

Writer delivers site information to reader, who receives it.

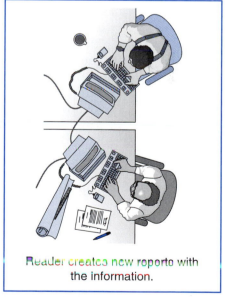

Reader creates new reports with the information.

Writing joins members together.

memo, grasp what I have done, and then use that material as you do your job. You in turn may have to rewrite this material into a report to give to your supervisor, thus enabling that person to act, and so on. The writing I do is deeply affected by my awareness of what members of my community need and expect. You need certain facts; you expect a certain format. You cannot know how to act on the facts I discover until I give them to you in a memo. Technical communication is based on this sense of community. "We write in order to help someone else act" (Killingsworth and Gilbertson, *Signs* 6).

Technical Communication Is Interactive

The key to all community exchanges is that they are interactive. Readers read the words in the document, but they also apply what they know or believe from past experiences. As the words and the experiences interact, the reader in effect recreates the memo so that it means something special to her, and that something may not be exactly what the writer intended.

Figure 1.3 shows how this interaction works. The writer presents a memo that tries to enable the reader to act. Acting on an awareness of community values, the writer chooses a form (memo), mentions certain facts ("too much gapping"), and interprets those facts ("the pillars are incorrect"). The reader interacts with the memo, using the document's words and format and her past experiences to make it meaningful to her. With her personal meaning, the reader may take a different course of action from the one that the writer may have intended. For instance, because of reading a previous report (knowledge from a prior experience) the excessive gapping allows the reader to conclude that the machine that built the mat and molding needs repair. The statements in the memo also tell her that the legal department needs to be informed because there is a potential contract problem. The memo is more than a report on a problem. Because the memo is read interactively, the reader constructs a meaning that tells her how to act in a situation that the writer in this case did not know about.

This interactive sense of writing and reading means that the document is like a blueprint from which the reader recreates the message (Green). The reader relates to certain words and presentation techniques from a framework of expectations and experiences and makes a new message (Rude; Schriver, *Dynamics*). Communication does not occur until the reader recreates the message.

Technical Communication Has Definite Purposes

Technical writers enable their readers to act in three ways: by informing, by instructing, and by persuading (Killingsworth and Gilbertson, "How Can"). Most writers use technical writing to inform. To carry out job responsibilities, people must supply or receive information constantly. They need to know or explain the scheduled time for a meeting, the division's projected profits, the physical description of a new machine, the steps in a process, or the results of an experiment.

Figure 1.3

Communication
Is Interactive

Writer composes memo based on facts.

Reader filters information, incorporates previous knowledge,
and anticipates future problems.

Writers instruct when they give readers directions for using equipment and for performing duties. Writing enables consumers to use their new purchase, whether it is a clock radio or a mainframe computer. Writing tells medical personnel exactly what to do when a patient has a heart attack.

Finally, with cogent reasons writers persuade readers to follow a particular course of action. One writer, for example, persuades readers to accept site A, not site B, for a factory. Another writer describes a bottleneck problem in a production process in order to persuade readers to implement a particular solution.

Technical Communication Is Appropriate

Appropriate can have two meanings in communication; the material needed in the situation is present (Schriver, "Foreword") or the material is socially acceptable (Sless).

The first meaning implies that the wording must be more than clear and well-structured. Suppose, for instance, that a reader consults a user manual to discover how to program a VCR that is connected to a digital TV. If that topic is not covered in the manual, or if the manual explains programming but does not deal with the particular steps needed to program the VCR/digital TV system—in other words, if the reader can't find the instructions that she needs—then the manual is useless, or inappropriate. Writers must learn to conceptualize the reader's needs in several situations and create the sections that help her or him to act.

The second meaning deals with what can be called *social appropriateness,* or accurately representing the relationships in the situation. One writer, for instance, gives the example of being called a *client* by a governmental tax agency: "I do not mind paying taxes, but to refer to me as a client in that context is unacceptable; it is a misrepresentation of the relationship. I am not a client of the tax office but a citizen contributing to society through their office" (Sless 64).

Social appropriateness also has ethical and global dimensions, which are discussed later in this chapter. The ethical dimension arises because writing affects relationships and empowers action. The global dimension arises because readers in other countries are members of other communities based in other cultures. Writers, aware of the role of writing to empower action, must learn to take into consideration the sometimes radically different needs of these other cultures.

Technical Communication Is Designed

Technical writers use design to help their readers both find information and understand it. Design has two ingredients—the appearance on the page and the structure of the content. Technical communicators design both the appearance *and* the content.

Design the Appearance

Designing the appearance means creating a page that helps readers locate information and see the relationship among various pieces of information. Figure

1.4 illustrates the use of basic design strategies. You can tell immediately by the design that the message has two main divisions, that the first division has two subdivisions, and that the text in the second division is supported by a visual aid. Technical writers use this kind of design to make the message easy to grasp (Cunningham; Hartley). The basic theory is that a reader can comprehend the message if he or she can quickly grasp the overall structure and find the parts (Rude; Southard). The basic design items that writers use are

- Headings
- Chunks
- Visual aids

Headings.   Headings, or heads, are words or phrases that name the contents of the section that follows. Heads are top-down devices. They tell the reader what will be treated in the next section. In Figure 1.4, page 14, the boldfaced heads clearly announce the topics of their respective units. They also indicate where the units begin and end. As a result, the readers always have a "map" of the message. They know where they are and where they are going.

Chunks.   A chunk is any block of text. The basic idea is to use a series of short blocks rather than one long block. Readers find shorter chunks easier to grasp.

Visual Aids.   Visual aids—graphs, tables, and drawings—appear regularly in technical writing. In Figure 1.4, the visual aid reinforces the message in the text, giving an example that would be impossibly long, and ineffective, as a piece of writing. Writers commonly use visual aids to present collections of numerical data (tables), trends in data (graphs), and examples of action (how to insert a disk into a computer). Documents that explain experiments or projects almost always include tables or graphs. Manuals and sets of instructions rely heavily on drawings and photographs. Feasibility reports often include maps of sites. More discussion of visual aids appears in Chapter 7.

Design the Content

Designing the content means selecting the sequence of the material and presenting it in ways that help the reader grasp it. Two common methods to use are

- Arrange the material top-down
- Establish a consistent visual logic

Arrange the Material Top-Down.   *Top-down* means putting the main idea first. Putting the main idea first establishes the context and the outline of the discussion. In Figure 1.4 (p. 14), the entire introduction is the top because it announces the purpose of the document. In addition, the list at the end of the introduction sets up the organization of the rest of the document. When the reader finishes the first paragraph, she or he has a clear expectation of what will

Figure 1.4

Sample Page

Top

List previous
organizations

Primary
subdivision

Secondary
subdivisions

Primary
subdivision

Technical writing is the practical writing that people do on their jobs. The goal of technical writing is to help people get work done. This memo explains two key characteristics of technical writing: audience centered and designed.

Audience Centered.   Writing is audience centered when it focuses on helping the audience. To help the audience, the writer must help the reader act and must remember community values.

   Help Act.   Writing helps readers get a job done or increase their knowledge so that they can apply it another time in their job.

   Community Values.   Everyone who belongs to any organization agrees with or lives by some of that organization's values. The writer must be sure not to offend those values.

Designed.   Technical writing appears in a more designed mode than many other types of writing. Design strategies help readers grasp messages quickly. Three key strategies are the top-down approach, the use of heads, and the use of chunks. Figure 1 illustrates the two methods. The first sentence is the top, or main, idea. The boldfaced words are the heads, which announce topics, and the x's represent the chunks or ideas.

There are two methods: heads and chunks.

Heads

xxxxxxxxxxxxxxxxxxxxxxxxxx
xxxxxxxxxxxxxxxxxxxxxxxx
xxxxxxxxxxxxxxxxxxxxxxxxx

Chunks

xxxxxxxxxxxxxxxxxxxxxxx
xxxxxxxxxxxxxxxxxxxx
xxxxxxxxxxxxxxxxxxxx

Figure 1.
Two Design Strategies

Remember, to be a good technical writer, always put your audience first and always design your material.

happen in the rest of the message. With this expectation established, the reader can grasp the writer's point quickly.

Establish a Consistent Visual Logic.   A consistent visual logic means that each element of format is presented the same as other, similar elements. Notice in Figure 1.4 that the heads that indicate primary subdivisions ("Audience Centered" and "Designed") look the same: boldfaced, the first letter of each word capitalized, and placed at the left margin. Notice that the heads that indicate the secondary subdivisions ("Help Act" and "Community Values") also look like each other, but differ from the primary heads because they appear indented five spaces. Notice the position of the visual aid, placed at the left margin, and the caption of the visual aid, italicized and in a smaller print size. If there were another visual aid, it would be treated the same way The key to this strategy is consistency. Readers quickly grasp that a certain "look" has a particular significance. Consistent treatment of the look helps the reader to grasp your meaning.

Technical Communication Is Responsible

Earlier sections focused on the audience and the text, but this section focuses on you, the writer. It is not enough just to help people act and to design your work to that end. Because readers count on you to be their guide, you must do what you can to fulfill their trust that you will tell them what—and all—they need to know. In other words, technical writing is an ethical endeavor (Griffin). The key principle here is to take responsibility for your writing (Mathes). In short, technical communicators must act ethically as they create and present documents. "What we write does have consequences, and we must accept responsibility for our words. The same is true for our images, sounds, web sites, or any other elements of communication" (Dombrowski 12).

The ethical dimension of writing is best expressed in this quote from Karen Schriver, an expert in document design:

> Since people rely on documents to make decisions that influence their safety, livelihood, health, and education, the highest ethical standards must be brought to bear in making textual choices—in deciding what to say and what not to say, in what to picture and what not to picture. Taking responsibility for these choices is central to the practice of document design. Expert practitioners distinguish themselves by skillfully selecting, structuring, and emphasizing content with the reader's needs in focus. (*Dynamics* 11).

You take responsibility because your readers, your employer, and society—also called "stakeholders"—rightfully expect to find in your document all the information necessary to achieve their goals, from assembling a tricycle to opening a factory (Harcourt). According to one expert, "Ethically it is the technical writer's responsibility to [ensure] that the facts of the matter are truly represented by the choice of words" (Shimberg 60).

In the text of your documents, then, you must tell the truth and you must do all you can to ensure that your audience understands your message. To help you with these important concepts, this chapter includes a definition of ethics and strategies to use for ethical presentation.

Definition of Ethics

Ethics deals with the question, What is the right thing to do? Philosophers since Plato have written extensively on the topic. It is a concern in daily life, in political life, in corporate life. Instances of its importance appear daily in our decisions about how to act and in news stories probing public actions. Ethics is a matter of judging both private and communal action. Individuals are expected to do the right thing, for their own personal integrity and for the well-being of their communities.

The issue, of course, is that the answer to the question, What is the right thing to do? is problematic. It is not always clear what to do or what value to base the decision on. Philosophers' answers to that dilemma have not always been consistent, but in relation to communication several common threads have emerged.

One major thread is that the communicator must be a good person who cares for the audience. Communicators must tell the full truth as convincingly as possible, because truth will lead to the good of the audience. Another thread is that the communicator must do what is right, regardless of possible outcomes. A third thread is that communicators must act for the greatest good for the greatest number of people (Dombrowski 16–18, 45–62). Of course, there are many ethical standards and writers on ethics, but it is commonly held that one must act not for self-gain but for the good of the community, or for the stakeholders in the situation.

Ethical Situations

The situations in which a person would have to make ethical decisions, and consequences from those decisions, vary dramatically. For instance, there are "this could cost me my job" situations, or *whistle-blowing,* a practice protected by federal law. In these situations, the employee becomes aware that the company is doing something illegal or that could cause great harm, perhaps because OSHA, FDA, or EPA standards are not being followed. For instance, before the terrible *Challenger* disaster, one employee had written a very clear memo outlining serious problems concerning the O-rings. This memo was subsequently used legally as the "smoking gun" to prove negligence on the part of those in charge. The writer subsequently lost his job, fought back and was reinstated under the law, only to leave the company because of challenges posed by remaining employed.(Dombrowski 132–140).

This kind of decision—and action—is incredibly intense, requiring more than just a sense of what is the right thing to do. It requires a courage to accept the negative consequences on self, and family, that losing employment entails.

Each person must ask him- or herself how to respond in a situation like this, but the ethical advice is clear—you should blow the whistle.

Much more common, however, are the everyday issues of communication. People rely on documents to act. These actions influence their well-being at all levels of their lives, from personal health, to large financial indebtedness, to accepting arguments for public policy. As a result, each document must be designed ethically.

Two examples from an ethics survey will give you a sense of the kind of daily decision that can be judged unethical. Sam Dragga interviewed several hundred technical communicators and asked them to evaluate these two issues, among others:

> You have been asked to design materials that will be used to recruit new employees. You decide to include photographs of the company's employees and its facilities. Your company has no disabled employees. You ask one of the employees to sit in a wheelchair for one of the photographs. Is this ethical?
>
> You are preparing materials for potential investors, including a 5-year profile of your company's sales figures. Your sales have steadily decreased every year for five years. You design a line graph to display your sales figures. You clearly label each year and the corresponding annual sales. In order to de-emphasize the decreasing sales, you reverse the chronology on the horizontal axis, from 1989, 1990, 1991, 1992, 1993 to 1993, 1992, 1991, 1990, 1989. This way the year with the lowest sales (1993) occurs first and the year with the highest sales (1989) occurs last. Thus the data line rises from left to right and gives the viewer a positive initial impression of your company. Is this ethical? (256–257)

Of the respondents, 85.6 percent found the first case and 71.8 percent found the second case "mostly" or "completely" unethical (260). Dragga found that the basic principle that the practitioners used was "The greater the likelihood of deception and the greater the injury to the reader as a consequence of that deception, the more unethical is the design of the document" (262–263).

Ethical considerations are integral parts of every project. In order to be a responsible member of the community, every communicator must investigate and find the principles—and courage—upon which to act ethically.

Strategies for Communicating Ethically

Many of the areas that require ethical decisions are listed in the code of ethics of the Society for Technical Communication. In the code, among other items, STC lists these five tenets:

My commitment to professional excellence and ethical behavior means that I will

- Use language and visuals with precision.
- Prefer simple direct expression of ideas.
- Satisfy the audience's need for information, not my own need for self-expression.
- Hold myself responsible for how well my audience understands my message.

- Respect the work of colleagues, knowing that a communication problem may have more than one solution.

This section explains some implications of this code. In applying high standards in the choice of words and images, communicators use unambiguous language; use design honestly; use visuals with precision; use simple, direct expression of ideas; and credit the ideas or work of others.

Use Unambiguous Language. Suppose, for instance, that you are writing a manual for a machine that has a sharp, whirling part under a protective cover. This dangerous part could slice off a user's fingers. When you explain how to clean the part, you inform the reader of the danger in a manner that prompts him or her to act cautiously. It would be unethical to write, "A hazard exists if contact is made with this part while it is whirling." That sentence is not urgent or specific enough to help a user prevent injury. Instead write, "Warning! Turn off all power before you remove the cover. The blade underneath could slice off your fingers!"

However, the need for unambiguous language appears in other much less dramatic situations. Take, for instance, the phrase "When I click on the hyperlink, nothing happens." Anyone familiar with hypertext knows that this message is not accurate. Something always happens—a message window appears, the cursor moves to a point on the screen that you did not expect it to, or the original screen reforms itself. The phrasing "nothing happens" is so imprecise that it does not allow another person to act in a helpful way. How can someone fix it if she does not know what is wrong? But that phrasing also indicates a moral stance—"I am not responsible. It is your job. I will not take the time and effort to right this, whatever inconvenience it may cause you." This kind of ambiguous use of language certainly is not dangerous, the way the previous example was, but it is a refusal to take responsibility in the situation. As such, the language does not help other people achieve their goals. It is wrong, not just because it is imprecise, but because it does not help the stakeholders.

Design Honestly. Suppose that in a progress report you must discuss whether your department has met its production goal. The page-formatting techniques you use could either aid or hinder the reader's perception of the truth. For instance, you might use a boldfaced head to call attention to the department's success:

> **Widget Line Exceeds Goals.** Once again this month, our widget line has exceeded production goals, this time by 18%.

Conversely, to downplay poor performance, you might use a more subdued format, one without boldface and a head with a vague phrase:

> Final Comments. Great strides have been made in resolving previous difficulties in meeting monthly production goals. This month's achievement is nearly equal to expectations.

If reader misunderstanding could have significant consequences, however, your use of "Final Comments" is actually a refusal to take responsibility for telling the stakeholder what he or she needs.

Create Helpful Visuals. Suppose readers had to know the exact location of the emergency stop button in order to operate a machine safely. To help them find the button quickly, you decide to include a visual aid. The two examples in Figure 1.5 indicate an imprecise way and a precise way of doing so.

Use Direct, Simple Expression. Say what you mean in a way that your reader will easily understand. Suppose you had to tell an operator how to deal with a problem with the flow of toxic liquid in a manufacturing plant. A complex, indirect expression of a key instruction would look like this:

> If there is a confirmation of the tank level rising, a determination of the source should be made.

A simple, direct expression of the same idea looks like this:

> Determine if the tank level is rising. Visually check to see if liquid is coming out of the first-floor trench.

Credit Others. Suppose a new coworker has found a way to modify a procedure and save the company money. You are assigned to write the internal proposal that suggests the change. Your obligation is to present the facts so that your manager understands who conceived the idea—and who gets the credit. To do otherwise would be to deny your coworker proper credit for the idea.

Throughout this book, you will learn strategies for the clear presentation of language, format, and visual aids. Use these communication devices responsibly

Figure 1.5

Imprecise Versus Precise Visual Aids

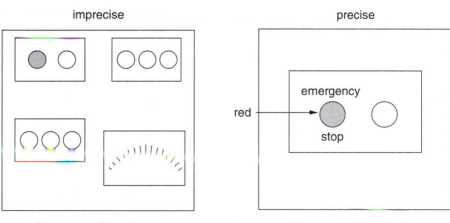

imprecise

precise

Emergency Stop Button

Emergency Stop Button

to ensure that your writing tells the audience everything it has a right to know. The audience trusts you because you are an expert. Be worthy of that trust.

Technical Communication Is Global

Today, business is international, and so too are writing and communication. As a result, people must now deal with the many languages and cultures throughout the world on a regular basis. For instance, since the passage of the North American Free Trade Act (NAFTA), many manuals for user products routinely appear in three languages—English, French, and Spanish. Workers, even at relatively small firms, indicate that they must e-mail colleagues across the globe. Websites, easily accessible to anyone in the world with a computer, must now be understandable to people who may speak many different languages and be of many different cultures. All of these factors mean that you as a technical communicator must understand the strategies of effective international communication.

While the goal of all communication is to use words and forms that enable the receiver to grasp your meaning (Beamer, "Learning"), in intercultural communication you need to give special consideration to cultural factors and to strategies for adapting communication for a variety of audiences.

The basic strategy for adapting writing and communication to other cultures is *localization*. Walter Bacah says, "Localization involves translating content and adapting it to local cultures—changing not only content, but also graphics, colors, time and date formats, units of measurement, currency, and symbols" (22). Nancy Hoft defines localization as "The process of creating or adapting an information product for use in a specific target country or a specific target market" (12). According to Hoft, there are two levels to localization: radical and general.

Radical Localization

Radical localization deals with those areas that affect the way users think, feel, and act (Hoft 13).These areas include rules of etiquette; attitudes toward time and distance; the rate and intensity of speech; the role of symbols; and local systems of economics, religion, and society—even the way people go about solving problems (60–77). In order to perform radical localization you must be able to look at social behavior from another culture's point of view, to understand the thinking patterns of the other person's culture, the role of the individual in the other person's culture, and the culture's view of direct and indirect messages (Beamer, "Teaching"; Martin and Chancey).

Another Point of View. Your ability to look at the meaning of behavior from a point of view other than your own is crucial to good communication. If you see the other point of view, you have no (or little) culture bias, and if you cannot see that point of view, you do have culture bias. When a person exhibits culture bias,

Globalization and Cultural Awareness

In our incredibly shrinking world, almost every kind of document produced has the potential to end up in the hands or on the screen of a person from another country and culture. This is both exciting and daunting. The potential for miscommunication is exponentially greater the farther from our home countries our communication moves, and not only because of translation issues. There are important cultural differences that must be considered. Establishing credibility has far-reaching implications. As more and more companies are themselves becoming international or partnering with multinational organizations, it is paramount that we become more culturally informed. To use language that is inappropriate and therefore misunderstood can have enormous negative consequences. Likewise, to commit a cultural faux pas can be understood as much more than just a social misstep—it can be interpreted as arrogance or elitism. We have a responsibility to our clients or coworkers to become acquainted with their customs and needs. Developing a good cultural sensibility is not only good business, it is the ethical thing to do.

he or she sends a "community" message, indicating that the recipient is not part of the sender's community and that furthermore the sender doesn't care. This subtext to any message makes communication much more difficult. In order to eliminate culture bias, you need to investigate what is important to the members of the other culture (Hoft). The associations commonly made by one culture about some objects, symbols, words, ideas, and the other areas mentioned above are not the same as those made by another culture for the same item—and remember, the differences do not indicate that one group is superior to the other.

For example, in China the color red is associated with joy and festivity; in the West red can mean stop, financial loss, or revolution (Basics). In the United States, *janitor* usually means a person who maintains a building, and is often associated with sweeping floors. But in Australia that same job is called a *caretaker*—a word that in the United States usually means someone who maintains the health of another person (Gatenby and McLaren). To take another example (Hoft 74, 94), conceptions of authority differ—the French prefer to come to conclusions after appeals to authority, but Scandinavians prefer more individual exploration. Levels of personal acquaintance differ in business relationships in other cultures. In the United States, people often conduct business, including very large sales, with people whom they hardly know, but in many countries people prefer to achieve some kind of personal relationship before entering into any significant business arrangement with them.

In order to communicate effectively, you must spend some time considering these differences and make changes in your documents accordingly.

Thinking Patterns. Much of U.S. thought focuses on cause–effect patterns and problem solving—identifying the causes of perceived effects and suggesting methods to alter the causes. In other cultures, however, a more common thought pattern is "web thinking." In Chinese tradition, for instance, everything exists not alone but in a relationship to many other things, so that every item is seen as part of an ever-larger web, but the web is as important as the individual fact. These thinking patterns become part of the way people structure sentences. In American English, one says "I go to lunch every day," but in Chinese, one says "Every day to lunch I go." The first sentence emphasizes the individual, and the second emphasizes the web or context (Beamer, "Teaching").

Role of the Individual. The individual is often perceived differently in a group dominated by web thinking, and web and group ideas can greatly affect the tone and form of communication. In the United States, long influenced by a tradition of individualism, many people feel that if they can just get their message through to the right person, action will follow. In other cultures, representatives of a group do not expect that same kind of personal autonomy or ease of identification from their readers.

Role of Direct and Indirect Messages. In the United States we teach that the direct method is best: State the main point right away and then support it with the facts. In some other cultures, that approach is unusual, even shocking. Although in the United States a writer would simply state in a memo that he or she needs a meeting, in a web culture, like China's, that request would come near the end of the letter, only after a context for the meeting had been established (Beamer, "Teaching").

General Localization

General localization deals with items, usually details of daily life, that change from country to country, for instance, date format, currency, and units of measurement. Much of the literature that contains advice for writing in a global context deals with these concerns. Expert writers change these when preparing documents for another country. These concerns fall into two broad areas—culture-specific references and style.

Culture-Specific References. Culture-specific items are those that we use everyday to orient ourselves. These items are often so ingrained that they are "invisible" to people in the culture—they are just the way things are done. The most common (based on Bacah; Hoft; Potsus; Yunker) of these are the following:

> *Time formats.* Countries configure the calendar date differently, some use Month/Day/Year, others use Day/Month/Year. However, a common prac-

tice is to present dates in numeral format, for instance, 01/03/04. Depending on the common configuration, these numbers could mean January 3, 2004, March 1, 2004, or even March 4, 2001. Be careful to use the appropriate configuration.

Weights and measurements. The United States is one of the few countries that does not use the metric system. Most of the world travels in kilometers, measures in grams and liters, and is hot or cold in degrees Celsius. While it is easy to interpret those weights and measurements you are familiar with, if you are not, the numbers can be very difficult to translate into common experience. Change miles to kilometers, Fahrenheit to Celsius. Americans know it takes about an hour to go 60 miles, but in Europe it would be better to say 96 kilometers. In the United States 86°F is hot, but in France the same temperature is 35°C. Switching between systems is difficult and you can help readers by performing the switch for them.

Currency. Try to express values in the country's money. Many websites and newspaper columns regularly report currency exchange values. Americans know that $8.50 is not a lot of money, but in Japan that figure is 1000 yen. (For help on the Web see, for instance, XE.Com The Universal Currency Converter at <www.xe.com/ucc/>)

Number formatting. In English, the comma divides a number into thousands, then millions, and so on. The decimal point divides the number into tenths or less—1,234,567.89. But in other countries, the same numbers use different punctuation. In Germany, that number is 1.234.567,89.

Telephone numbers and addresses. In the United States, telephone numbers are grouped in threes and fours—715-444-9906, but in other countries they are often grouped by two's—33 (0)1 23 34 76 99. In the United States, it is common practice to address an envelope with the name at the top and lists in descending order the street address and city. In some countries, Russia, for instance, the address list is reversed; the country is placed on the first line and the name of the person on the bottom line.

Page size. In the United States, the standard paper size is 8.5 × 11 inches; most documents are designed with these basic dimensions in mind. In many other parts of the world, however, the basic size is called A4 (8.25 × 11.66 inches). The difference in size can cause difficulties in copying material.

Style. Style items are the subjects of many articles on globalization. The goals of managing style are to make English easier to understand and to make it easier to translate. Many of the style tips are simply calls for good, clear unambiguous writing. Here are a few common style items (based on Hoft 214–236; Locke; Potsus) to consider:

Avoid using slang, idioms, and jargon. Most of these are simply impossible to translate: He is a brick. She hit a home run with that presentation.

Avoid using humor. When a joke fails to get a laugh, the lame excuse is often, "You had to be there. "Much humor is so culture dependent that what is hilarious to people of one culture is nearly incomprehensible to people from another culture. Humor often just does not work except in very small communities. Good writers generally avoid humor in their writing for other cultures.

Avoid puns, metaphors, and similes. Metaphors and similes compare items to indicate worth or appearance. These devices are helpful, but only if the reader gets the point of the comparison. Puns are plays on words, often used in ads. But puns are virtually untranslatable. Use these devices only if you are sure the reader would understand them.

Use glossaries. If you do use jargon, be sure to include a glossary of definitions.

Don't omit little words: a, an, the, of, these. Often, they are omitted to save space and to get to the point, but just as often they obscure the exact nature of the phrase. Compare "Click down arrow to bring up menu" with "Click on the down arrow to bring up the menu."

Include relative pronouns. The relative pronouns are *who, whom, whose, which,* and *that. That* is often the problem. A sentence like "A fire alarm losing power will beep" can be changed to "A fire alarm that is losing power will beep," or "The switches found defective were replaced" can become "The switches that were found defective were replaced" or "Maintenance personnel replaced the defective switches."

Don't use long noun phrases. Often English speakers string together a series of nouns." Damage recovery results," for instance, could be the results of damage recovery or the act of damaging those results. To avoid misinterpretation, rewrite the phrase for the non-native speaker: "results of the damage inspection."

Avoid using homophones. Homophones are two or more words that sound alike but have different meanings, and may have different spellings—like damage, which can be a noun or a verb. "Damage results" can mean "to damage the results" or "the results of the inspection of damage." To native speakers, the context often makes the meaning of these phrases clear, but non-natives often have trouble with the meaning.

Use clear modifier strings. Consider the phrase "black ergonomic keyboards and mouse pads." Does this mean that both the keyboards and mouse pads are black and ergonomic? Or just the keyboards? To help non-natives, you need to express the material in a more precise, though longer, form: "mousepads and black ergonomic keyboards" or "keyboards and mousepads that are black and ergonomic."

Write in clear subject–verb–object order. If speakers are not familiar with the rhythms of English language speech, they can become lost in the quickness and turns that sentences in English can take. Use the sentence order that it is likely non-native speakers learned in textbooks. Use "The director of the lab ordered new computers," rather than something like "Ordering lab computers was taken care of by the director."

If your text is to be translated, also be aware of these concerns: Leave space for expansion (Locke; Potsus). English phrases often expand in translation. Translated text can be as much as 30 percent longer in other languages. Even a simple Canadian highway sign in English and French illustrates this: Chain-up area. Attachez vos chaines ici. (13 spaces vs. 24 spaces). If you have pages designed so that text should fall at a certain spot, leave extra room in your English original so that after the translation and subsequent expansion, the text will still be relatively at the same spot.

Choose a simple font and avoid text effects (like boldface, italics, underlining)(Hoft; Locke). Many languages that use roman letters have diacritical marks that are not used in the United States (like Å or Ç). Custom fonts often do not include these letters, though "common" fonts, like New York, do. Many languages do use text effects and they are simply not recorded in the translation, thus losing any emphasis they may have originally carried.

A Final List

Here is a helpful synopsis of many of the points made in this section (Gillette 17).

When designing a site for a professional, international audience, you must follow most of the standard international communication guidelines commonly used for printed documents, online help, and other forms of software design. In brief:

- Keep sentences short and to the point.
- Use simple subject–verb–object sentence structure.
- Avoid the use of embedded or dependent clauses.
- Use short paragraphs to allow for easier paragraph-by-paragraph interpretation.
- Avoid regional idioms or turns of phrase.
- Avoid any visual, textual, or interactive metaphors based on a specific national or social context (e.g., mailboxes and envelopes vary from country to country, so a mailbox icon that indicates "send mail" in the United States may just look like a blue box to the international visitor).
- Define technical terms as directly as possible, avoiding elaborate metaphor whenever possible.
- If you have any doubt about users' knowledge of a specific term, define it.
- Accompany all graphical buttons with a verb-based identifier (e.g., left-pointing arrow with "Go Back").

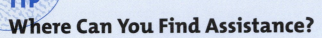

Where Can You Find Assistance?

Many resources exist to help communicators in the international arena, including many easily accessible on the Web (based on St. Amant):

CIA World Factbook On-line. 15 Aug. 2003 <www.odci.gov/cia/publications/>. "The World Factbook is produced by CIA's Directorate of Intelligence. The Factbook is a comprehensive resource of facts and statistics on more than 250 countries and other entities."

Culturegrams. 15 Aug. 2003 <www.culturegrams.com/>. "We go beyond mere facts and figures to deliver an insider's perspective on daily life and culture, including the history, customs, and societies of the world's people." This site contains both a free and a subscriber section.

Getting Through Customs. 15 Aug. 2003 <www.getcustoms.com>. A site that prints numerous articles and tips on international cultural issues.

Lone Writer SIG. 15 Aug. 2003 <www.stcsig.org/lw/resources.htm>. This site has numerous links to helpful resources, including <www.stcsig.org/lw/Translation_Project_Planning_Checklist.pdf>, a succinct guide to communication issues in international work.

Nielsen, Jakob. "International Web Usability." August 1996. 15 Aug. 2003 <www.useit.com/alertbox/9608.html>. Explains international concerns for use on the Web.

Try Google <www.google.com>. A recent search of Google, using the term "international communication" turned up 3,820,000 sites.

Self-help texts are also available:

Axtell, Roger E., Tami Briggs, Margaret Corcoran, and Mary Beth Lamb. *Do's and Taboos Around the World for Women in Business*. New York: Wiley, 1997.

Hager, Peter J., and H. J. Scheiber, eds. *Managing Global Communication in Science and Technology*. New York: Wiley, 2000.

Hofstede, Geert. *Culture and Organizations: Software of the Mind*. New York: McGraw-Hill, 1997.

Morrison, Terri, Wayne A. Conaway, and George A. Borden. *Kiss, Bow or Shake Hands: How to Do Business in Sixty Countries*. Avon, MA: Adams Media, 1995.

Morrison, Terri, Wayne A. Conaway, and Joseph J. Douress. *Dun & Bradstreet's Guide to Doing Business Around the World*. New York: Prentice-Hall, 1996.

Niemeier, Susanne, Charles P. Campbell, and René Dirven, eds. *The Cultural Context in Business Communication*. Philadelphia: Hon Benjamins, 1998.

See also Works Cited for this chapter for many helpful items, especially Nancy Hoft, *International Technical Communication*.

Exercises

▶ You Create

1. Make a list of several communities to which you belong (for instance, university students, this class, X corporation). Write a paragraph that explains how you used writing as a member of one of those communities to enable another member or members of the community to act. Specifically explain your word, format, and sequencing (which item you put first, which second, etc.) choices.

2. Write a paragraph that persuades a specific audience to act. Give two reasons to enroll in a certain class, to purchase a certain object, to use a certain method to solve a problem, or to accept your solution to a problem.

3. Write a paragraph that gives an audience information that they can use to act. For example, give them information on parking at your institution.

4. Draw a visual aid to enable a reader to act. Choose one of these goals: show the location of an object in relation to other objects (machines in a lab; rooms in a building); show someone how to perform an act (how to insert a disk into a computer; how to hold a hammer); show why one item is better than another (cost to purchase an object like a stereo or a TV or class notebooks; features of two objects).

5. Interview a professional in your field of interest. Choose an instructor whom you know or a person who does not work on campus. Ask questions about the importance of writing to that person's job. Questions you might ask include

   • How often do you write each day or week?
   • How important is what you write to the successful performance of your job?
   • Is writing important to your promotion?
   • What would be a major fault in a piece of writing in your profession?
   • What are the features of writing (clarity, organization, spelling, and so on) that you look for in someone else's writing and strive for in your own?

   Write a one-page memo in which you present your findings. Your instructor may ask you to read your memo to your classmates.

▶ You Analyze

6. Explain a situation in which you would write to a member of a community to enable him or her to act. Identify the community and detail the kind of writing you would do and what the reader would do as a result of your writing.

7. Bring to class a piece of writing that clearly assumes that you (or the reader) belong to a particular community (good sources include newspaper stories on social issues like taxes, editorials, letters that ask for contributions). Point out the

words and design devices that support your analysis. Alternate assignment: For a piece of writing given to you by your instructor, determine the community to which the writer assumed the reader belongs.

8. Choose one of the models at the end of a chapter in this book or a sample of writing you find in your daily life. Write a paragraph that describes how you interact with that piece of writing to gather some meaning. Describe your expectations about the way this kind of writing should look or be organized, what features of the writing led you to the main point, and any reactions to presentation language, visual aids, or context.

9. Research your library's electronic card catalog. Write a paragraph that alerts your instructor to commands, screens, or rules that will give students trouble if they are not aware of them ("use ^u to go up on the screen"; "the log-in word must be typed in lowercase"; "the library closes at 9:00 p.m. on Fridays").

10. Research a database available through your library's electronic catalog. Tell students about at least two types of material in the database (abstracts of articles, U.S. demographic information), and explain how that material will help them.

11. Analyze the following paragraph to decide who the audience is and what their need is; then rewrite it for a different audience with a different need. For instance, you might recount it as a set of instructions or use it to tell a person what objects to buy for this step and why.

> The fixing solution removes any unwanted particles that may still be in the paper. This process is what clears the print and makes the image more "crisp." The photographer slips the print into the fixing solution, making sure it is entirely submerged. He or she will agitate the print occasionally while it is in the fixer. After two minutes, he or she may turn the room lights on and examine the print. The total fixing time should be no less than 2 minutes and no more than 30 minutes. After the fixing process is over, the print then needs to be washed.

▶ You Revise

12. Arrange the following block of information into meaningful chunks. Some chunks may contain only several sentences.

> You have expressed an interest in the process used to carve detailed feathers on realistic duck decoys. The tools used are the same ones needed to prepare the carving up to this point: flexible-shaft grinder, stone bits, soft rubber sanding disc, pencil, and knife. My intention is to explain the ease with which mastery of this process can be achieved. There are five steps involved: drawing, outlining, and concaving the feathers, stone carving the quill, and grinding the barbs. The key to drawing the feathers is research. Good-quality references have been used in getting the carving to this stage and they will prove invaluable here. When comfortable with the basic knowledge of placement and types of feathers, drawing can begin.

As with the actual carving, drawing should be done in a systematic manner. All feathers should be drawn in from front to back and top to bottom. Drawing should be done lightly so that changes can be made if necessary. All of the other steps are determined by what is done here, so the carver must be satisfied before beginning. Outlining creates a lap effect similar to fish scales or the shingles on a house. The carver uses the flexible-shaft grinder and a tapered aluminum-oxide bit to achieve the fish scale effect. The carver starts at the front and works toward the back using the lines that were drawn as guides. Each feather is tapered from a depth of about 1/32 inch at the lines up to the original height of the carving. This step also works toward a "shingle effect." Concaving by the artist is nothing more than using the soft rubber sanding disc and gently cupping each feather toward the quill (center) area. The carver only uses slight sweeping motions with the disc to achieve good results at this phase. Concaving starts at the back outside edge of each feather and proceeds toward the tips. In stone carving the quill, the carver uses the aluminum-oxide bit and flexible-shaft grinder to raise and outline the quill area. As with all of the other steps, none of the procedures should be exaggerated. The goal is to make everything as life-like as possible. The carver grinds the barbs to match as closely as possible the hair-like structures of the feather using the same stone and grinder that were used in carving the quill. Actual feathers are used to get the exact angles needed for realism. Gentle sweeping motions are used, starting at the quill and moving toward the outside edge. Using only the tip of the stone creates the desired effects. When finished, the carver uses a loose wire wheel to remove any unwanted hair-like matter on the surface of the carved areas.

▶ Group

13. In groups of three, ask each other if the writing you do as a student or as an employee enables other people to do something. As a group, create a paragraph in which you list the kinds of people and actions that your writing affects. Use the Magolan memo (pp. 37–38) as a guide.

14. Your instructor will assign groups of three or four to read any of the following documents that appear later in this book: Instructions (four examples, pp. 277–283), IMRD (Examples 12.1 or 12.2, pp. 306–310), or Informal Recommendation (pp. 290–299, 311–313). After reading it, explain what made it easy or hard to grasp. Consider all the topics mentioned in this chapter. Compare notes with other people. If your instructor so requires, compose a memo that explains your results.

15. In groups of three or four, analyze the following sample memo. Explain how the memo enables a reader to act, demonstrates its purpose, and uses specific practices to help the reader grasp that purpose. If your instructor so requires, create a visual aid that would encourage a reader to agree with the recommendation (perhaps a table that reveals all the results at a glance) and/or create another memo to show how the memo is ethical. Your instructor will ask one or two groups to report to the class.

DATE: April 1, 2005
TO: Isaac Sparks
FROM: Keith Munson
SUBJECT: Recommendation on whether we should issue bicycles to the maintenance department

Introduction

As you requested, I have investigated the proposal about issuing bicycles to the maintenance department and have presented my recommendation in this memo. I consulted with a company that has already implemented this idea and with our maintenance department. The decision I made was based on five criteria:

- Would machine downtime be reduced?
- Is the initial cost under $5000?
- Will maintenance actually use them?
- Will maintaining them be a problem?
- Are the bicycles safe?

Recommendation

Through my investigation, I have found that the company could realize substantial savings by implementing the proposal and still sufficiently satisfy all the criteria. Therefore, I fully recommend it.

Would Machine Downtime Be Reduced?   Yes. There would be less machine downtime if bicycles were used because maintenance could get to the machines faster and have an average of 2 hours more per day to work on them. This could save the company approximately $500 a week by reducing lost production time.

Is the Initial Cost Under $5000?   Yes. The initial cost of approximately $1500 is well within our financial limitations.

Will Maintenance Actually Use Them?   Yes. I consulted with the maintenance department and found that all would use the bicycles if it became company policy. The older men felt that biking, instead of walking, would result in their fatiguing more slowly.

Will Maintaining Them Be a Problem?   No. The maintenance required is minimal, and parts are very cheap and easy to install.

Are the Bicycles Safe?   Yes. OSHA has no problem with bicycles in the plant as long as each is equipped with a horn.

16. Perform this exercise individually or in a group, as your instructor requires. Assume that you work for a manufacturer of one of the following items: (a) electric motors, (b) industrial cranes, (c) microprocessing chips, or (d) a product typical of the kind of organization that employs you now or that will when you graduate. Assume that you have discovered a flaw in the product. This flaw will eventually cause the product to malfunction, but probably not before the warranty period has expired. The malfunction is not life threatening. Write a memo recommending a course of action.

17. In groups of three or four, react to the memos written for Exercise 16. Do not react to your own memo. If all individuals wrote memos, pick a memo from someone not in your group. If groups wrote the memos, pick the memo of another group. Prepare a memo for one group (customers, salespeople, manufacturing division) affected by the recommendation. Explain to them any appropriate background, and clarify how the recommendation will affect them. Your instructor will ask for oral reports of your actions.

18. You have just learned that the malfunction discussed in Exercises 16 and 17 *is* life threatening. Write new memos. Your instructor will ask for oral reports of your actions.

Web Exercise

Analyze a website to determine how it fills the characteristics of technical communication explained in this chapter. Use any site unless your instructor directs you to a certain type (e.g., major corporation, research and development site, professional society). Write a memo or IMRD (see Chapter 12) in which you explain your findings to your classmates or coworkers.

Works Cited

Allen, Nancy J. "Community, Collaboration, and the Rhetorical Triangle." *Technical Communication Quarterly* 2.1 (1993): 63–74.

Bacah, Walter. "Trends in Translation." *intercom* (May 2000): 22–23.

The Basics of Color Design. Cupertino, CA: Apple, 1992.

Beamer, Linda. "Learning Intercultural Communication Competence." *Journal of Business Communication* 29.3 (1992): 285–303.

———. "Teaching English Business Writing to Chinese-Speaking Business Students." Bulletin of the Association for Business Communication LVII.1 (March 1994): 12–18.

Business Roundtable. "The Rationale for Ethical Corporate Behavior." *Business Ethics* 90/90. Ed. John E. Richardson. Guilford, CT: Dushkin, 1989. 204–207. Originally published in *Business and Society Review* 20 (1988): 33–36.

Carliner, Saul. "An Introduction to the Work of Technical Communicators." 15 Aug. 2003 <http://saulcarliner.home.att.net/idbusiness/workoftcs.htm>.

Collins, Robert C. Letter to Art Muller, packaging concentration coordinator. The Dial Corp. Scottsdale, AZ. 8 March 1994.

Cunningham, Donald. Presentation. CCC Convention. Minneapolis. 20 March 1985.

Currie, Cynthia. "International Technical Publications Competition." *intercom* (July/August 2003): 21.

Dombrowski, Paul. *Ethics in Technical Communication.* Needham Heights, MA: Allyn & Bacon, 2000.

Dragga, Sam. "'Is This Ethical?' A Survey of Opinion on Principles and Practices of Document Design." *Technical Communication* 43.3 (Third Quarter 1996): 255–265.

Gatenby, Beverly, and Margaret C. McLaren. "A Comment and a Challenge." *Journal of Business Communication* 29.3 (1992): 305–307.

Gillette, David. "Web Design for International Audiences." *intercom* (Dec. 1999): 15–17.

Green, Georgia M. "Linguistics and the Pragmatics of Language Use" *Poetics* 11 (1982): 45–76.

Griffin, Jack. "When Do Rhetorical Choices Become Ethical Choices?" *Technical Communication and Ethics.* Ed. R. John Brockman and Fern Rook. Arlington, VA: Society for Technical Communication, 1989. 63–70. Originally published in *Proceedings of the 27th International Technical Communication Conference* (Washington, DC: STC, 1980).

Hall, Dean G., and Bonnie A. Nelson. "Integrating Professional Ethics into the Technical Writing Course." *Journal of Technical Writing and Communication* 17.1 (1987): 45–61. Exercises 16–18 are based on material from this article.

Harcourt, Jules. "Developing Ethical Messages: A Unit of Instruction for the Basic Business Communication Course." *Bulletin of the Association for Business Communication* 53 (1990): 17–20.

Hartley, Peter. "Writing for Industry: The Presentational Mode versus the Reflective Mode." *Technical Writing Teacher* 18.2 (1991): 162–169.

Hoft, Nancy L. *International Technical Communication: How to Export Information About High Technology.* New York: Wiley, 1995.

Jeske, Holly. "Writing and Me." E-mail to Dan Riordan. 20 Aug. 2003.

Killingsworth, M. Jimmie, and Michael Gilbertson. "How Can Text and Graphics Be Integrated Effectively?" *Solving Problems in Technical Writing.* Ed. Lynn Beene and Peter White. New York: Oxford, 1988. 130–149.

———. *Signs, Genres, and Communities in Technical Communication.* Amityville, NY: Baywood, 1992. The material on community and interaction is based on ideas developed in this book.

Locke, Nancy A. "Graphic Design with the World in Mind." *intercom* (May 2003): 4–7.

Martin, Jeanette S., and Lillian H. Chancey. "Determination of Content for a Collegiate Course in Intercultural Business Communication by Three Delphi Panels." *Journal of Business Communication* 29.3 (1992): 267–284.

Mathes, J. C. "Assuming Responsibility: An Effective Objective in Teaching Technical Writing." *Technical Communication and Ethics.* Ed. R. John Brockman and Fern Rook. Arlington, VA: Society for Technical Communication, 1989. 89–90. Originally published in *Proceedings of the Technical Communication Sessions at the 32nd Annual Meeting of the Conference on College Composition and Communication* (Dallas, TX: NASA Publication 2203, 1981).

Morrison, Terri, and Wayne A. Conaway. "Your Cultural IQ." *American Way* (15 March 1997): 140.

Pinelli, Thomas E., Myron Glassman, Rebecca 0. Barclay, and Walter E. Oliu. *Technical Communications in Aeronautics: Results of an Exploratory Study—An Analysis of Managers' and Non-Managers' Responses.* NASATM-101625. Washington, DC: National Aeronautics and Space Administration, August 1989 (Available from NTIS, Springfield, VA).

Potsus, Whitney Beth. "Is Your Documentation Translation Ready?" *intercom* (May 2001): 12–17.

Redish, Janice. "What Is Information Design?" *Technical Communication* 47.2 (May 2000): 163–166.

Rude, Carolyn D. "Format in Instruction Manuals: Applications of Existing Research." *Journal of Business and Technical Communication* 2 (1988): 63–77.

Schriver, Karen. *Dynamics in Document Design: Creating Texts for Readers.* New York: Wiley, 1997.

———. "Foreword." *Content and Complexity: Information Design in Technical Communication.* Ed. Michael J. Albers and Beth Mazur. Mahwah, NJ: Lawrence Erlbaum, 2003. ix–xii.

Selzer, Jack. "Arranging Business Prose." *New Essays in Technical and Scientific Communication: Research, Theory, Practice.* Ed. Paul V. Anderson, R. John Brockman, and Carolyn Giller. Farmingdale, NY: Baywood, 1983. 37–54.

Shimberg, Lee H. "Ethics and Rhetoric in Technical Writing." *Technical Communication and Ethics.* Ed. R. John Brockman and Fern Rook. Arlington, VA: Society for Technical Communication, 1989. 54–62. Originally published in *Technical Communication* 25.4 (Fourth Quarter 1988): 173–178.

Sless, David. "Collaborative Processes and Politics in Complex Information Design." *Content and Complexity: Information Design in Technical Communication.* Ed. Michael J. Albers and Beth Mazur. Mahwah, NJ: Lawrence Erlbaum, 2003. 59–80.

Society for Technical Communication. *Code for Communicators.* Washington, DC: n.d.

Southard, Sherry. "Practical Considerations in Formatting Manuals." *Technical Communication* 35.3 (Third Quarter 1988): 173–178.

St. Amant, Kirk R. "Resources and Strategies for Successful International Communication." *intercom* (Sept./Oct. 2000): 12–14.

Yunker, John, E. "A Hands-On Guide to Asia and the Internet." *intercom* (May 2000): 14–19.

Focus on
Codes of Ethical Conduct

Many companies and most professional associations—Johnson & Johnson and the American Marketing Association, for instance—publish codes of conduct for their employees or practitioners. These codes provide guidelines for ethical action. They include a variety of topics, but several are typically addressed: fundamental honesty, adherence to the law, health and safety practices, avoidance of conflicts of interest, fairness in selling and marketing practices, and protection of the environment (Business Roundtable).

As a technical writer, you should be aware of the guidelines presented in the code of the Society for Technical Communication:

> As a technical communicator, I am the bridge between those who create ideas and those who use them. Because I recognize that the quality of my services directly affects how well ideas are understood, I am committed to excellence in performance and the highest standards of ethical behavior.
>
> I value the worth of the ideas I am transmitting and the cost of developing and communicating those ideas. I also value the time and effort spent by those who read or see or hear my communication. I therefore recognize my responsibility to communicate technical information truthfully, clearly, and economically. My commitment to professional excellence and ethical behavior means that I will

- Use language and visuals with precision.
- Prefer simple, direct expression of ideas.
- Satisfy the audience's need for information, not my own need for self-expression.
- Hold myself responsible for how well my audience understands my message.
- Respect the work of colleagues, knowing that a communication problem may have more than one solution.
- Strive continually to improve my professional competence.
- Promote a climate that encourages the exercise of professional judgment and that attracts talented individuals to careers in technical communication.

Source: Guidelines presented in the Code of the Society for Technical Communication. Reprinted with permission from The Society for Technical Communication, Arlington, Virginia, USA.

Chapter
2
Profiling Audiences

Chapter 2
In a Nutshell

You write a different document based on how you define your audience. Because your understanding of your audience controls so many of your writing decisions, analyze the audience before you write. Create an audience profile by answering these questions:

▶ Who are they?
▶ How much do they know?
▶ What do they expect?

Find out who your audience is. Is it one person or a group or several groups? Are you writing a memo to a specific individual or instructions for "typical" workers?

Estimate how much they know. If they are advanced, they know what terms mean, and they understand the implications of sentences. If you are addressing beginners, you have to explain more.

Determine expectations. Expectations are the factors that affect the way in which the audience interprets your document. Will it conform to their sense of what this kind of document should look and sound like? Will it help them act in the situation? Does it reflect a sense of the history of the situation or the consequences of acting?

Chapter 1 showed how audience is a major concern in technical communication. Every piece of writing has an intended audience—the intended reader or readers of the document. Because your goal is to enable those readers to act, you must analyze the intended readers in order to discover the facts and characteristics that will enable you to make effective decisions as you write. The facts and characteristics that you discover will affect planning, organizing, and designing all aspects of the document, from word choice to overall strategy and structure.

This chapter explains the factors that writers investigate in order to analyze audiences. The chapter begins with an example of a technical memo, and then presents sections that help you answer these key questions:

- Who is the audience?
    - What are their demographic characteristics?
    - What is the audience's role?
    - How does the reader feel about the subject?
    - How does the reader feel about the sender?
    - What form does the reader expect?
- What is the audience's task?
- What is the audience's knowledge level?
- What factors influence the situation?
- How do I create an audience profile?

An Example of Technical Writing

The following memo illustrates how writers communicate. Todd Magolan performs routine inspections of thick rubber cargo mats that fit into the beds of special hauling equipment in manufacturing facilities. He reports on their performance to Marjorie Sommers, his supervisor. Sommers uses the reports to determine whether or not her company has met the conditions of its contract and to decide whether or not to change manufacturing specifications.

As you read the memo, note the following points:

1. The writer names the audience (Marjorie Sommers) and states the purpose of the memo for the audience (to deliver information on his impressions).

2. The writer uses unambiguous language to focus on the specific parts of the cargo mat (trim lines, holes, kick plate) and to point out specific problems ("the front edge of vinyl/maratex still needs to be evaluated with a base kick plate"). Note that the writer uses the word "good" to mean "implements the specification exactly," knowing that his audience understands that usage.

3. The information is designed, appearing in easy-to-scan chunks set off by heads. The writer sets up the document in the first paragraph by naming the

three items—the 410, 430, and 480 mats—that he discusses in the body of the memo. He repeats these key words as section headings, presents information in a consistent pattern (vinyl/maratex, hole location, concerns) for each section, and numbers individual points within sections.

4. The writer uses a visual aid—the drawing—to convey a problem discussed in the 480 section (Figure 1).

5. The writing is responsible. The writer tells the stakeholder (Sommers) all that she needs to know to be able to do her job. In addition, the memo treats other stakeholders properly. Magolan's company, for instance, has informed individuals handling its affairs. The customer has received honest treatment of the problem, allowing them to interact with Magolan's company in an informed manner.

Audience named

Date:	3-15-05
To:	Marjorie Sommers
From:	Todd Magolan
Subject:	Review of Mats at Oxbow Creek Plant

Introduction "sets up" discussion

Purpose

After seeing the 410 and 430 Cargo mats, as well as the 480 Front mat, installed in the vehicles at Oxbow Creek, my impressions of each are as follows:

Heading

410 CARGO

Unemotional presentation

Assumes audience knowledge

1. The rear and side vinyl/maratex both fit. (The rear kick plate fits perfectly.)
2. All hole locations were good.
3. The front edge of vinyl/maratex was not evaluated because there was no base kick plate for the front.

Overall, I feel that the 410 Cargo mat fit was very good. However, the front edge of vinyl/maratex still needs to be evaluated with a base kick plate.

Words repeated as section heads

430 CARGO

Chunking of information

1. With our revised vinyl and maratex trim lines, I feel the mat fit is excellent. There was no pull out of the kick plate such as we noticed before.
2. All trim lines and holes are now good.

It is my feeling that our proposed design is much more functional than the original design and should be incorporated if feasible.

Words repeated as section heads

480 FRONT

1. All trim lines seemed good.
2. Hole locations were good. There was a little concern/suggestion that the rear group of holes (for the rear seat) be moved outward a few millimeters. The added lytherm seemed to bring them inward slightly.

Chunking of
information

3. The major concerns came in the B- and C-pillar areas (see attached sketch). There is much gapping between our mat and the molding. It is most evident in the C-pillar area. It is my opinion that our mat is correct in being molded to the sheet metal contour (in the C pillar) and that the pillars themselves are incorrect.

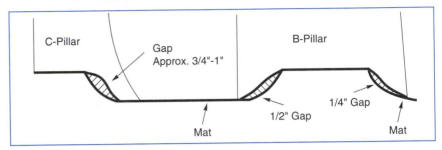

Figure 1
480 B-Pillar and C-Pillar Gap Problems

If you have any questions or would like to discuss these findings further, please let me know.

This brief memo illustrates the skills and attitudes that technical writers employ. Although the memo is a straightforward report of a site visit, it is nonetheless a well-crafted document that effectively conveys the information that writer and reader need to fulfill their roles in the organization.

Who Is the Audience?

The answer to this basic question dictates much of the rest of what you do in the document.

The audience is either someone you know or a generalized group, such as "college freshmen" or "first-time cell phone users" (Coney and Steehouder). The audience could be a single person (your supervisor, a coworker), a small group (members of a committee), or a large group (the readers of a set of instructions). Sometimes the audience is multiple, that is, a primary audience who will act on the contents of your document and a secondary audience who read the document for information—to keep them in the loop—but who will not act on the information.

In order to communicate effectively with your audience, you have to engage them, that is, write in a way that makes it clear to them "that their knowledge and values are understood, respected, and not taken for granted" (Schriver 204). To create that engagement (see Schriver 152–163), you must answer these questions:

▸ What are the audience's demographic characteristics?
▸ What is the audience's role?

> How does the reader feel about the subject?
> How does the reader feel about the sender?
> What form does the reader expect?

What Are the Audience's Demographic Characteristics?

Demographic characteristics are basically objective items dealing with common ways of classifying people—age, ethnicity, and gender. The answers to these questions provide a base from which to act, but, in order to prevent stereotyping, have to be used with the answers to the other questions. For instance, in the Magolan memo, the audience Marjorie Sommers is a 45-year-old white woman with a number of years' experience as a manager. However, the memo is not written for a generic 45 year old, it is written to Marjorie, an audience much more specific than a member of an age group. The memo can be as individualized as it is because the writer also personally knows the audience and thus knows her attitudes and expectations.

Sometimes, however, if an audience is treated stereotypically, the message can fail. One researcher discovered that a brochure urging African-American teens to stay away from drugs failed because the writers used an image of a person who had an outdated hairstyle. The teens who read the brochure felt that the outmoded image indicated that the writer was someone who was out of touch; thus, they dismissed the brochure. In other words, the writers used demographic data but did not find answers to the other audience questions, and as a consequence failed to engage their audience (Schriver 171–189).

Exercises

▶ You Create

1. Write three different sentences in which you use a technical concept to explain three situations to a person who knows as much as you do. (Example: "You can't print that because the printer doesn't have that font in memory." In this sentence, the writer assumes that the reader understands font, memory, font in memory, and the relationship of memory to printing.)

▶ You Revise

2. Rewrite one of the three sentences in Exercise 1 into a larger paragraph that makes the same idea clear to someone who knows less than you do.

What Is the Audience's Role?

In any writing situation, your audience has a role. Like actors in a drama, audience members play a part, using the document as a "script." In the script, you "write a part" for department managers or tricycle assemblers or parents or students. The reader assumes that role as he or she reads. If the role in the document reflects the role the reader has in the real-life situation, the document will then engage the reader and thus make effective communication more likely. Their willingness to assume the role in the document often depends upon the way the role depicted in the writing is similar to the role they play in a real-life situation (Coney and Steeholder).

To create an effective role in the script, you have to understand the audience's role in real life. Because of their role, they have specific tasks, that is, specific responsibilities and actions. For instance, Marjorie Sommers is department manager. She evaluates data in relation to other aspects of the company. Her decisions have consequences because the corporation chooses to fund different actions based on her evaluations. She is professionally concerned that her clients be satisfied, that the company sell quality products. Personally, she is concerned that her work be judged as effective both by clients and by her supervisors. This matrix of characteristics defines her role. A writer must understand those characteristics in order to design an effective memo or any other document for her.

Role and task, in turn, are connected to need, those items necessary to fulfill the role. In order to fulfill her role in the organization, Marjorie Sommers must be assured that the mats serve the purpose for which they were sold. She must know of possible problems so that she can keep the customer happy with the product and with the company's service. She must be able to explain to her supervisor how her department is functioning. She has to decide whether she should talk to people in manufacturing about the fabrication of the part. In short, she needs the information to help her carry out her job responsibilities.

You can easily see the different effects of need by considering two audiences: operators of a machine and their department managers. Both groups need information but of different kinds. Operators need to know the sequence of steps that make the machine run: how to turn it on and off, how to set it to perform its intended actions, and how to troubleshoot if anything goes wrong. Managers need to know whether the machine would be a useful addition to the workstation and thus to purchase it. They need to know whether the machine's capabilities will benefit staff and budget. They need to know that the machine has a variable output that can be changed to meet the changing flow of orders in the plant; that the personnel on the floor can easily perform routine maintenance on the machine without outside help; and that problems such as jamming can be easily corrected.

Because the audiences' needs differ, the documents directed at each are different. For the operator, the document would be a manual, with lots of numbered how-to-do-it steps, photos or drawings of important parts, and an index to help the operator find relevant information quickly. For the manager, the document would contain explanatory paragraphs rather than numbered how-to-do-it steps. Instead of photos, you might use a line graph that shows the effect of the variable rate of production or a table that illustrates budget, cost, or savings.

How Does the Reader Feel About the Subject?

The reader's feelings can be described as positively inclined, neutral, or negatively inclined toward the topic or the writer. If the audience is positively inclined, a kind of shared community can be set up rather easily. In such a situation, many of the small details won't make as much difference; the form that is chosen is not so important, and the document can be brief and informal. Words that have some emotional bias can be used without causing an adverse reaction. Marjorie Sommers is positively inclined toward the subject. Knowing about mats is part of her job; she is responsible for seeing that clients are satisfied with the product they purchased.

Much the same is true of an audience that is neutral. A writer who has to send a neutral audience a message about a meeting or the results of a meeting might choose a variety of forms, perhaps a memo or just a brief note. As long as the essential facts are present, the message will be communicated.

However, if the audience is negatively inclined, the writer cannot assume a shared community. The small details must be attended to carefully. Spelling, format, and word choice become even more important than usual because negatively inclined readers may seize upon anything that lets them vent their frustration or anger. Even such seemingly trivial documents as the announcement of a meeting can become a source of friction to an audience that is negatively inclined.

Heather Lazzaro, a technical communicator writing about the design of manuals, has suggested some questions to help you discover the audience's attitudes toward the subject. Although these questions are phrased for the audience of a manual, they are easily generalizable to any situation:

> Do your users hold any biases against the technology that will prevent them from using the product or reading documentation? Are they experts in other technologies who are reluctant to develop new specializations?
>
> Are they forced to use the product to accomplish their jobs? Would they opt to use it if given a choice?
>
> How long have your users been in their current jobs? What kinds of positions did they have previously? (19)

How Does the Reader Feel About the Sender?

A writer must establish a relationship with the reader. Readers feel positively about a message if they feel that it is organized around their needs and if the writer "has taken the time to speak clearly, knowledgeably, and honestly to them" (Schriver 204). To create this positive sense, writers must create the belief that they are credible and authoritative and they must create documents that are "inviting" (Coney and Steeholder) and "seductive" (Horton). These documents tend to motivate the reader both to read and act, a key requirement for effective communication (Carliner).

Credibility means that you are a person who can be listened to. Credibility (see "Focus on Credibility," p. 54) arises because of your role or your actions.

If readers know that you are the quality control engineer, they will believe what you write on a quality issue. If readers know that you have followed a standard or at least a clear method of investigating a topic, they will believe you.

Authority means that you have the power to present messages that readers will take seriously (Lay). Basically, you have the right to speak because you have expertise, gained by either your role or your actions. Naturally, this authority is limited. Your report is authoritative enough to be the basis for company policy, even though you might not be the one who actually sets the policy.

Inviting documents cast the writer in a helpful role toward the reader. For instance, the writer could assume a role of guide who shows visitors the path through the forest of instructions in assembling a tricycle, a librarian who leads users through the information to find what they need. Often how inviting the document is depends on the wording. For instance, Coney and Steeholder suggest that on Web pages the phrase "e-mail us" is much more seductive than a link to an unnamed webmaster (332).

Seductive, in this usage, means "to attract our readers' attention and win their sympathy" (Horton 5). To create a seductive document is partly a matter of your attitude (and partly, as Horton explains, the way you design and present the information, concepts that will be dealt with in later chapters). Horton suggests that the key attitude is to present yourself as a person who will "guide and protect the reader" in order to stimulate them into action.

In our example, Sommers is positively inclined toward Magolan. Magolan knows that Sommers likes and trusts him because the two have worked together for a while. His past actions have generated a sense of authority. He is a person who has the right to speak because he has acted well in the past. Todd knows that Sommers is the supervisor and expects clear information without much comment.

Todd, however, creates a credible, inviting document. He establishes credibility in the first paragraph by explaining his methodology—he inspected the site. He clearly feels that he has the authority to make evaluative comments that suggest future actions (incorporate the proposed design, not the original one). His tone is informal, using "I" and "you"; notice, too, that he is comfortable enough to structure his communication as a short, no-nonsense list. In short, Todd presents himself as a person who will guide and protect the reader. He has created an easy-to-follow document that invites further action, should it be necessary. He is a person whom the reader can trust.

What Form Does the Reader Expect?

Many audiences expect certain types of messages to take certain forms. To be effective, you must provide the audience with a document in the form they expect. For instance, a manager who wants a brief note to keep for handy reference may be irritated if he gets a long, detailed business letter. An electronics expert who wants information on a certain circuit doesn't want a prose discussion because it is customary to convey that information through schematics and specifications. If an office manager has set up a form for reporting accidents, she expects reports in that form. If she gets exactly the form that she specified, her attitude may easily turn from neutral to positively inclined. If she gets a different form, her attitude may change from neutral to negatively inclined.

Marjorie Sommers expects an informal memo that she can skim over easily, getting all the main points. She expects that this memo, like all that she receives reporting on site visits, will have the usual lines (Date, To, From, Subject) at the top and heads to break up the text. Todd knows that the message must be brief (one to two pages), that its method of production must be a word processor, and that it must appear on 8½-by-11-inch paper.

What Is the Audience's Task?

What will the reader do after reading the document? Although fulfilling a need is why the audience is involved in the situation, the task is the action they must accomplish (Rockley). Tasks vary greatly, and can be nearly anything—to assemble a tricycle or a workstation, to say no to drugs, to agree to build a retail outlet at a site, to evaluate the sanitary conditions of a restaurant. The document must enable the reader to perform that task.

Marjorie Sommers's task is to act to protect the interests of the company. As a result, she will alert her superiors to the problem with the pillars. Because Magolan feels the problem is the customer's, Sommers will not ask manufacturing to change their process. But because she has been informed, she will have the facts she needs if she must act at a later date.

What Is the Audience's Knowledge Level?

Every audience has a knowledge level, the amount they know about the subject matter of the document. This level ranges from expert to layperson (or nonexpert). An expert audience understands the terminology, facts, concepts, and implications associated with the topic. A lay audience is intelligent but not well informed about the topic. Knowing how much the audience knows helps you choose which information to present and in what depth to explain it.

Adapting to Your Audience's Knowledge Level

You adapt to your audience's knowledge level by building on their schemata—that is, on concepts they have formed from prior experiences (Huckin). The basic principles are "add to what the audience knows," and "do not belabor what they already know." If the audience knows a term or concept (has a schema for it), simply present it. But if the audience does not know the term or concept (because they have no schema), you must help them grasp it and add it to their schemata.

Suppose for one section of a report you have to discuss a specific characteristic of a digitized sound. If the reader has a "digitized sound schema," you can use just the appropriate terminology to convey a world of meaning. But if the reader does not have this schema, you must find a way to help him or her develop it. The following two examples illustrate how writers react to knowledge level.

For a More Knowledgeable Audience

For a more knowledgeable audience, a writer may use this sentence:

That format allows only 8-bit sampling.

The knowledgeable reader knows the definitions of the terms "format" and "8-bit." He or she also understands the implication of the wording, which is that the sound will not reproduce as accurately if it is sampled at 8 bits, but that the file will take up less disk space than, say, 16-bit sampling.

For a Less Knowledgeable Audience

A less knowledgeable audience, however, grasps neither the definitions nor the implications. To develop a schema for such readers, the writer must build on the familiar by explaining concepts, formatting the page to emphasize information, making comparisons to the familiar, and pointing out implications. You might convey the same information about sampling to a less knowledgeable audience in the following manner (Stern and Littieri 146):

Explanation of concept

Highlighted text

Implication of explanation

Analogy

Highlighted text
Implications

In many ways, it helps to think of digitized sound as being analogous to digitized video.

Digitized sound is actually composed of a sequence of individual sound samples. The number of samples per second is called the *sample rate* and is very much like a video track's frame rate. The more sound samples per second, the higher the quality of the resulting sound. However, more sound samples also take up more space on disk and mean that more data need to be processed during every second of playback. (The amount of data that must be processed every second is called the *data rate*.)

Sound samples can be of different sizes. Just as you can reproduce a photograph more faithfully by storing it as a 24-bit (full-color) image than as an 8-bit image, 16-bit sound samples represent audio more accurately than 8-bit sound samples. We refer to the size of those samples as a sound's *sample size*. As with the sample rate, a larger sample size increases the accuracy of the sound at the expense of more storage space and a higher data rate.

To see more about writing for a less knowledgeable audience, refer to Chapter 9, "Defining."

Finding Out What Your Audience Knows

Discovering what the audience knows is a key activity for any writer. It complements and is as important as discovering the audience's role. To estimate an audience's knowledge level, you can employ several strategies (Coney; Odell et al.; Selzer).

Ask Them Before You Write

If you personally know members of the audience, ask them in a phone call or brief conversation how much they know about the topic.

Ask Them After You Write

Ask the audience to indicate on your draft where the concepts are unfamiliar or the presentation is unclear.

Ask Someone Else

If you cannot ask the audience directly, ask someone who knows or has worked with the audience.

Consider the Audience's Position

If you know what duties and responsibilities the audience members have, you can often estimate which concepts they will be familiar with.

Consider Prior Contacts

If you have had dealings with the audience before, recall the extent of their knowledge about the topic.

Exercises

▶ You Create

3. Write a brief set of instructions (three to six steps) to teach an audience how to use a feature of a machine—for instance, the enlarge/reduce feature on a photocopier or the proper method of moving the cursor on a computer. Then exchange your instructions with a partner. After interviewing the partner to learn the procedure, rewrite the instructions as a paragraph that explains the value of the feature to a manager.

▶ You Analyze

4. Analyze either the memo in Exercise 15, Chapter 1 (p. 30), or the report in Example 15.2 (pp. 397–399) in order to determine the audience's need and task. Be prepared to report your findings orally to the class.

▶ **You Revise**

5. The Magolan memo on pages 37–38 is aimed at an audience of one. Rewrite it so that it includes a secondary audience who is interested in customer relations. If the subject matter of that memo is too unfamiliar to you, use a subject you know well (in-line fillers or a client's computer network).

. . . .

What Factors Influence the Situation?

In addition to the personal factors mentioned above in relation to role, many business or bureaucratic situations have factors external to both the reader and writer that will affect the design of a document. These factors can powerfully affect the way readers read. Common questions (based on Odell et al.) to answer are

▶ What consequences will occur from this idea?
▶ What is the history of this idea?
▶ How much power does the reader have?
▶ How formal is the situation?
▶ Is there more than one audience?

What Consequences Will Occur from This Idea?

Consequences are the effects of a person's actions on the organization. If the effect of your suggestion would be to violate an OSHA standard, your suggestion will be turned down. If the effect would be to make a profit, the idea probably will be accepted.

What Is the History of This Idea?

History is the situation prior to your writing. You need to show that you understand that situation; otherwise, you will be dismissed as someone who does not understand the implications of what you are saying. If your suggestion to change a procedure indicates that you do not know that a similar change failed several years ago, your suggestion probably will be rejected.

How Much Power Does the Reader Have?

Power is the supervisory relationship of the author and the reader. Supervisors have more power. Orders flow from supervisors to subordinates, and suggestions move in the reverse. The more powerful the reader, the less likely the document is to give orders and the more likely it is to make suggestions (Driskill; Fielden; Selzer).

How Formal Is the Situation?

Formality is the degree of impersonality in the document. In many situations, you are expected to act in an official capacity rather than as a personality. For an oral presentation to a board meeting concerning a multimillion-dollar planning decision, you would simply act as the person who knows about widgets. You would try to submerge personal idiosyncrasies, such as joking or sarcasm, for example. Generally, the more formal the situation, the more impersonal the document.

Is There More Than One Audience?

Sometimes a document has more than one audience. In these situations, you must decide whether to write for the primary or the secondary audience. The primary audience is the person actually addressed in the document. A secondary audience is someone other than the intended receiver who will also read the document. Often you must write with such a reader in mind. The secondary reader is often not immediately involved with the writer, so the document must be formal. The following two examples illustrate how a writer changes a document to accommodate primary and secondary audiences.

Suppose you have to write a memo to your supervisor requesting money to travel to a convention so that you can give a speech. This memo is just for your supervisor's reference; all he needs is a brief notice for his records. As an informal memo intended for a primary audience, it might read like this:

Informal use of name

No formatting of document

March 19, 2007
John

This is my formal request for $750 in travel money to give my speech about widgets to the annual Society of Manufacturing Engineers convention in San Antonio in May. Thanks for your help with this.

Fred

If this brief note is all your supervisor needs, neither a long, formal proposal with a title page and table of contents nor a formal business letter would be appropriate. The needs of the primary audience dictate the form and content of this memo.

Suppose, however, that your supervisor has to show the memo to his manager for her approval. In that case, a brief, informal memo would be inappropriate. His manager might not understand the significance of the trip or might need to know that your work activities will be covered. In this new situation, your document might look like this:

Date:	March 19, 2005
To:	John Jones
From:	Fred Johnson
Subject:	Travel money for speech to Society of Manufacturing Engineers convention

A more official format, including formal use of names

As I mentioned to you in December, I will be the keynote speaker at the Annual Convention of the Society of Manufacturing Engineers in San Antonio. I would like to request $750 to defray part of my expenses for that trip.

Orients reader to background and makes request

Explains background of request

This group, the major manufacturing engineering society in the country, has agreed to print the speech in the conference Proceedings so that our work in widget quality control will receive wide readership in M.E. circles. The society has agreed to pay $250 toward expenses, but the whole trip will cost about $1000.

Adds detail that primary audience knows

I will be gone four days, May 1–4; Warren Lang has agreed to cover my normal duties during that time. Work on the Acme Widget project is in such good shape that I can leave it for those few days. May I make an appointment to discuss this with you?

As you can see, this document differs considerably from the first memo. It treats the relationship and the request much more formally. It also explains the significance of the trip so that the manager, your secondary audience, will have all the information he or she needs to respond to the request.

Creating Audience Profiles

Before you begin to write, you need to use the concerns outlined in this chapter to create an audience profile, a description of the characteristics of your audience. The profile is an image of a person who lives in a situation. You use that description as the basis for decisions you make as you create your document (Lazzaro; Rockley).

In order to create a profile, you need to ask specific questions and use an information-gathering strategy.

Questions for an Audience Profile

To create the profile, ask the questions discussed in this chapter:

▶ Who is the audience?
   What are their demographic characteristics?
   What is the audience's role?

> How does the reader feel about the subject?
> How does the reader feel about the sender?
> What form does the reader expect?
> ◗ What is the audience's task?
> ◗ What is the audience's knowledge level?
> ◗ What factors influence the situation?

Information-Gathering Strategies

To create these answers, use one of two methods—create a typical user or involve the actual audience at some point in your planning (Schriver 154–163).

Create a Typical User

Creating a typical user means to imagine an actual person about whom you answer all the profile questions. Suppose that you have to write a set of instructions for uploading a document to a webserver. You could create a typical user whom you follow in your mind as she enacts the instructions.

Here is such a creation: Marie Williamson, a sophomore (*demographics*) arrives in the lab (*situation*) with the assignment to place several files on the Web (*task*). She is taking her first course that deals completely with computers and needs to get the files on the Web so that she receives credit (*role*). She intensely dislikes computers (*attitude toward the subject*), has never really liked using manuals (*expectation about sender and form*). She has never done this before by herself (*knowledge level*). She is stressed because she has only 20 minutes in the lab to do this before she has to go to work, and she still has no baby sitter lined up for her child (*other factors*) (see Garret 54–56).

To write the manual, you try to accommodate all the "realities" that you feel will affect "Marie's" or any user's ability to carry out the instructions. Your goal is to write an inviting, seductive manual, one that entices her to read and to act. If you can write a manual that would help "Marie," it is a good guess that it will help other people also. Using "Marie" you can make decisions that will help create a document that enables any reader to accomplish his/her task and enhance his/her relationships with teachers, coworkers, and children.

Involve the Actual Audience

You could interview actual members of the target audience. Instead of creating "Marie," you interview several people who have to upload files but have no experience. You ask them the profile questions and, in the best practice, later ask them to review the manual before the final draft is published. When you interview them, you ask the same profile questions as you would if you were creating a typical user, but of course you get their answers rather than your imagined ones. While this method is slower than creating an imaginary user, it is often more accurate in gauging real users' needs because it clarifies what the attitudes and experiences really are.

Exercises

▶ You Create

6. Write a brief memo in which you propose a change to a more powerful, positively inclined audience. Use an emotional topic, such as eliminating all reserved parking at your institution or putting a child-care room in every building. Before you write, answer the questions on pages 48–49. In a group of three or four, explain how your memo reflects the answers you gave to those questions.

▶ You Analyze

7. Analyze this brief message to determine whether you should rewrite it. Assume it is written to a less powerful, negatively inclined audience (Kostelnick). If you decide it needs revision, rewrite it. If your instructor so requires, discuss your revisions in groups of three.

   Thank you for your recent proposal on the 4-day, 10-hour-a-day week. We have rejected it because

   It is too short.

   It is too narrow-minded.

   It has too many errors.

▶ Group

8. In class, set up either of these two role-playing situations. In each, let one person be the manager, and let two others be employees in a department. In the first situation, the employees propose a change, and the manager is opposed to it. In the second, the employees propose a change, and the manager agrees but asks pointed questions because the vice-president disagrees. In each case, plan how to approach the manager, and then role-play the situation. Suggestions for proposed changes include switching to a 4-day, 10-hour-a-day week; instituting recreational free time for employees; and having a random drawing to determine parking spaces instead of assigning the spaces closest to the building to executives. After you complete the role playing, write a memo to the manager requesting the change. Take into consideration all you found out in the role playing.

9. In groups of three or four, agree on a situation in which you will propose a radical change, perhaps that each class building at your college contain a child-care room. Plan how to approach an audience that feels positive about this subject. For instance, which arguments, formats, and visual aids would be persuasive? Then write the memo as a group. For the next class, each person should bring a memo that requests the same change but addresses an audience that is neutral or negative (you choose) about this subject. Discuss the way you changed the original plan to accommodate the new audience. As a group, select the best memo and read it to the class.

Worksheet for Defining Your Audience

☐ **Who will read this document?**

- Name the primary reader or readers.
- Name any secondary readers.

☐ **Determine the audience's level of knowledge.**

- What terms do they know?
- What concepts do they know?
  To find your reader's knowledge level, you must (1) ask them directly, (2) ask someone else who is familiar with them, or (3) make an educated guess.
- Do they need background?

☐ **Determine the audience's role.**

- What will they do as a result of your document?
- Have you presented the document so they can take action easily?
- Why do they need your document? for reference? to take to someone else for approval? to make a decision?

☐ **Determine the audience's community attitudes.**

- What are the social factors in the situation?
- Is the audience negatively or positively inclined toward the message? toward you?
- What format, tone, and visuals will make them feel that you are focused on their needs?

☐ **What form does the audience expect?**

Writing Assignments

1. Interview one or two professionals in your field whose duties include writing. Find out what kinds of audiences they write for by asking a series of questions. Write a memo summarizing your findings. Your goal is to characterize the audiences for documents in your professional area. Here are some questions that you might find helpful:

   - What are two or three common types of documents (proposals, sets of instructions, informational memos, letters) that you write?
   - Do your audiences usually know a lot or a little about the topic of the document?
   - Do you try to find out about your audience before you write or as you write?
   - What questions do you ask about your audience before you write?

- Do you change your sentence construction, sentence length, or word choice to suit your audience? If so, how?
- Do you ever ask someone in your intended audience to read an early draft of a document?
- Does your awareness of audience power or inclination affect the way you write a document?
- Does your awareness of the history of a situation affect the way you write a document?
- Do you ever write about the same topic to different audiences? If yes, are the documents different?
- Do you ever write one document aimed at multiple audiences? If yes, how do you handle this problem?

2. Write two different paragraphs about a topic that you know thoroughly in your professional field. Write the first to a person with your level of knowledge. Write the second to a person who knows little about the topic. After you have completed these two paragraphs, make notes on the writing decisions you made to accommodate the knowledge level of each audience. Be prepared to discuss your notes with classmates on the day you hand in your paragraphs. Your topic may describe a concept, an evaluating method, a device, or a process. Below are some suggestions; if none applies to your field, choose your own topic or ask your instructor for suggestions.

Using a Web search engine	A photo cell phone
A machining process used in wood	The food pyramid
Finding an address on the Internet	The relation of calories to grams
Speakers for an audio system	Registering for a class
Cranking amps of a battery	Antilock brakes
The dots-per-inch feature of a laser printer	Using a PDA to receive and send e-mail
Creating a document using HTML code	Just-in-time manufacturing
	Securing an internship

3. Form groups of three or four. If possible, the people in each group should have the same major or professional interest. Decide on a short process (four to ten steps) that you want to describe to others. As a group, analyze the audience knowledge and then write the description.

   Alternate: After writing the memo, plan and write the same description for an audience with different characteristics than your first audience.

Web Exercise

Analyze a website to determine how the authors of the site "envision" their readers. What have they assumed about the readers in terms of knowledge and needs, role, and community attitudes? Write an analytical report to your classmates or coworkers in which you explain strategies they should adopt as they develop their websites.

Works Cited

Coney, Mary. "Technical Communication Theory: An Overview." *Foundations for Teaching Technical Communication: Theory, Practice, and Program Design*. Ed. Katherine Staples and Cezar Ornatowski. Greenwich, CT: Ablex, 1997. 1–16.

Coney, Mary, and Michael Steeholder. "Role Playing on the Web: Guidelines for Designing and Evaluating Personas Online" *Technical Communication* 47.3 (Third Quarter 2000): 327–340.

Driskill, Linda. "Understanding the Writing Context in Organizations." *Writing in the Business Professions*. Ed. Myra Kogen. Urbana, IL: NCTE, 1989. 125–145.

Fielden, John S. "What Do You Mean You Don't Like My Style?" *Harvard Business Review* 60 (1982): 128–138.

Garret, Jesse James. *The Elements of User Experience: User-Centered Design for the Web*. Indianapolis, IN: American Institute of Graphic Arts, 2003.

Horton, William. *Secrets of User-Seductive Documents*. Arlington, VA: Society for Technical Communication, 1997.

Huckin, Thomas. "A Cognitive Approach to Readability." *New Essays in Technical and Scientific Communication: Research, Theory, and Practice*. Ed. Paul V. Anderson, R. John Brockman, and Carolyn Miller. Farmingdale, NY: Baywood, 1983. 90–108.

Kostelnick, Charles. "The Rhetoric of Text Design in Professional Communication." *Technical Writing Teacher* 17.3 (1990): 189–203.

Lay, Mary. *The Rhetoric of Midwifery: Gender, Knowledge, and Power*. New Brunswick, NJ: Rutgers University Press, 2000.

Lazzaro, Heather. "How to Know Your Audience." *intercom* (November 2001): 18–20.

Magolan, Todd. Personal memo, 15 March 1991. Used by permission.

Odell, Lee, Dixie Goswami, Ann Herrington, and Doris Quick. "Studying Writing in Non-Academic Settings." *New Essays in Technical and Scientific Communication: Research, Theory, and Practice*. Ed. Paul V. Anderson, R. John Brockman, and Carolyn Miller. Farmingdale, NY: Baywood, 1983. 17–40.

Rockley, Ann. "Single Sourcing and Information Design." *Content and Complexity: Information Design in Technical Communication*. Ed. Michael J. Alebers and Beth Mazur. Mahwah, NJ: Lawrence Erlbaum, 2003. 307–335.

Schriver, Karen. *Dynamics in Document Design: Creating Texts for Readers*. New York: Wiley, 1997.

Selzer, Jack. "Composing Processes for Technical Discourse." *Technical Writing: Theory and Practice*. Ed. Bertie E. Fearing and W. Keats Sparrow. New York: MLA, 1989. 43–50.

Stern, Judith, and Robert Littieri. *QuickTime and Moviemaker Pro for Windows and Macintosh*. New York: Peachpit, 1998. 146.

Focus on
Credibility

When you deliver information, a key ingredient of your message is your credibility. I will tend to accept your message if I feel you are a credible person in the situation. Credibility grows out of competence and method.

Competence is control of appropriate elements. If you act like a competent person, you will be perceived as credible. The items discussed in this text will all improve your credibility—attention to formatting, to organization, to spelling and grammar, and to the audience's needs. Competence is also shown by tone. You will not seem credible if you sound casual when you should sound formal or facetious when you should sound serious.

Method includes the acts you have taken in the project. Simply put, audiences will view you as credible—and your message as believable—if they feel you have "acted correctly" in the situation. If I can be sure that you have worked through the project in the "right way," I will be much more likely to accept your requests or conclusions. If you have talked to the right people, followed the right procedures, applied the correct definitions, and read the right articles, I will be inclined to accept your results. For instance, if you tell me that package design A is unacceptable because it failed the Mullen burst test (an industry-wide standard method of applying pressure to a corrugated box until it splits), I will believe you, because you arrived at the conclusion the right way.

In all the informative memos and reports you write, you should try to explain your methodology to your reader. The IMRD report (Chapter 12) provides a specific section for methodology, and in other types of reports you should try to present the methodology somewhere, often in the introduction, as explained in this chapter. Sometimes one sentence is all you need: "To find these budget figures I interviewed our budget analyst." Sometimes you need to supply several sentences in a paragraph that you might call "background." However you handle it, be sure to include methodology in both your planning (what do I need to do to find this information?) and in your writing (be sure to add a reference or a lengthier section).

Chapter 3

The Technical Communication Process

Chapter 3
In a Nutshell

You have to plan, draft, and edit your document, either by yourself or in a group.

Plan by establishing your relationship with your audience and your document. Situate yourself by determining your knowledge of the topic and your goal for the audience. Also determine whether legal, ethical, or global issues are involved. Create a clean document design, both for physical appearance and for content strategy. Develop a realistic production schedule. You want your audience to accept what you tell them. They have to accept you as credible—because they know "who you are" or because you have performed the "right action" to familiarize yourself with the topic.

Draft by carrying out your plan. Find your best production method. Some people write a draft quickly, focusing on "getting it out," whereas others write a draft slowly, focusing on producing one good sentence after another. Keep basic strategies—for instance, the top-down method of first announcing the topic and then filling in the details—in mind as you write. If the writing causes you to see a new, better way to present the material, change.

Edit by making the document consistent. Look for surface problems, such as spelling, grammar, and punctuation. Make sure all the presentation elements—heads, captions, margins—are the same. Set and meet quality benchmarks.

Work in a group by expanding your methodology. For groups, add into your planning a method to handle group dynamics—set up a schedule, assign responsibilities, and, most important, select a method for resolving differences.

Like all processes, document production proceeds in stages. This chapter explains each of these stages and introduces you to writing as a member of a group, a practice common in industry and business.

An Overview of the Process

The goal of technical communication is to enable readers to act. To do so, you need to create a document that is helpful and appropriate, one in which the audience can find what they need, understand what they find, and use what they understand appropriately (Redish). In addition, you need to create a document that engages readers so that they focus their attention on the patterns in the document. In order to achieve those goals, you need to follow a process. This chapter explains that process, including special items you need to consider if you are composing a document as part of a group.

Many technical writers have discussed the process that they use to produce documents. These discussions have a remarkable consistency. Almost all writers feel that the process is both linear (following the sequence step by step) and recursive (returning to previous steps or skipping ahead as necessary).

Almost all use some version of this sequence:

▶ Plan by discovering and collecting all relevant information about the communication situation.
▶ Draft, test, and revise by selecting and arranging the elements in the document.
▶ Finish by editing into final form.

Figure 3.1, a flow chart of the process (adapted from Goswami et al. 38 and Redish 164), indicates the steps and nature of the process. The blue arrows indicate the linear sequence: first plan, then write, then edit. This path is the standard logical sequence that most people try to follow in any project. The black arrows, however, indicate the recursive nature of this process—you must be ready to return to a previous stage or temporarily advance to a subsequent stage to generate a clear document.

Planning Your Document

During the planning stage, you answer a set of questions concerning your audience, your message, your document's format, and the time available for the project. The answers to those questions will give you important information about your audience and a general idea of the document you will create.

Figure 3.1

Process Chart

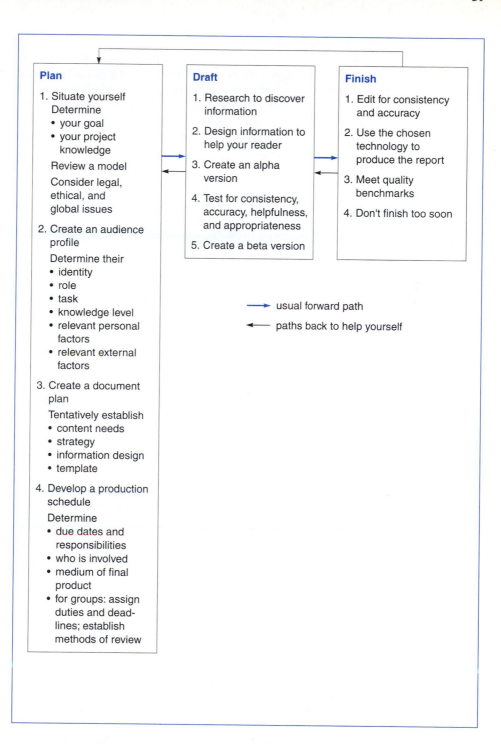

Plan

1. Situate yourself
   Determine
   • your goal
   • your project knowledge
   Review a model
   Consider legal, ethical, and global issues

2. Create an audience profile
   Determine their
   • identity
   • role
   • task
   • knowledge level
   • relevant personal factors
   • relevant external factors

3. Create a document plan
   Tentatively establish
   • content needs
   • strategy
   • information design
   • template

4. Develop a production schedule
   Determine
   • due dates and responsibilities
   • who is involved
   • medium of final product
   • for groups: assign duties and dead-lines; establish methods of review

Draft

1. Research to discover information

2. Design information to help your reader

3. Create an alpha version

4. Test for consistency, accuracy, helpfulness, and appropriateness

5. Create a beta version

Finish

1. Edit for consistency and accuracy

2. Use the chosen technology to produce the report

3. Meet quality benchmarks

4. Don't finish too soon

→ usual forward path

← paths back to help yourself

Depending on the situation, planning can be either brief or lengthy. For a short memo or e-mail, the planning step might be just making a few mental notes that guide you as you compose. For instance, to send out a routine meeting announcement via e-mail, you might only have to select the recipients from your address list; confirm the date, place, and agenda; and review previous routine announcements to get the correct page design. But for a lengthy proposal or manual, the planning sessions could yield a written document that specifies the audience and explains the way they will use the planned document, a detailed style sheet for format, and a realistic production schedule.

Regardless of the amount of time spent on this step, planning is a key activity in writing any document. Better writing results from better planning (Dorff and Duin; Flower).

To plan effectively, situate yourself, create an audience profile, create a document plan, design your information, design your template, and create a production schedule.

Situate Yourself

To situate yourself, you need to determine what you are trying to do. Ask these questions:

- What is your goal?
- How much project knowledge do you have?
- Is there a model to help you focus your thinking?
- Are global issues involved?
- Are legal or ethical issues involved?

What Is Your Goal?

Your goal is to direct the audience to a specific experience after reading your document. What should they know or be able to do after reading your document, and how should they be helped to that new condition? (See Screven; Shedroff .) The answer depends on the situation, and often seems obvious, but specifically stating your goal will give you a mission statement that will guide your work. Examples include, "Through reading this document the audience will assemble the tricycle." "Through reading this document the audience will find compelling reasons to fund my research project." "Through reading this document the audience will understand the results and implications of this experiment." (See Horton, Chapter 1.)

How Much Project Knowledge Do You Have?

You need to decide how much you know. The answer to this question depends on the relation of the project and the document. If the project is completed, you have the knowledge and need to perform little if any research. You know what happened in the experiment; you know what you found out when you investigated sites for a new retail store. On the other hand, in many situations you are handed an assignment that results in a document, but concerns a topic about which you know nothing. In this case, you need to plan how you will discover all the relevant information. If you are assigned to create a set of instructions for software that you are unfamiliar with, then you will have to undergo some kind of training and perform research to find the information. If you have to write a proposal for funding a program, you will have to research all the information that the granting agency requires.

This information will eventually find its way into your production schedule, discussed below.

Is There a Model to Help You Focus Your Thinking?

Most people find a document easier to create if they have some idea of what it ought to look like. As a college student, you no doubt have asked a teacher for specifications on a research paper's length and whether you need to include formal footnotes so you know how to shape the material. To find a model, consider any of these methods. Is there a look imposed by a genre? Many technical documents belong to a *genre*, a formalized way of handling recurring writing tasks. For instance, sets of instructions are a genre; most of them look roughly alike, with a title, introduction, and chronologically sequenced steps. A memo is a broader genre, with To, From, Subject, and Date lines. Is there a company style sheet? Many companies have style guidelines to follow when creating a report—the title page, for instance must contain specified information such as title, author, recipient, date, and project number; chapters must begin on a new page, using a font of a certain size. Look for an earlier document of the same kind to serve as a model. Many times, authors who have to write a report, say, a personnel evaluation or committee minutes, will find an example of previous work done acceptably, and use that approach.

All these are examples of the author finding out in the planning stage the "vessel" into which he or she will "pour" his or her ideas. Having this look in mind allows you to plan other elements effectively.

Are Global Issues Involved?

If the document will be translated for or used by people in other countries, there are global issues to be dealt with. In that case, you need to review the type of localization (as explained in Chapter 1, pp. 20–25) you will use—radical or general. You will then have to adjust the document's form and wording to include the audience who does not share your cultural background.

Are Legal or Ethical Issues Involved?

Ethical issues, as discussed in Chapter 1, deal with doing the right thing for the stakeholders involved. These issues can be very large, such as whether this project and the report that supports it actually break the law, or they can deal with issues of treating information clearly for all stakeholders. At times, issues may need to be reviewed by legal counsel, for instance, phrasing sentences so that you remain in compliance with terms of a contract.

Create an Audience Profile

To create an audience profile you need to answer the audience questions listed in Chapter 2. Rephrased slightly, those questions are

▶ Who is the audience?
▶ What is their role?
▶ What is their task?
▶ What is their knowledge level?
▶ What personal factors influence the situation(feelings about subject, sender, expected form of communication)?
▶ What external factors influence the situation?

As was explained in Chapter 2, planning for your audience is a key method for writing helpful and appropriate documents. Since these questions have all been dealt with there, they will not be discussed here. If the audience is located in another country, use ideas from Chapter 1 (see "Technical Communication Is Global") in your planning.

Create a Document Plan

To create a document plan, establish your content needs, establish your strategy, decide whether to use a genre, decide whether to use an established pattern, and consider using a metaphor.

Establish Your Content Needs

To establish your content needs you must determine what you know about the topic and what the audience knows and needs to learn. If you know quite a bit, then you understand the content and have little need for more research. If you know little or nothing about the topic, then you have to set up a plan to discover all the information before you can decide what the audience needs.

To deal with the audience knowledge and needs, refer again to the schemata discussion (Chapter 2, pp. 43–44) where you will find the basic principle: Add to what the audience knows and don't belabor the obvious. If they know a lot, then you can assume that they know broad terminology and implications of actions (e.g., if you say, "Open a Word file," they will perform all the actions necessary to open a new file on the screen without further instruction). If they do not know a lot, then build on what they do know (if your audience is composed of computer novices, you would instruct them to turn on the computer, find the Word icon, double-click on it, and choose New from the File menu).

Establish Your Strategy

To establish your strategy, you determine how to carry out your goal. Strategy is your "creative concept," the way you present the material so that your reader can easily grasp and act upon it. What design will help the reader to do that? Essentially, you must create an experience for your reader, helping him or her to build a unique model of the information (Albers). In order to build the model, the reader must be able to see all the parts and then work to join them together into a final product. It is as if every writing situation were a box filled with parts that have to be assembled. As the writer, you have to lay out those parts and then give the reader what is needed to assemble them. For instance, you lay out the problem and then help the reader understand (and agree to) your solution. Or, you lay out the goal and then take the reader step by step through a process that leads to the goal. While establishing a strategy and building a model are often difficult to do, it is also a fun way to think about relating to your topic (Garret 40–59).

The idea of strategy is to help readers grasp the big picture so that they can interrelate all the details that you present to them. The organization of your document helps them find that big picture. Three common ways to organize are to follow a genre, to use established patterns, and to use a metaphor.

Decide Whether to Use a Genre

A genre is a standardized way to present information (Carliner 49). Various genres include documents such as sets of instructions, proposals, trip reports, or meeting agendas, documents so common that just by looking at them the reader knows what to expect. He or she understands what they are, what they are likely to include, and how they are likely to be structured. In other words, they know what experience they will have as they read the document. For instance, in instructions, lists of necessary materials appear in the introduction, and the steps appear in the body in sequential order. Readers know that they have to find the parts and then follow the steps to reach the goal.

Decide Whether to Use an Established Pattern

If no genre exists, you could try several other options. You could use a common rhetorical sequence: for instance, definition followed by example and analogy. You could establish an organizational progression, such as most vital to least important or top to bottom. To treat the components of a computer system from most to least important, you might start with the central processing unit and end with the electrical cord. To discuss it from top to bottom, you might start with the screen and end with the keyboard.

Consider Using a Metaphor

A metaphor is a figure of speech in which a word or phrase is used symbolically: Recall our earlier example of the author "pouring" ideas into the "vessel," or genre. It is often helpful to a reader when one thing is used to stand for the organization of the material. For instance, you can compare your document's parts to those of a building, so that the introduction is the entryway and the various parts are hallways that lead to rooms, which are the subsections of the report. Or you could organize in terms of a wheel, talking about the important information as the hub and various implications of the information as the spokes. Remember that metaphors tend to be culture specific, so choose carefully if your audience is a global one. Metaphors are most useful if they fit into the user's experience. For example, if the audience knows what an automobile's stick shift is, dividing a topic into gears will make a lot of sense, but if such a vehicle is unknown, the reference only serves to confuse (Shedroff 280).

Design Your Information

In addition to choosing an overall organization, plan how you will present information within that organization. Several methods will help you. The key items are choosing an arrangement and using various cueing devices to help the reader pick out what is important and to "assemble" relationships. Since many of these devices become clearer to a writer as he or she actually writes, they are discussed at length in the sections on drafting, testing, and finishing your document.

Design Your Template

A *template* is a general guide for the look of your page. It is closely related to a style sheet, a list of specifications for the design of the page. The goal of a style sheet is to create a visual logic—a consistent way to visually identify parts of a paper. The parts of a style sheet are the width of margins, the appearance of several layers of heads, the treatment of visuals and lists, the position of page numbers, and the typeface. These items are described in detail in Chapter 6. Most sophisticated word processing programs (Microsoft Word and WordPerfect, for instance) allow you to set these formats before you begin.

Here is a sample template. The style sheet for this template would read like this

Title—24 point bold Arial, set up and down (i.e., capitalize all important words)
Level 1 Heading—12 point Arial bold, set flush left, up and down
Level 2 Heading—12 point Arial bold, set flush left, up and down, followed by a period
Text font—12 point Arial
Spacing—single-space within paragraphs and double-space between

Visual aids—set flush left
Caption—10 point Arial italicized, set flush left, up and down; include the word *Figure* followed by a period, then the title and no period

---

Sample Report Template

Level 1 Heading

This is text font. It is Arial 12 point.

Level 2 Heading. This is more text font. Below is a visual aid with its caption.

Figure 1. Caption

An Example of Developing a Style Sheet

Let's return again to the memo on pages 37–38 to see how the author developed a style sheet. He used headings to make the three sections of the report easy to locate, and he used numbering to set off each item. His lists use the "hanging indent" method, the second line starting under the first word of the first line. The visual aid is set at the left margin with the caption above it.

Create a Production Schedule

A production schedule is a chronological list of the activities required to generate the document and the time they will consume. Your goal is to create a realistic schedule, taking into consideration the time available and the complexity of the document. You need to answer these questions: How much time do I have? Who is involved in producing the document? What constraints affect production?

How Much Time Do I Have?

You have from the present to the deadline, be it one hour, two weeks, or four months. Determine the end point and then work backward, considering how long it will take to perform each activity. How long will it take to print the final document? How much time must you allot to the revision and review stage? How long will it take to draft the document? How much time do you need to discover the "gist" of your document?

A major problem with time management is "finishing too soon." Many people bring hidden time agendas to projects. They decide at the beginning that they have only so many hours or days to devote to a particular project. When that time is up, they must be finished. They do not want to hear any suggestions for change, even though these suggestions are often useful and, if acted upon, could produce a much better document. Another time management problem often occurs in research projects. A fascinated researcher continually insists on reading "just one more" article or book, consuming valuable time. When he or she begins to write the report, there is not enough time left to do the topic justice. The result is a bad report.

As your ability to generate good documents increases, you will get better at estimating the time it will take to finish a writing project. You will also develop a willingness to change the document as much as necessary to get it right. Developing these two skills is a sure sign that you are maturing as a writer. The worksheet on page 78 provides a useful editing checklist.

Also consider the time it will take to produce the final document. Who will type it? What kind of system will you use? word processor? professional typesetter and printer? e-mail? How well do you know the system? The less familiar you are with your production tools, the more time you are likely to spend.

For a short document, of a type you have created before, the answers are obvious. But for long, complex documents, these questions are critical. The creation of such a document is far easier if you answer these questions realistically and accurately in the planning stage.

Who Is Involved in Producing the Document?

The number of people involved in the procedure varies from one (you) to many especially if there is a review process. If it is only you who is involved, you need to consider only your own work habits as you schedule time to work through the document. If many people are involved—such as reviewers for technical accuracy legality and internal consistency—you will have to schedule deadlines for them to receive and return your document.

An Example of Scheduling.   To produce the site-visit memo was an easy task for Todd. Because he had noted all the facts and had drawn a rough sketch while on the site, he saw that he could produce the memo in less than an hour on his word processor. He also knew that drawing would require another half an hour at most. No one else had to review the document, so Todd scheduled an hour of time and easily completed the document.

What Constraints Affect Production?

Constraints are factors that affect production of the document. They include time, length, budget, method of production, method of distribution, and place of use (Goswami et al.).

> ▶ *Time* is the number of hours or weeks until the date the document is due. *Length* is the number of pages in the final document.
> ▶ The *budget* specifies the amount of money available to produce the document. A negligible concern in most brief documents (those one to three pages in length), the budget can greatly influence a large, complex document. For instance, a plastic spiral binding could increase costs slightly but a glued, professional-type binding could be prohibitively expensive.
> ▶ *Method of production* is the type of equipment used to compose the document. If you plan to use two word processors, one in a public lab and one at home, both must run the same software so that you can use your disk in either system.
> ▶ *Method of distribution* is the manner in which the document is delivered to the reader. If it is to be mailed, for instance, it must fit into an envelope of a certain size.
> ▶ *Place of use* is the physical surroundings in which the audience reads the document. Many manuals are used in confined—even dirty—spaces that could require smaller paper sizes and bindings that can withstand heavy use.

Worksheet for Planning— Short Version

☐ **What is the situation?**
   a. Who will do what? When? Where?
   b. Are there any special issues to be aware of? History? Personal ties? Physical limitations?
   c. Do you need to do any work to be comfortable with the topic?

☐ **Who is the audience?**
   a. How much do they know about the topic?
   b. What is their goal in reading your document?
   c. What do they need in order to accomplish their goal?

☐ **What will the final document look like?**

☐ **How will you present the information?**

☐ **How will you design your template?**

☐ **What is a realistic schedule for completing the document?**

Worksheet for Planning— Long Version

☐ **What is true of my audience?**
  a. Who will read this document?
  b. Why do they need the document?
  c. What will they do with it or because of it?
  d. How much do they know about the topic?
  e. What is their personal history with the topic?
  f. What expectations exist for this kind of document's appearance or structure?
  g. Will other people read this document? What do they need to know?
  h. What tone do they expect? Should I use personal pronouns and active voice?

☐ **What is true of me?**
  a. Why am I credible? because of my role? because of my actions? Do I need to tell the reader that role or those actions?
  b. What authority do I feel I have? Do I believe that I have the right to speak expertly and be taken seriously?
  c. What should I sound like? How formal or informal should I make myself appear?

☐ **What is my goal in this writing situation?**
  a. What basic message do I want to tell my audience?
  b. Why do I need to convey that message? to inform? to instruct? to persuade?

☐ **What constraints affect this situation?**
  a. How much time do I have?
  b. How long should the document be?
  c. Is budget an issue? If so, how much money is available? Can I produce the document for that amount?
  d. How will I produce the document? Hard copy? Web document? Do I really understand the potential problems with that process? (For example, will the library printers be working that day?)
  e. Is distributing the document an issue? If so, do I understand the process?
  f. Where will the reader read the document? Does that affect the way I create the document?

☐ **What are the basic facts?**
  a. What do I already know?
  b. Could a visual aid clarify essential information?
  c. Where can I read more?
  d. With whom can I discuss the matter further?
  e. Where can I observe actions that will reveal facts?

☐ **What is an effective strategy?**
  a. Should I choose a standard genre (set of instructions, technical report)?

    b. Should I follow one example throughout the document or should I use many examples?

    c. Should I develop a central metaphor?

    d. Should I use definition followed by example and analogy?

    e. Which organizational principle (such as top to bottom) is best?

☐ **What format should I use?**

    a. What margins, heads, and fonts do I want?

    b. What will my template for the document look like?

☐ **What is my production schedule?**

    a. What is the due date of the document?

    b. By what date can I complete each of the stages (research, alpha version, review, beta version—see p. 57)?

Exercises

▶ You Create

1. Following the format of the on-site review (pp. 37–38), write a one-page plan for one of the following situations. Your instructor may ask you to form groups of three. In each situation, assume that you have completed the research; that is, the plan is for the document, not the entire project.

    a. Recommend a change in the flow of work in a situation. For example, change the steps in manufacturing a finished part or in moving the part from receiving to point of sale.

    b. Think of a situation at your workplace or at your college that you want changed. Write the plan for a document proposing that change.

    c. You have discovered a new opportunity for manufacturing or retail at your workplace. Plan a report to obtain permission to use this opportunity.

    d. You have been assigned to teach coworkers or fellow students how to create a simple webpage. Plan the set of instructions you will write. Your goal is to enable your audience to create a page with four or five paragraphs, heads, one hyperlink to another document, and one visual (optional).

▶ Group

2. In groups of three, critique the plans you wrote in Exercise 1. Your goal is to evaluate whether a writer could use the plan to create a document. Your instructor will ask for oral reports in which your group explains helpful and nonhelpful elements of plans.

Drafting and Finishing Your Document

Drafting is often part of planning because, as you write, you discover more about the topic and how to present it. You may suddenly think of new ideas or new ways to present your examples. Or you may discover an entirely new way to organize and approach the whole topic, and so you discard much of your tentative plan. In the finishing stage, you produce a consistent, accurate document. This section explains strategies for these stages.

Research to Discover Information

If you do not know the topic, obviously you must learn about it. To do so requires time spent researching. The basic methods used are keyword searching in a library or on the Web, interviewing users, and interviewing experts. These topics are covered in depth in Chapter 5.

Design Your Information to Help Your Reader

To design your information, remember that you are creating an experience for your readers. You are helping them build a model. An easy way to envision this concept is to think about how information is given along the interstate highway system. Large green signs, placed at crucial junctures, indicate such information as exit numbers and distances between points. You receive the information you need at the point where you have to act: Signage tells you where to exit, or enter, the system. As a result of the placement of the information, you perform the correct action. Your information design should accomplish the same thing for your readers because you engineer an experience for them (Shedroff).

The following sections contain many hints for engineering interactions and experiences. All of them are ways to design the kinds of "signs" needed to arrive at the "final destination." However, you must place them at the key points, depending on what you know about the audience and their goal in the situation.

General Principles

Researchers (Duin; Huckin; Slater; Spyridakis) have developed some specific guidelines to help you create the kind of interactive document your reader needs. For more detailed information on these topics, see Chapters 4 and 9.

1. For an audience with little prior knowledge about a topic, use the familiar to explain the unfamiliar. Provide examples, operational definitions, analogies, and illustrations. These devices invite your reader to become imaginatively involved with the topic and make it interesting.
2. For readers familiar with a topic, don't belabor the basics. Use accepted terminology.

3. For all readers, do the following:

- State your purpose explicitly. Researchers have found that most readers want a broad, general statement that helps them comprehend the details.
- Make the topic of each section and paragraph clear. Use heads. Put topic sentences at the beginning of paragraphs. This top-down method is very effective.
- Use the same terminology throughout. Do not confuse the reader by changing names. If you call it *registration packet* the first time, don't switch to *sign-up brochure* later.
- Choose a structuring method that achieves your goal. If you want your readers to remember main ideas, structure your document hierarchically; if you want them to remember details, use a list format. Figure 3.2 shows the same section of a report arranged both as a hierarchy and as a list. Writers typically combine both methods to structure an entire document or smaller units such as paragraphs (see Spyridakis). The list provides the details that fill out the hierarchy.

Figure 3.2

Hierarchical and List Formats

Hierarchy	**List**
Site A has two basic problems.	Site A has these problems.
Terrain	Water drains into the basement
Water in the basement	Hills block solar heating
Hills block solar heating	Roof leaks
Disrepair	Windows are broken
Roof leaks	Air conditioning system is
Windows are broken	broken
Air conditioning system is	
broken	

- Write clear sentences. You should try to write shorter sentences (under 25 words), rely on the active voice, employ parallelism, and use words the reader understands.
- Make your writing interesting (Duin; Slater). Use devices that help readers picture the topic. Include helpful comparisons, common examples, brief scenarios, and narratives. Include any graphics that might help, such as photographs, drawings, tables, or graphs.

Use Context-Setting Introductions

Your introduction should supply an overall framework so that the reader can grasp the details that later explain and develop it. You can use an introduction to orient readers in one of three ways: to define terms, to tell what caused you to write, and to explain the document's purpose.

Define Terms

You can include definitions of key terms and concepts, especially if you are describing a machine or a process.

> A closed-loop process is a system that uses feedback to control the movement of hydraulic actuators. The four stages of this process are position sensing, error detecting, controlling the flow rate, and moving the actuator.

Tell What Caused You to Write

Although you know why you are writing, the reader often does not. To orient the reader to your topic, mention the reason you are writing. This method works well in memos and business letters.

> In response to your request at the June 21 action group meeting, I have written a brief description of the closed-loop process. The process has four stages: position sensing, error detecting, controlling the flow rate, and moving the actuator.

State the Purpose of the Document

The purpose of the document refers to what the document will accomplish for the reader.

> This memo defines the basic concepts related to the closed-loop system used in the tanks that we manufacture. Those terms are position sensing, error detecting, controlling the flow rate, and moving the actuator.

Place Important Material at the Top

Placing important material at the top—the beginning of a section or a paragraph—emphasizes its importance. This strategy gives readers the context so that they know what to look for as they read further. Put statements of significance, definitions, and key terms at the beginning.

The following two sentences, taken from the beginning of a paragraph, illustrate how a writer used a statement of significance followed by a list of key terms.

> A bill of materials (BOM) is an essential part of every MRP plan. For each product, the BOM lists each assembly, subassembly, nut, and bolt.

The next two sentences, also from the beginning of a paragraph, illustrate how a writer used a definition followed by a list of key terms.

> The assets of a business are the economic resources that it uses. These resources include cash, accounts receivable, equipment , buildings, land, supplies, and the merchandise held for sale.

Use Preview Lists

Preview lists contain the key words to be used in the document. They also give a sense of the document's organization. You can use lists in any written communication. Lists vary in format. The basic list has three components: an introductory sentence that ends in a "control word," a colon, and a series of items. The *control word* (*parts* in the sample that follows) names the items in the list and is followed by a colon. The series of items is the list itself (italicized in this sample).

> A test package includes three parts: *test plans, test specifications,* and *tests.*

A more informal variation of the basic list has no colon, and the control word is the subject of the sentence. The list itself still appears at the end of the sentence.

> The three parts of a test are *test plans, test specifications,* and *tests.*

Lists can appear either horizontally or vertically. In a horizontal list, the items follow the introductory sentence as part of the text. In a vertical list, the items appear in a column, which gives them more emphasis.

> A test package includes three parts:
> - Test plans
> - Test specifications
> - Tests

Use Repetition and Sequencing

Repetition means restating key subject words or phrases from the preview list; *sequencing* means placing the key words in the same order in the text as in the list. The author of the following paragraph first lists the three key terms— *test plans, test specifications,* and *tests.* She repeats them at the start of each sentence in the same sequence as in the list.

> A test package includes three parts: test plans, test specifications, and tests. *Test plans* specify cases that technicians must test. *Test specifications* are the algorithmic description of the tests. The *tests* are programs that the technicians run.

Use Coordinate Structure

Coordinate structure means that each section of a document follows the same organizational pattern. Readers react positively once they realize the logic of the structure you are presenting. The following paragraphs have the same structure: first a definition, followed by details that explain the term.

CUTTING PHASE

The cutting phase is the process of cutting the aluminum stock to length. The aluminum stock comes to the cutoff saw in 10-foot lengths and the saw cuts off 6-inch lengths.

MILLING PHASE

The milling phase is the process of shaving off excess aluminum from the stock. The stock comes to the milling machine from the cutoff saw in exact lengths. The milling machine shaves the stock down to exact height and width specifications of 5 inches.

DRILLING/THREADING PHASE

The drilling/threading phase has two steps: drilling and threading. The drilling phase is the process of boring holes into the aluminum stock in specified positions. After the stock comes from milling, the drilling machine bores a hole in each of the four sides, creating tunnels. These tunnels serve as passageways for oil to flow through the valve.

The threading phase is the process of putting threads into the tunnels. After a tool change to enable threading, the drilling machine cuts threads into the tunnels. The threads allow valve inlets and outlets to be screwed into the finished stock.

Testing

Testing is asking other people to interact with your document in order to discover where it is effective and where it needs revision. The goal is to turn your alpha version into a finished, smooth beta version. There are two types of testing—formal and informal.

Formal Testing

Formal testing is conducted in a Usability Lab, which contains enough technology (e.g., one-way mirrors, video cameras, scan converters, banks of monitors) to allow an empirical rendering of what occurred as the reader interacted with the document (Barnum 14–15). This type of testing can occur only in a formal laboratory setting and is usually beyond the normal resources of a writer.

Informal Testing

Informal testing involves the writer asking members of the target audience, or sometimes people familiar with the target audience, to review the document (Barnum; Lazzaro; Schriver). Methods for performing informal tests vary, ranging from the author creating various questions for the tester, to the author simply getting from a tester feedback on what is effective. Two common methods of testing are soliciting opinions of quality (Schriver) and soliciting comments about the five dimensions of usability (the 5 e's) (Quesenbery).

The quality information that the writer hopes to find often falls into two broad areas: information on the quality of the written prose and on the quality of the page design and use of visual aids. Questions on quality can generate opinions on such items as weak sentence structure, poorly designed tables, and use of jargon. Questions on page design can generate comments on whether the design is consistent, whether the information is easy to scan, and whether text and visual aids act together to send meaning (Schriver 448–449).

The 5 e questions generate a slightly different kind of information. These questions ask, Is the document

effective? Does it help readers achieve their goals?
efficient? Can users complete their tasks with speed and accuracy?
engaging? Is the document pleasant to use?
easy to learn? How well does the document facilitate the readers' interaction with the process? (This question and the following are often used with sets of instructions.)
error tolerant? How well does the document help the reader avoid or recover from mistakes? (Quesenbery 83–89)

In other words, you can easily create a test by creating key questions, either of quality or related to the 5 e's, and asking one or several members of the target audience to read the document. After you receive the results, you need to revise your document. If you have time, you could ask for a second round of tests before you finalize the document.

Worksheet for Drafting

☐ **Choose strategies that help your readers.**
    a. For unspecialized readers, use comparison, example, and brief narrative to make the unfamiliar familiar.
    b. For specialized readers, do not overexplain.
    c. For all readers:
       State your purpose.
       Make topic sentences and headings clear.
       Use consistent terminology.
       Organize material hierarchically to emphasize main ideas.
       Use a list format to emphasize details.
       Write clear sentences by employing the active voice and parallelism.
       Note: If these strategies inhibit your flow of ideas, ignore them in the first draft and use them later.

(continued)

(continued)

☐ Be sure your introductory material accomplishes one of these three purposes:
- Defines terms.
- Tells why you are writing.
- States the purpose of the document.

☐ Use preview lists as appropriate in the body.

☐ Repeat keywords from the preview list in the body.

☐ Use coordinate structure to develop the paragraphs in the body.

☐ Test your draft.
- Ask a member of the target audience or of people familiar with the target audience to read your draft.
- Give them a list of specific items to check (from spelling to legal, contextual considerations)
- Have them "talk through" the document with you. Record each time they have a question or comment. Revise those spots.

Exercises

▶ You Create

3. Using the plan that you constructed in Exercise 1 (p. 67), draft the document.

▶ You Revise

4. Revise the following paragraph. Use the strategies listed in the Worksheet for Drafting (p. 73 under "Choose strategies"). The new paragraph should contain sections on reasons for writing and on format. Revise individual sentences as well.

Proposals are commonly used in the field of retail management. These recommendations are written in a standard format which has a number of company parts that we often use. These are stating the problem, a section often used first in the proposal. Another one is providing a solution, of course, to the problem that you had. You write a proposal if you want to knock out a wall in your department. You also write a recommendation if you want to suggest a solution that a new department be added to your store. Another section of the proposal is the explanation of the end results. You could also write a proposal if in the situation you wanted to implement a new merchandise layout. A major topic involved with writing proposals is any major physical change throughout the store. A proposal can be structured so as to enable implementation of any major physical changes to the upper management level, who would be the audience for which the proposal is intended.

· ·

Editing
· · · · · ·

Editing means developing a consistent, accurate text. In this stage, you refine your document until everything is correct. You are looking for surface, consistency problems. You check spelling, punctuation, basic grammar, format of the page, and accuracy of facts. You make the text agree with various rules of presentation. When you edit, ask, Is this correct? Is this consistent? In general, you edit by constructing checklists.

A key item to create for yourself is a set of quality benchmarks. Widely used in industry, benchmarks are quality standards used to judge a product. In order to edit effectively, you must set similar benchmarks for your work. Typically, benchmarks are divided into categories with statements of quality levels. Below is a simple benchmark set for a Web document created early in a college technical writing course.

STYLE DESIGN

- No spelling or grammar errors.

INFORMATION DESIGN

- Title appears.
- Introduction appears. Introduction tells point of document and, if it is long, sections of the document.
- Body sections are structured similarly in type and depth of points.

PAGE DESIGN

- Fonts are standard roman, and large enough to be easily readable.
- Heads appear and indicate subject of their section.

VISUAL DESIGN

- Visuals appear to support a point in the text or provide a place for the text to begin.
- Visuals are effectively sized, captioned, and referenced.

NAVIGATION DESIGN

- Every link works.
- Link size and placement indicate the type of content (e.g., return to homepage or major section of document?) that it connects to.
- Links provide helpful paths through the work.

Constructing checklists of typical problems is a helpful strategy. The key is to work on only one type of problem at a time. For example, first read for apostrophes, then for spelling errors, then for heading consistency, then for consistency in format, and so forth. Typical areas to review include paragraph indicators (indented? space above?); heads (every one of each level treated the same?); figure captions (all treated the same?); and punctuation (for instance, the handling of dashes).

The following paragraphs demonstrate the types of decisions that you make when you edit. The goal is to correct errors in spelling, punctuation, grammar, and consistency of presentation. Version 1 is the original; Version 2 is edited.

Version 1

TECHNICAL REPORTS

Unclear topic sentence

Misused semicolon

List elements are inconsistent

Vertical list misemphasizes content

Sentence fragment

"An" indicates one example, but two appear

Misused semicolon

The detailed technical report to upper management will be submitted at the end of the project. It must explain;

1. the purpose of the machine,
2. its operation,
3. and the operation of its sub systems.
4. Assembly methods will also be presented.

It will also include all design calculations for loads, stresses, velocities, and accelerations. Justification for the choice of materials of subsystems. An example might be; the rationale for using plastic rather than steel and using a mechanical linkage as compared to a hydraulic circuit. The report also details the cost of material and parts.

Version 2

TECHNICAL REPORTS

Clear topic sentence

List consistent

Two examples suggested

Complete sentence

The technical report to upper management, submitted at the end of the project, contains several sections. The report explains

1. the machine's purpose and operation.
2. the operation of its subsystems and methods to assemble it.

It also includes all design calculations for loads, stresses, velocities, and accelerations, as well as justification for the choice of materials in subsystems. Examples of this justification include the rationale for using plastic rather than steel and a comparison between a mechanical linkage and a hydraulic circuit. The report also details the cost of material and parts.

Editing with a Word Processor

As you work to achieve the consistency that is the goal of the finishing stage, you can use style aids.

Style aids are word processing features that highlight errors in standard usage. The two most common are spell checkers and grammar checkers (Krull).

A spell checker indicates any words in your document that are not in its dictionary. If you have made a typo, such as typing *wtih*, the checker highlights the word and allows you to retype it. Most spell checkers have an autocorrect feature. Once you engage it (check your program's instructions), it will automatically change every mistake, such as *teh* to *the*. However, these programs have problems. If your typo happens to be another word—such as *fist* for *first*—the program does not highlight it. Also, if you misuse a word—such as *to* instead of *too*—the program does not detect the error.

Grammar checkers indicate many stylistic problems, not just faulty grammar. Checkers detect such problems as subject-verb disagreement, fragments, and comma splices. Checkers also can detect features of your writing. For instance, the checker might highlight all the forms of *to be* in your paper, thus pointing out all the places where you may have used the passive voice. Checkers can also highlight words that could be interpreted as sexist or racist, that are overused, or that are easily confused. Thus the checker will highlight every *your* and *you're*, but you must decide whether you have used the correct form.

Follow these guidelines:

Use your spell checker.
Set your spell checker to "AutoCorrect."
Use your grammar checker for your key problems. (If you have trouble, for example, with fragments or passive voice, set the checker to find only those items.)
After using the spelling and grammar checkers, proofread to find any "OK, but wrong" words, for example, *their* used for *there* or *luck* used for *link*.

Producing the Document

Producing a document involves the physical completion—the typing or printing—of the final document. This stage takes energy and time. Failure to plan enough time for physical completion and its inevitable problems will certainly cause frustration. Many people have discovered the difficulties that can plague this stage when their hard drive crashes or their printer fails. Although physical completion is usually a minor factor in brief papers, in longer documents it often takes more time than the drafting stage.

Worksheet for Editing

☐ **Make a checklist of possible problems.**
  a. Head format (does each level look the same?)
  b. Typographical items (for example, are all dashes formed the same way?)
  c. Handling of lists
  d. Handling of the beginning of paragraphs (for example, are they all indented five spaces?)
  e. References to figures
  f. Spelling
  g. Grammar
  h. Consistent word use

Exercises

▶ You Create

5. Produce the final version of Exercise 1 (p. 67): the change in work flow, the situation change, the new opportunity, or the creation of a webpage. Then write a memo in which you (1) describe the process you used, step by step, and (2) evaluate the strength of your process.

6. Create a brief top-down document. Choose a topic that you can easily break into parts (e.g., computer memory includes RAM and ROM). Use a preview list, coordinate structure, and repetition and sequencing. Use a highlighter or some other method to indicate the devices you used to achieve the top-down structure.

▶ You Revise

7. Edit the following sections for consistency. Correct errors in spelling and grammar, as well as such details of presentation as indentation, capitalization, and treatment of heads.

For whom will I write?

For quick correspondence, memos will be sent to entry level employees, co-workers, Immediate Supervisors and, occasionally, to chief executives. Proposals will be sent to potential clients in hopes of attracting their business. Response request forms are to be sent out following the Preposals in order to obtain a response regarding the proposal.

When I Will Write:

Memos will be written on a daily basis. Proposals and response request letters will be written on demand—once or twice a week.

Importance of Writing

Memos are crucial for effective inter-office communication. Proposals are the key for attrcating business to the organiztion—with successful proposals come revenue. The Response Request Forms are important in communication because they complete the process begun by the proposal. Because of all these forms are written, good writing is the core of good communion.

Importance to promotion—

Since writing is tangible, it is easily alleviated. Hence, the quality of writing is the primary source for promotional evaluation.

LONGER DOCUMENTS

I would be required to write a longer document if a proposal is excepted by a client. Once a proposal is excepted, I will be required to write longer documents --   workbooks, reasearch reports, and needs analysis evaluations.

Workbooks which consist of guidelines for a workshop and appropriate information on the subject.

RESEARCH REPORTS--a closer and more detailed look at a particular topic or organization

Needs Analysis Elevations. These are documents we use in order to determine the needs of the organization and to assist in setting up a workshop

8. Edit this paragraph for effective structure.

Additional Writing

Additional types of writing that may be encountered by a software developer include quality tips memos, trip reports, and development proposals. Other types of proposals are infrequently submitted since there are no moving parts or assembly lines in software development. In addition, since hardware is so expensive, changes in hardware are generally management directed. Consequently, proposals for a change in machinery are also rarely submitted. These types of suggestions and requests are generally incorporated into weekly status reports. Development proposals rise from recognizing the need for a software tool to do work more efficiently. These proposals are directed at management in areas that could utilize this tool. It describes the need for the tool and the advantages using it. The quality tips memos are usually for the department and are a summary of a quality roundtable. Trip reports contain an overview of the conference materials and the points that the attender found most interesting. It also makes the conference literature available to interested parties.

9. Analyze and revise the following paragraph. Create several paragraphs, and revise sentences to position the important words first.

The problem solving process is started by the counselor and client working together to define the problem. The goal of defining the problem is getting the client to apply the problem in concrete and measurable terms. The client may have several problems, so the counselor should allow the client to speak on the problem he or she may want to talk about first. The client usually speaks about smaller problems first until the client feels comfortable with the counselor. Determining the desired goal of the client is the second step in the process. The main point of determining the desired goal of the client is making sure the client and counselor feel the patient's goal is realistic. A bad example of this would be "I don't want to be lonely anymore," but a measurable example that would be realistic is getting the client to spend two hours a day socializing with other people. A realistic goal is considered to be one in which the client can reach without overcoming huge obstacles. The next step is for the counselor to discover the client's present behavior. The goal of this step is for the counselor to rationalize where the client is in relation to the problem. The present behavior of the client is what they are currently doing in concrete terms. Following the previous example, the lonely client could state, "I spend about a half hour a week with other people outside of working." Working out systematic steps between the present behavior and the goal is the final step in the problem solving process. The object is for the counselor to break the process into small parts so the client can work toward their goal and to give the client concrete, real work. Making schedules is a common formality clients use to follow their goals. Counselors are to be supportive and give feedback when the client is not working toward the goal. Feedback would be pointing out to the client that he or she is procrastinating. The client is always open to redefine their goals and the counselor is to stay open and to renegotiate the process.

10. Exchange the document you created in Exercise 6 with a classmate, rewrite that document, and then compare results and report to the class.

▶ Group

11. From any document that explains technical information to lay readers, photocopy a page that illustrates several of the organization strategies explained on pages 68–72. In groups of three or four, read one another's photocopies to identify and discuss the strategies; list them in order of how frequently they occur. Give an oral report of your findings to the class or, as a group, prepare a one-page memo of your findings for your instructor. In either case, base your report on the list you construct and use specific examples.

12. Your instructor will divide you into groups of three or four. Each member interviews three or four people on a relevant topic in your community. (Are you getting a good education at this college? Does the school district meet your children's needs?) Combine the results into a report to the relevant administrator or official, using one visual aid.

First, as a group, create a memo report in which you answer the questions in the Worksheets for Drafting and Editing (pp. 73–74, 78). Then give each member of your group and your instructor a copy, conduct the interviews, and create the final report. Your instructor will provide you with more details about constraints. After you hand in the report, be prepared to give an oral report describing the process you used and evaluating the strengths and weaknesses of the process.

13. In groups of three or four, develop the document you planned in Exercise 1 (p. 67). Plan the entire project, from collecting data and assessing audiences to producing a finished, formatted report. First, create a memo report in which you complete the Worksheet for Planning (pp. 65–67, choose either the long or short version) and for Drafting (pp. 73–74). Give each member of your group and your instructor a copy. Then produce the report. Your instructor will provide you with more detailed comments about constraints.

Writing Assignments

1. Assume that you have discovered a problem with a machine, process, or form that your company uses. Describe the problem in a memo to your supervisor. She will take your memo to a committee that will review it and decide what action to take. Before you write the description, do the following:

   - Write your instructor a memo that contains your plan, that is, the answers to all the planning questions about your document. Include a schedule for each stage of your process.
   - Hand in your plan with your description of the problem.

   Here are some suggestions for a problem item. Feel free to use others.

   oscilloscope

   oxyacetylene torch

   condenser/enlarger

   photoelectric sensing device

   speed drive system for a conveyor

   deposit slip for a bank

   application form for an internship

   black-and-white plotter

   tool control software

   bid form

   layout of a facility (a computer lab, a small manufacturing facility)

   instruction screen of a database

   plastic wrap to cover a product

   method of admitting patients

   method of deciding when to order

   desktop publishing program, an older version

   method for testing X (you pick the topic)

   tamper-evident cap to a pill bottle

   compliance with an OSHA (Occupational Safety and Health Administration) or a DOT (United States Department of Transportation) regulation

   method for inspecting X (you pick the topic)

   disposal of styrofoam drinking cups

   posting your resume on Monster.com

2. Interview three people who write as part of their academic or professional work to discover what writing process they use. They should be a student whose major is the same as yours, a faculty member in your major department, and a working professional in your field. Prepare questions about each phase of their writing process. Show them the model of the process (see Figure 3.1), and ask whether it reflects the actual process they undergo. Then prepare a one- to two-page memo to your classmates summarizing the results of your interviews.

3. Your instructor will assign you to groups of three or four on the basis of your major or professional interest. As a group, perform Writing Assignment 2. Each person in the group should interview different people. Your goal is to produce a two-page memo synthesizing the results of your interviews. Follow the steps presented in the "Focus on Groups" box of this chapter (pp. 84–86). In addition to the memo, your teacher may ask you to hand in a diary of your group activities.

4. Write a one- to two-page paper in which you describe the process you use to write papers. Include a process chart. Then, as part of a group, produce a second paper in which your group describes either one optimal process or two competing models.

Web Exercise

Explain the strategy of a website that you investigate. Establish the ways in which the authors develop a starting point, a presentation "map," and an identity. Write an analytical report to your classmates or coworkers in which you explain strategies they should adopt as they develop their own websites.

Works Cited

Albers, Michael J. "Complex Problem Solving and Content Analysis." *Content and Complexity: Information Design in Technical Communication*. Ed. Michael J. Albers and Beth Mazur. Mahwah, NJ: Lawrence Erlbaum, 2003. 263–283.

Barnum, Carol. *Usability Testing and Research*. New York: Longman, 2002.

Carliner, Saul. "Physical, Cognitive, and Affective: A Three-Part Framework for Information Design." *Content and Complexity: Information Design in Technical Communication*. Ed. Michael J. Albers and Beth Mazur. Mahwah, NJ: Lawrence Erlbaum, 2003. 39–58.

Dorff, Diane Lee, and Ann Hill Duin. "Applying a Cognitive Model to Document Cycling." *Technical Writing Teacher* 16.3 (1989): 234–249.

Duin, Ann Hill. "How People Read: Implications for Writers." *Technical Writing Teacher* 15.3 (1988): 185–193.

Flower, Linda. "Rhetorical Problem Solving: Cognition and Professional Writing." *Writing in the Business Professions*. Ed. Myra Kogen. Urbana, IL: NCTE, 1989. 3–36.

Garret, Jesse James. *The Elements of User Experience: User-Centered Design for the Web*. Indianapolis, IN: American Institute of Graphic Arts, 2003.

Goswami, Dixie, Janice C. Redish, Daniel B. Felker, and Alan Siegel. *Writing in the Professions: A Course Guide and Instructional Materials for an Advanced Composition Course*. Washington, DC: American Institute for Research, 1981.

Horton, William. *Secrets of User-Seductive Documents*. Arlington, VA: Society for Technical Communication, 1997.

Huckin, Thomas. "A Cognitive Approach to Readability." *New Essays in Technical and Scientific Communication: Research, Theory, Practice*. Ed. Paul V. Anderson, R. John Brockman, and Carolyn R. Miller. Farmingdale, NY: Baywood, 1983. 90–108.

Krull, Robert. "Using Electronic Writing Aids." *Word Processing for Technical Writers*. Ed. Robert Krull. Amityville, NY: Baywood, 1988. 61–71.

Lazzaro, Heather. "How to Know Your Audience." *intercom* (November 2001): 18–20.

Quesenbery, Whitney. "The Five Dimensions of Usability." *Content and Complexity: Information Design in Technical Communication*. Ed. Michael J. Albers and Beth Mazur. Mahwah, NJ: Lawrence Erlbaum, 2003. 81–102.

Redish, Janice. "What Is Information Design?" *Technical Communication* 47.2 (May 2000): 163–166.

Schriver, Karen. *Dynamics in Document Design*. New York: Wiley, 1997.

Screven, C. G. "Information Design in Informal Settings: Museums and Other Public Spaces." *Information Design*. Ed. Robert Jacobson. Cambridge, MA: MIT, 2000. 131–192.

Shedroff, Nathan. "Information Interaction Design: A Unified Field Theory of Design Information Design." *Information Design*. Ed. Robert Jacobson. Cambridge, MA: MIT, 2000. 267–292.

Slater, Wayne H. "Current Theory and Research on What Constitutes Readable Expository Text." *Technical Writing Teacher* 15.3 (1988): 195–206.

Spyridakis, Jan H. "Guidelines for Authoring Comprehensible Web Pages and Evaluating Their Success." *Technical Communication* 47.3 (Third Quarter 2000): 359–382.

Focus on Groups

Once they are in business and industry, many college graduates discover that they must cowrite their documents. Often a committee or a project team of three or more people must produce a final report on their activities (Debs). Unless team members coordinate their activities, the give and take of a group project can cause hurt feelings and frustration, and result in an inferior report. The best way to generate an effective document is to follow a clear writing process. For each of the writing stages, not only must you perform the activities required to produce the document, you must also facilitate the group's activities.

Before you begin planning about the topic, you must:

Select a leader
Understand effective collaboration
Develop a method to resolve differences
Plan the group's activities
Choose a strategy for drafting and revising
Choose a strategy for editing
Choose a strategy for producing the document

Select a Leader

The leader is not necessarily the best writer or the person most informed about the topic. Probably the best leader is the best "people person," the one who can smooth over the inevitable personality clashes, or the best manager, the one who can best conceptualize the stages of the project. Good leadership is an important ingredient in a group's success (Debs).

Understand Effective Collaboration

Group members must understand how to collaborate effectively. Two effective methods are goal sharing and deferring consensus (Burnett).

Goal sharing means that individuals cooperate to achieve goals. As any individual strives to achieve a goal, he or she must simultaneously try to help someone else achieve a goal. Thus, if two members want to use different visual aids, the person whose visual is used should see that the other person's point is included in the document. This key idea eliminates the divisiveness that occurs when people see each issue competitively, so that someone wins and someone loses.

Deferring consensus means that members agree to consider alternatives and voice explicit disagreements. Groups should deliberately defer arriving at a consensus in order to explore issues. This process initially takes more time, but in the end it gives each member "ownership" of the project and document.

According to one expert, group success is greatly helped by the "ability to plan and negotiate through difficulties; failures may be caused by a group's inability to resolve conflict and to reach consensus" (Debs 481).

Develop a Method to Resolve Differences

Resolving differences is an inevitable part of group activity. Your group should develop a reasonable, clear method for doing so. Usually the group votes, reaches a consensus, or accepts expert opinion. Voting is fast but potentially divisive. People who lose votes often lose interest in the project. Reaching a consensus is slow but affirmative. If you can thrash through your differences without alienating one another, you will maintain interest and energy in the project. Accepting expert opinion is often, but not always, an easy way to resolve differences. If one member who has closely studied citation methods says that the group should use a certain format, that decision is easy to accept. Unfortunately, another group member may disagree. In that case, your group will need to use one of the other methods to establish harmony.

Plan the Group's Activities

Your group must also develop guidelines to manage activities and to clarify assignments and deadlines.

To *manage activities*, the group must make a work plan that clarifies each person's assignments and deadlines. Members should use a calendar to set the final due date and to discuss reasonable time frames for each stage in the process. The group should put everything in writing and should schedule regular meetings. At the meetings, members will make many decisions; for example, they will create a style sheet for head and citation format. Write up these decisions and distribute them to all members. Make—and insist on—progress reports. Help one another with problems. Tell other group members how and when they can find you.

To *clarify assignments* and *deadlines,* answer the following questions:

What is the exact purpose of this document?
Must any sections be completed before others can be started?
What is each person's research and writing assignment?
What is the deadline for each section?
What is the style sheet for the document?

Each member should clearly understand the audience, the intended effect on the audience, any constraints, and the basic points that the document raises. This sense of overall purpose enables members to write individual sections without getting off on tangents (McTeague).

Worksheet for Group Planning

• *What considerations are necessary if this is a group planning situation?*
a. Who will be the leader?
b. How will we resolve differences?
c. What are the deadlines and assignments?
d. Who will keep track of decisions at meetings?
e. Who will write progress reports to the supervisor?

Choose a Strategy for Drafting and Revising

To make the document read "in one voice," one person often writes it, especially if it is short. A problem with this method is that the writer gets his or her ego involved and may feel "used" or "put upon," especially if another member suggests major revisions. The group must decide which method to use, considering the strengths, weaknesses, and personalities of the group members.

Once drafting begins, groups should review each other's work. Generally, early reviews focus on large matters of content and organization. Reviewers should determine whether all the planned sections are present, whether each section contains enough detail, and whether the sections achieve their purpose for the readers. At this stage, the group needs to resolve the kinds of differences that occur if, for instance, one person reports that the company landfilled 50,000 square yards of polystyrene and another says it was 5 tons. Later reviews typically focus on surface-level problems of spelling, grammar, and inconsistencies in using the style sheet.

Worksheet for Group Drafting and Revising

• *Ask and answer planning questions.*
a. What is the purpose of this document?
b. Who is the audience of this document?
c. What is the sequence of sections?
d. Must any sections be completed before others can be started?
e. What is each person's writing assignment?
f. Does everyone understand the style sheet?
g. What is the deadline for each section?

• *Select a method of drafting.*
a. Will everyone write a section?
b. Will one person write the entire draft?

Choose a Strategy for Editing

Groups can edit in several ways. They can edit as a group, or they can designate an editor. If they edit as a group, they can pass the sections around for comment, or they can meet to discuss the sections. However, this method is cumbersome. Groups often "overdiscuss" smaller editorial points and lose sight of larger issues. If the group designates one editor, that person can usually produce a consistent document and should bring it back to the group for review. The basic questions that the group must decide about editing include:

Who will suggest changes in drafts? one person? an editor? the group?
Will members meet as a group to edit?
Who will decide whether to accept changes?

In this phase, the conflict-resolving mechanism is critical. Accepting suggested changes is difficult for some people, especially if they are insecure about their writing.

(continued)

(continued)

Choose a Strategy for Producing the Document

The group must designate one member to oversee the final draft. Someone must collect the drafts, oversee the inputting, and produce a final draft. In addition, if the document is a long, formal report, someone must write the introduction and attend to such matters as preparing the table of contents, the bibliography, and the visual aids. These tasks take time and require close attention to detail.

Questions for the group to consider at this stage include:

Who will write the introduction?
Who will put together the table of contents?
Who will edit all the citations and the bibliography?
Who will prepare the final version of the visual aids?
Who will oversee production of the final document?

The group writing process challenges your skills as a writer and as a team member. Good planning helps you produce a successful report and have a pleasant experience. As you work with the group, it is important to remember that people's feelings are easily hurt when their writing is criticized. As one student said, "Get some tact."

Worksheet for Group Editing

- *Select a method of editing.*
  a. Will the members meet as a group to edit?
  b. Who will determine that all sections are present?
  c. Who will determine that the content is complete and accurate?
  d. Who will check for conflicting details?
  e. Who will check for consistent use of the style sheet?
  f. Who will check spelling, grammar, punctuation, etc.?
- *Oversee the production process.*
  a. Who will write the introduction?
  b. Who will put together the table of contents?
  c. Who will edit the citations and bibliography?
  d. Who will prepare the final versions of the visual aids?
  e. Who will direct production of the final document?

Note: Each project generates its own particular editing checklist. Formulate yours on the basis of the actual needs of the document.

Works Cited

Burnett, Rebecca E. "Substantive Conflict in a Cooperative Context: A Way to Improve the Collaborative Planning of a Workplace Document." *Technical Communication* 38.4 (1991): 532–539.

Debs, Mary Beth. "Recent Research on Collaborative Writing in Industry." *Technical Communication* 38.4 (1991): 476–484.

McTeague, Michael. "How to Write Effective Reports and Proposals." *Training and Development Journal* (Nov. 1988): 51–53.

Chapter 4

Technical Communication Style

Chapter 4
In a Nutshell

Develop strategies to use as you write and revise documents. Mastering just a few key strategies makes your ideas seem clearer to your audience and causes them to trust you.

The basic principles are

▸ Write in the active voice.

▸ Use parallelism.

▸ Write 12- to 25-word sentences.

▸ Use *there are* sparingly.

▸ Avoid "garden path" sentences.

▸ Change tone by changing word choice.

The first five principles are common writing advice. The last principle helps create an identity that is appropriate for the situation.

When you draft and edit, it is helpful if you have developed a repertoire of strategies to guide the choices you have to make. Having a sense of effective ways to present sentences, paragraphs, and tone helps you produce a document that your readers find clear and easy to grasp. The basic concept is to arrange material top-down: Put the most important idea at the beginning (the "top"), and follow with the explanatory details (the "down").

As you revise drafts, look for language that might cause confusion, making your writing harder to understand. The key is to learn to recognize these constructions as "flags," indicators that the spot needs to be evaluated. For instance, whenever you write a *there are* or whenever you repeat a word, decide whether to change the sentence. The following sections help you develop an awareness of many constructions that cause imprecise, difficult-to-read sentences. Although you may produce some of these constructions in your early drafts, learn to identify and change them. For more information, review Appendix A, which outlines basic problems in sentence structure.

Write Clear Sentences for Your Reader

Following these guidelines for composing sentences will make your writing clear:

- Place the sentence's main idea first.
- Use normal word order.
- Use the active voice.
- Employ parallelism.
- Write sentences of 12 to 25 words.
- Avoid "garden path" systems.
- Use *there are* sparingly.
- Avoid nominalizations.
- Avoid strings of choppy sentences.
- Avoid wordiness.
- Avoid redundant phrases.
- Avoid noun clusters.
- Use *you* correctly.
- Avoid sexist language.

Place the Main Idea First

To put the main idea first (at the "top") is a key principle for writing sentences that are easy to understand. Place the sentence's main idea—its subject—first. The subject makes the rest of the sentence accessible. Readers relate subjects to their own ideas (their schema) and thus orient themselves. After readers know the topic, they are able to interact with the complexities you develop.

Note the difference between the following two sentences. In the first, the main idea, "two types of professional writing," comes near the end. The sentence is difficult to understand. In the second, the main idea is stated first, making the rest of the sentence easier to understand.

Main idea is last — The writing of manufacturing processes, which explain the sequence of a part's production, and design specifications, which detail the materials needed to produce an object, are two types of professional writing I will do.

Main idea is first — Two types of professional writing that I will do are writing manufacturing processes, which explain the sequence of a part's production, and design specifications, which detail the materials needed to produce an object.

Use Normal Word Order

The normal word order in English is subject-verb-object. This order makes reading easier because it reveals the topic first and then develops the idea. It also usually produces the clearest, most concise sentences.

Normal — The ASTM definition describes the process by which polymers break down.

Inverted — Polymers break down in a process described by the ASTM definition.

Use the Active Voice

The active voice emphasizes the performer of the action rather than the receiver. The active voice helps readers grasp ideas easily because it adheres to the subject-verb-object pattern and puts the performer of the action first. When the subject acts, the verb is in the active voice ("I wrote the memo"). When the subject is acted upon, the verb is in the passive voice ("The memo was written by me").

Change Passive to Active

To change a verb from the passive to the active voice, follow these guidelines:

▶ Move the person acting out of a prepositional phrase.

Passive — The test was conducted *by the intern.*

Active — *The intern* conducted the test.

▶ Supply a subject (a person or an agent).

Passive — This method was ruled out.

Active — *The staff* ruled out this method.

Active — *I* ruled out this method.

▶ Substitute an active verb for a passive one.

Passive — The heated water *is sent* into the chamber.

Active — The heated water *flows* into the chamber.

Use the Passive If It Is Accurate

The passive voice is sometimes more accurate; for instance, it is properly used to show that a situation is typical or usual or to avoid an accusation.

Typical situation needs no agent

Robots are used in repetitive activities.

Active verb requires an unnecessary agent (companies)

Companies use robots in repetitive activities.

Active accuses

You violated the ethics code by doing that.

Passive avoids accusing

The ethics code was violated by that act.

The passive voice can also be used to emphasize a certain word.

Use passive to emphasize *milk samples*

Milk samples are preserved by the additive.

Use active to emphasize *additive*

The additive preserves the milk samples.

Employ Parallelism

Using parallelism means using similar structure for similar elements. Careful writers use parallel structure for *coordinate elements,* elements with equal value in a sentence. Coordinate elements are connected by coordinating conjunctions (*and, but, or, nor, for, yet, so*) or are words, phrases, or clauses that appear in a series. In the following sentence, the italicized words make up a series.

Technical writers create *memos, proposals,* and *manuals.*

If coordinate elements in a sentence are not treated in the same way, the sentence is awkward and confusing.

Faulty

My duties included *coming in early in the morning and doing preparation work, cooking on the front line, trained new employees,* and *took inventory.*

Parallel

My duties included *coming in early to do preparation, cooking on the front line, training new employees,* and *taking inventory.*

Faulty

Typical writing situations include *proposals, the sending of electronic mail,* and *how to update the system.*

Parallel

Typical writing situations include *editing proposals, sending electronic mail,* and *updating the system.*

Write Sentences of 12 to 25 Words

An easy-to-read sentence is 12 to 25 words long. Shorter and longer sentences are weaker because they become too simple or too complicated. However, this is only a generality. Longer sentences, especially those exhibiting parallel construction, can be easy to grasp. The first of the following sentences is harder to understand, not just because it is long, but also because it ignores the dictum of putting the main idea first. The revision is easier to read because the sentences are shorter and the main idea is introduced immediately.

One sentence, 40 words long	The problem is the efficiency policy, which has measures that emphasize producing as many parts as possible, for instance, 450 per hour, compared to a predetermined standard, usually measured by the machine's capacity, say, 500, for a rating of 90%.
Two sentences, 19 and 21 words long	The problem is the efficiency policy, which calls for as many parts as possible compared to a predetermined standard. If a machine produces 450 per hour and if its capacity is 500 per hour, it has a rating of 90%.

Avoid "Garden Path" Sentences

People read by making guesses about what will come next in a sentence. They look for cues that allow them to analyze sentence structure or identify parts of speech. If, however, the writer muddles the cue, the sentence leads the reader down the "garden path," to a place that is unexpected (Kohl; Schriver).

Garden path	If the gauge fails the process will end.
Proper cue (a comma) added	If the gauge fails, the process will end.
Garden path	The material machines are made from is purchased locally.
Proper cue (*that*) added	The material that machines are made from is purchased locally.
Garden path	In this system there's complete sharing of information between individuals with more detailed information preferable to the less detailed information.
Proper cue (comma) added	In this system, there's complete sharing of information between individuals with more detailed information preferable to less detailed information.
Corrected	In this system, individuals share all information; they prefer more detailed information to less detailed information.
Garden path	The task of reconciling a checkbook with a statement from the bank has two steps of checking subtraction and cross checking entries.
Corrected by complete revision	Reconciling a checkbook and bank statement has two steps: cross checking entries and checking subtraction.

Use *There Are* Sparingly

Overuse of the indefinite phrase *there are* and its many related forms (*there is, there will be,* and so on) weakens meaning by "burying" the subject in the middle of the sentence. Most sentences are more effective if the subject is placed first.

Ineffective   *There is* a change in efficiency policy that could increase our profits.

Effective   Our profits will increase if we change our efficiency policy.

Use *there are* for emphasis or to avoid the verb *exist*.

Weak   Three standard methods exist.

Stronger   There are three standard methods.

Avoid Nominalizations

Avoid using too many *nominalizations,* verbs turned into nouns by adding a suffix such as *-ion, -ity, -ment,* or *-ness.* Nominalizations weaken sentences by presenting the action as a static noun rather than as an active verb. These sentences often eliminate a sense of agent, thus making the idea harder for a reader to grasp. Express the true action in your sentences with strong verbs. Almost all computer style checkers flag nominalizations.

Static   The training policy for most personnel will have the requirement of the completion of an initial one-week seminar.

Active   The training policy will require most personnel to complete a one-week seminar.

Static   There will be costs for the installation of this machine in the vicinity of $10,000.

Active   We can install this machine for about $10,000.

             The machine will cost $10,000 to install.

Exercises

❯ Passive Sentences

1. Make the following passive sentences active.
    - When all work is completed, turn the blueprint machine off.
    - A link, such as products or specifications, was checked by me.
    - The OK button was clicked and the correct link was scrolled down to.
    - After these duplicate sites were eliminated there were still too many sites to be looked at.

- Profits can be optimized by the manufacturer with the use of improved materials and the result can be better product value for the customer.
- Revise the passive voice sentences in these paragraphs.

   Numerous problems with flexographic printing have to be considered. On very short runs of 1000 or less the cost can be excessive. In addition dot gain is another issue that can be a problem. Small print and reverse print can be other problems with flexography.

   The page is entitled features. All of the features of the Netmeeting software are listed down the center of this page. Under each feature a brief discussion of the feature is given so that the user is told what the feature is for. The uses of each feature in detail can be found by clicking on the icon.

▶ Parallel Structure

2. Revise the following sentences to make their coordinate elements parallel.

- It serves the purpose of pulling the sheet off the coil and to straighten or guide the sheet through the rest of the machine.
- The main uses of the data are to supply estimates of the workforce and estimating unemployment.
- There were clear headings to each section, some links for navigating around the site, and it provided a glossary of terms.
- During the internship I was responsible for merchandising, creating floor plans, and to train new employees.
- Along with a case, the laptop package includes 18 feet of top-quality UGA cable, a surge protector with circuit checker, and there is also an extra battery.

3. Write a sentence in which you give three reasons why a particular Web search engine is your favorite. Alternate: Your instructor will provide topics other than Web search engines.

4. Create a paragraph in which your sentence from Exercise 6 is the topic sentence. Write each body sentence using active voice and parallelism.

▶ "Garden Path" Sentences

5. Correct the following "garden path" sentences.

- When the cycle ends the life of the battery is nearly over.
- When I called the manager was not in.
- The writing papers are composed of should be clear.
- One bad thing about this is the cost of it is going to be considerably higher.
- The audience for the brochure will be people who want information on sexual assault and services and support will provide all the information.

- If you don't want one of those other colors are available.
- I have been working with people who have developmental disabilities with more than three and a half years of experience to offer you.
- Place your mouse cursor over the cell in your table that you would like to adjust the properties on.

▶ Use of *There Are*

6. Eliminate *there are,* or any related form, from the following sentences.

- There are several methods of research that I will use.
- There are slowness and inefficiency in the data processing and information retrieval.
- There is a need for some XML code to be learned by me.
- There is a reliance on business to provide product information on the Web.
- There is the necessity to modify packaging methods so that there can be fewer contaminants released into the environment.

▶ Nominalization

7. Correct the nominalizations in the following sentences.

- Insertion of the image into the document occurs when you click OK.
- Specification of the file as a Word document was necessary to make the conversion of the file from text to Word.
- The manipulation of the layout in order to cause the transformation of it into film is the process of color separation.
- The use of my two keywords resulted in the best results for me.
- The conversion of the screen captures to JPEGs was easy to perform.

8. Write a paragraph about a concept you know well. Use as many nominalizations, *there are*'s, and passive voice combinations as you can. Then rewrite it eliminating all those constructions.

· ·

Avoid Strings of Choppy Sentences

A string of short sentences results in choppiness. Because each idea appears as an independent sentence, the effect of such a string is to deemphasize the more important ideas because they are all treated equally. To avoid this, combine and subordinate ideas so that only the important ones are expressed as main clauses.

Choppy

Both models offer safety belts. Both models have counterbalancing. Each one has a horn. Each one has lights. One offers wing-sided seats. These seats enhance safety.

Clear   Both models offer safety belts, counterbalancing, a horn, and lights. Only one offers wing-sided seats, which enhance safety.

Avoid Wordiness

Generally, ideas are most effective when they are expressed concisely. Try to prune excess wording by eliminating redundancy and all unnecessary intensifiers (such as *very*), repetition, subordinate clauses, and prepositional phrases. Although readers react positively to the repetition of keywords in topic positions, they often react negatively to needless repetition.

Unnecessary subordinate clause   I found the site by the use of keywords that are nanotechnology and innovation

Revised   I found the site using *the keywords* "nanotechnology" and "innovation."

Redundant intensifiers plus unnecessary subordinate clause   It is made of *very* thin glass *that is milky white in color*.

Revised   It is made of thin, milky white glass.

Redundant   The tuning handle is a metal protrusion that can be easily grasped *hold of by the hand* to turn the gears.

Revised   The tuning handle is a metal protrusion that can be easily grasped to turn the gears.

Unnecessary repetition plus overuse of prepositions   *This search* was done by a *keyword search* of the *same words* using the *search function* of different *search engines*.

Revised   This investigation used the same keywords in different search engines.

Avoid Redundant Phrases

Here is a list of some common redundancies. A better way to express the idea follows each.

Redundant Phrase	More Concise Word or Phrase
due to the fact that	because
employed the use of	used
basic fundamentals	fundamentals
completely eliminate	eliminate
alternative choices	alternatives
actual experience	experience
connected together	connected
final result	result
prove conclusively	prove
in as few words as possible	concisely

Avoid Noun Clusters

Noun clusters are three or more nouns joined in a phrase. They crop up everywhere in technical writing and usually make reading difficult. Try to break them up.

Noun cluster

Allowing *individual input variance* of *data process entry* will result in higher morale in the keyboarders.

Revised

We will have higher morale if we allow the keyboarders to enter data at their own rate.

Use *You* Correctly

Do not use *you* in formal reports (although writers often use *you* in informal reports). Use *you* to mean "the reader"; it should not mean "I," or a very informal substitute for "the" or "a" (e.g., "This is your basic hammer.").

Incorrect as "I"

I knew when I took the training course that *you* must experience the problems firsthand.

Correct

I knew when I took the training course that I needed to experience the problems firsthand.

Avoid Sexist Language

Language is considered sexist when the word choice suggests only one sex even though both are intended. Careful writers rewrite sentences to avoid usages that are insensitive and, in most cases, inaccurate. Several strategies can help you write smooth, nonsexist sentences. Avoid such clumsy phrases as *he/she* and *s/he*. Although an occasional *he or she* is acceptable, too many of them make a passage hard to read.

Sexist

The clerk must make sure that *he* punches in.

Use an infinitive

The clerk must make sure *to punch* in.

Use the plural

The clerks must make sure that *they* punch in.

Use the plural to refer to "plural sense" singulars

Everyone will bring their special dish to the company potluck.

In the last example, *their,* which is plural, refers to *everyone,* which is singular but has a plural sense.

Exercises

▶ Choppiness

1. Eliminate the choppiness in the following sentences.

   • XYZ has introduced an LCD monitor. The monitor is 17 inches. The monitor has a Web camera. Web conferencing applications can use the Web camera.

- Numerical control exists in two forms. CNC is one form. CNC is Computer Navigated Control. DNC is Distributed Numerical Control.
- On-line registration is frustrating. It should make it faster to register. The words on the screen are not self-explanatory. Screen notices say things like "illegal command." Retracing a path to a screen is difficult.

▶ Wordiness

2. Revise the following sentences, removing unnecessary words.
   - Due to the fact that we have two computer platforms connected together, we must pay attention to basic fundamentals when we send a document such as an e-mail attachment.
   - This project will be presented in Web format with links to the resource sites as well as other links that are associated with the sites and links that are associated with the topic of researching a report.
   - In fact many sites are available on the Web where the viewer can have the actual experience of purchasing equipment, new and used, from the site.
   - In the printing business there are two main ways of printing. The first is by using offset and the second is by using flexography.

▶ Use of *You*

3. Correct the use of *you* in the following sentences.
   - To judge the site's accuracy I read the tutorial and evaluated the lessons. This is where you look for your clear technical explanations and your computer code examples.
   - This evaluation is different than the change from your old remote control to the new laser ones.
   - Two methods exist to enhance fiber performance. In the first you orient the fibers. The second is the alkali process.

▶ Sexist Language

4. Correct the sexist language in the following sentences.
   - Each presenter must bring his own laptop.
   - Every secretary will hand in her timecard on Friday.
   - If he understands the process, the machinist can improve production.
   - Details were very good up to a point, but then the author seemed to lose her focus.
   - I am not sure who Dr. Jones is, but I am sure he is a good doctor.

Write Clear Paragraphs for Your Reader

A paragraph consists of a topic sentence followed by several sentences of explanation. The *topic sentence* expresses the paragraph's central idea, and the remaining sentences develop, explain, and support the central idea. This top-down arrangement enables readers to grasp the ideas in paragraphs more quickly (Slater).

Put the Topic Sentence First

Putting the topic sentence first, at the top of the paragraph, gives your paragraphs the direct, straightforward style most report readers prefer (Slater). Consider this example:

Topic sentence

Supporting details

The second remarkable property of muskeg is that, like a great sponge, it absorbs and accumulates water. Water enters a muskeg forest through precipitation (rain and snow) and through the ground (rivers, streams, seeps). It leaves by evaporation (chiefly of vapor transpired by the plants) and by outflow through or over the ground. However, input and output are not in balance. Water accumulates and is held absorbed in the accumulating peat. One of the commonest plants of muskeg is sphagnum moss, otherwise known as peat moss; alive or dead, its water-holding capacity is renowned, and is what gives peat its great water-retaining power. (Pielou 97)

Structure Paragraphs Coherently

In a paragraph that exhibits coherent structure, each sentence amplifies the point of the topic sentence. You can indicate coherence by repeating terms, by placing key terms in the dominant position, by indicating class or membership, and by using transitions (Mulcahy). You can also arrange sentences by level.

Repeat Terms

Repeat terms to emphasize them. In the following example, notice that the new information, *collided* (sentence 1), becomes old information, *collision* (sentence 2), and that *mountain range* is new information in sentence 3 but old information in sentence 4.

Subduction stopped when the continent collided with the island arc along its northern margin. This collision resulted in extensive deformation of the island arc as well as deformation of the sedimentary rocks on the continental margin described earlier. The collision produced a mountain range across northern Wisconsin. This ancient mountain range is called the Penokean Mountains. The eroded remnants of these mountains constitute much of the bedrock of Wisconsin, Minnesota, and Michigan. (LaBerge 111)

Use the Dominant Position

Placing terms in the dominant position means to repeat a key term as the subject, or main idea, of a sentence. As a result, readers return to the same topic and find it developed in another way. In the following short paragraph, *hyperfocal distance* is the dominant term.

> The depth-of-field scale is used for presetting the lens to the hyperfocal distance for any given f-stop. The hyperfocal distance is the point of focus where everything from half that distance to infinity falls within the depth of field. To find this distance for any f-stop, just prefocus your lens by placing the infinity mark on the focusing collar over the f-stop mark for the aperture you're using. (Shaw 31)

Maintain Class or Membership Relationships

To indicate class or membership relationships, use words that show that the subsequent sentences are subparts of the topic sentence. In the following sentences, *distributed media* and *online systems* are members of the class *paradigm*.

> Interactive multimedia follows one of two paradigms. Distributed media, such as CDs, are self-contained and circulated to audiences in the same manner as books or audio recordings. Online systems, such as intranets and the web, resemble broadcasting in that the content originates from one central location and the use accesses it from a distance. (Bonime and Pohlman 177)

Provide Transitions

Using transitions means connecting sentences by using words that signal a sequence or a pattern. Common examples include:

Sequence	first . . . second . . . then . . . next . . .
Addition	and also furthermore
Contrast	but however
Cause and effect	thus so therefore hence

Arrange Sentences by Level

You can also develop coherence by the way you place sentences in a paragraph. In almost all technical paragraphs, each sentence has a level. The first level is that of the topic sentence. The second level consists of sentences that support or explain the topic sentence. The third level consists of sentences that develop

Globalization and Style

When writing for an audience whose first language is not English, you need to be as clear and concise as possible. You do not need to "dumb down" your style, but in order for your document to be effective, you will need to think about the way in which you write (Dehaas; Dixon; Thrush).

- Use the active voice rather than the passive voice. Use "This machine has three functions" rather than "There are three functions to this machine" or "The manager accepted the proposal" instead of "The proposal was agreed to by the manager."
- Focus on making the subject and the action clear—you can accomplish this by keeping your sentences in natural order (subject first, verb in the middle, and predicate last) and by putting only one idea in each sentence.
- Try not to use contractions, especially in negative statements; say "Do not use" rather than "Don't use"; the use of the separate word *not* gives the statement more emphasis.
- Stay away from examples that require readers to understand metaphors outside of their experience. For instance, the sentence "That approach hit a home run in the States" depends for its meaning on the ability to translate "home run" into "very successful."
- Use consistent vocabulary. Once you use a word for an object, continue to use the same word. If you use "screen," don't suddenly switch to a synonym, for example, to "page" or "webpage." English often has many words for the same thing and many writers like to vary to avoid monotony, but variation often confuses a non-English reader.

Create a style guide when you write for localization. This type of document helps you regularize your language so that you treat items consistently. One expert (Myer) suggests that you create a guideline for how you will handle all of the following:

Rules for capitalization
Rules for translating titles and subtitles
Rules for handling conversions of measurements, dates, currencies, addresses, and phone numbers
Rules for abbreviations and punctuation

In order to create these rules, you will have to investigate the usages common in the culture for which you are writing.

One interesting aspect of style is the font you choose. Many languages have diacritical marks not used in English—for instance Ä, ê, ü, å. While all the common fonts (e.g., Times, Helvetica) can use these symbols, many of the more unusual fonts cannot. Choose a font that will accommodate these symbols, because the inability to reproduce them amounts to a spelling error, and perhaps even an offense.

For further reference, check out these websites:

IBM's Globalization site at <www-306.ibm.com/software/globalization/topics/writing/index.jsp> has articles on writing, communicating, and presenting for an international community. You'll find tips on writing style as well as articles on cultural sensitivity issues.

Erin Heximer and Lisa Wu's article, "Going Global: The Challenges of Writing for International Audiences," at the Society for Technical Communication's website <http://216.239.39.104/search?q=cache:KSojMhBYRZUJ:www.stc.org/confproceed/2001/PDFs/STC48-000110.PDF> goes into great detail on writing style and international audiences. They include many tips and several examples, as well as an extensive references list for further research.

Works Cited

Dehaas, David. "Say What You Mean." *OHS Canada.* 31 Jan. 2004 <www.ohscanada.com/Training/saywhatyoumean.asp>.

Dixon, Duncan. "Clear Writing for International English Readers." *Encyclopedia of Educational Technology.* 1 Feb 2004 <http://coe.sdsu.edu/eet/Articles/clearwriting/index.htm>.

Myer, Thomas. "Localization Gotchas for Documentation." Triple Dog Dare Media. 31 Jan. 2004 <www.tripledogdaremedia.com/content/l10ngotchas.pdf.> or <http://216.329.37.104/search?q=cache:aK0EqN9A_7kj:www.tripledogdaremedia.com/content/l10ngotchas.pdf+localization+and+%22word+choice%22&hl=en&ie=UTF-8>.

Thrush, Emily. "Writing for an International Audience, Part II." Suite 101.com. 7 Feb. 2004 <www.suite101.com/article.cfm/5381/32971>.

one of the second-level ideas. Four sentences, then, could have several different relationships. For instance, the last three could all expand the idea in the first:

1 First level
   2 Second level
   3 Second level
   4 Second level

Or sentences 3 and 4 could expand on sentence 2, which in turn expands on sentence 1:

1 First level
   2 Second level
     3 Third level
     4 Third level

Putting the topic idea in the first sentence makes it possible for readers to get the gist of your document by skimming over the first sentences.

As you write, evaluate the level of each sentence. Decide whether the idea in the sentence is level 2, a subdivision of the topic, or level 3, which provides details about a subdivision. Consider this example:

> (1) Hydraulic pumps are classified as either nonpositive or positive displacement units. (2) Nonpositive displacement pumps produce a continuous flow. (3) Because of this design, there is no positive internal seal against leakage, and their output varies as pressure varies. (4) Positive displacement pumps produce a pulsating flow. (5) Their design provides a positive internal seal against leakage. (6) Their output is virtually unaffected as system pressure varies.

The sentences of this paragraph have the following structure:

> (1) Hydraulic pumps are classified as either nonpositive or positive displacement units.
> > (2) Nonpositive displacement pumps produce a continuous flow.
> > > (3) Because of this design, there is no positive internal seal against leakage, and their output varies.
> > (2) Positive displacement pumps produce a pulsating flow.
> > > (3) Their design provides a positive internal seal against leakage.
> > > (3) Their output is virtually unaffected as the system pressure varies.

Choose a Tone for the Reader

The strategies discussed thus far in this chapter produce clear, effective documents. Using them makes your documents easy to read. These strategies, however, assume that the reader and the writer are unemotional cogs in an information-dispensing system. It is as though reader and writer were computers and the document were the modem. One computer emotionlessly activates the modem and sends out bits of information. The other, activated, receives and stores that information.

In fact, situations are not so predictable. The tone, or emotional attitude implied by the word choice, can communicate almost as much as the content of a message (Fielden). To communicate effectively, you must learn to control tone. Let's consider four possible tones:

- Forceful
- Passive
- Personal
- Impersonal

The *forceful* tone implies that the writer is in control of the situation or that the situation is positive. It is appropriate when the writer addresses subordinates or when the writer's goal is to express confidence. To write forcefully,

Ethics and Style

Clarity is the gold standard for all communication. Jargon, shop talk, or techno-babble that marginalizes or excludes the reader or audience is not only confusing, it is unethical. It is both reasonable and desirable to create prose, be it technical or otherwise, that is written for its intended audience. Unfortunately, sometimes it is all too easy to slide into a vernacular that is common among those in-house. To use terms that are unique to a particular discursive community can create a boundary between the document and its intended audience. If you must use jargonistic terms, include a glossary, or define the term the first time you use it in the text. If your language can be misconstrued, it can cause problems. A good general rule is to guard against any use of terms that are common in-house when the audience for the document is "out of house." This is not only good practice, it is the ethical thing to do.

> Use the active voice.
> Use the subject-verb-object structure.
> Do not use "weasel words" (*possibly, maybe, perhaps*).
> Use imperatives.
> Clearly indicate that you are the responsible agent.

I have decided to implement your suggestion that we supply all office workers with laptops and eliminate their towers. This suggestion is excellent. You have clearly made the case that this change will reduce eyestrain issues and will greatly enhance the flow of information in the department. Make an appointment with me so we can start to implement this fine idea.

The *passive* tone implies that the reader has more power than the writer or that the situation is negative. It is appropriate when the writer addresses a superior or when the writer's goal is to neutralize a potentially negative reaction. To make the tone passive,

> Avoid imperatives.
> Use the passive voice.
> Use "weasel words."
> Use longer sentences.
> Do not explicitly take responsibility.

The proposal to implement laptops in our department has not been accepted. The ergonomic benefits of the screens are not seen as offsetting the potential disruption that will be caused by the migration of files to the new machines. The large footprint of the docking station has also been suggested as a possible problem for our employees due to their already restricted desk space. Because the need for action on computer replacement is necessary, a meeting will be scheduled next week to discuss this.

Compare this to a forceful presentation:

> The steering committee and I reject the laptop proposal. You have not included enough convincing data on morale or work flow and you have not dealt with work flow disruption and the large size of the docking station. Make an appointment to see me if necessary.

The *personal* tone implies that reader and writer are equal. It is appropriate to use when you want to express respect for the reader. To make a style personal,

- Use the active voice.
- Use first names.
- Use personal pronouns.
- Use short sentences.
- Use contractions.
- Direct questions at the reader.

> Ted, thanks for that laptop suggestion. The steering committee loved it. Like you, we feel it will solve the eyestrain issue and will facilitate data flow. And we think it will also raise morale. I'd like you to begin work on this soon. Can you make an appointment to see me this week?

This tone is also appropriate for delivering a negative message when both parties are equal.

> Ted, thanks for the laptop suggestion, but we can't do it this cycle. The steering committee understands the ergonomic issue you raise, but they are very concerned about the disruption that migrating all those files will cause. In addition, they feel that we need to work out the entire issue of footprint—the model you suggested would cause a number of problems with current desk configurations. I know that this is a disappointment. Could we get together soon to discuss this?

The *impersonal* tone implies that the writer is not important or that the situation is neutral. Use this tone when you want to downplay personalities in the situation. To make the tone impersonal:

- Do not use names, especially first names.
- Do not use personal pronouns.
- Use the passive voice.
- Use longer sentences.

> A decision to provide each employee with a laptop has been made. Laptops will reduce the eye fatigue that some employees have experienced and the laptops will increase data flow. Ted Baxter will chair the implementation committee. Donna Silver and Robert Sirabian will assist. The committee will hold its initial meeting on Monday, October 10, at 3:00 P.M. in Room 111.

Worksheet for Style

☐ Find sentences that contain passive voice. Change passive to active.

☐ Look for sentences shorter than 12 or longer than 25 words. Either combine them or break them up.

☐ Check each sentence for coordinate elements. If they are not parallel, make them so.

☐ Find "garden path" sentences and revise them.

☐ Read carefully for instances of the following potential problems:
- Nominalizations
- Sexist language
- Too frequent use of *there are*
- Choppiness
- Incorrect use of *you*
- Wordiness

☐ *Sentences.* Look for four types of phrasing. Change the phrasing as suggested here or as determined by the needs of your audience and the situation.
- The word *this.* Usually you can eliminate it (and slightly change the sentence that is left), or else you should add a noun directly behind it. ("By increasing the revenue, this will cause more profit" becomes "Increasing the revenue will cause more profit.")
- The words *am, is, are, was, were, be,* and *been.* If these are followed by a past tense (*was written*) the sentence is passive. Try to change the verb to an active sense (*wrote*).
- Lists of things or series of activities. Put all such items, whether of nouns, adjectives, or verb forms, in the same grammatical form (*to purchase, to assemble, and to erect*—not *to purchase, assembling, and to erect*). This strategy will do more to clarify your writing than following any other style tip.
- The phrases *there are* and *there is.* You can almost always eliminate these phrases and a *that* which appears later in the sentence. ("There are four benefits that you will find" becomes "You will find four benefits.")

☐ Use the top-down principle as your basic strategy.

☐ Make sure that each paragraph has a clear topic sentence.

☐ Check paragraph coherence by reviewing for
a. Repetition of terms.
b. Placement of key terms in a dominant position.
c. Class or membership relationships.
d. Transitions.

☐ Evaluate the sentence levels of each paragraph. Revise sentences that do not clearly fit into a level.

Exercises

▶ You Analyze

1. Analyze Example 4.1. Either in groups or individually, revise the document using the concepts outlined in this chapter. You may revise both wording and tone.

Example 4.1

Methods Statement

> In this section of the report, I will discuss *www.flipdog.com*. I used the job search titles of Retail Analyst, Construction Project Manager, and Packaging Engineer. For each of these job search titles, I researched the entire United States, Minnesota, and Wisconsin.
>
> When I first got to the website, I clicked on "Find Jobs." My first step was to then choose the area I wanted to search. I decided to start with the entire United States so I just left "Search: All of U.S." as the default search. With this option, there was a maximum of 264,941 jobs searched. I then had the choice to choose a category of search. You may search all job categories, for a category individually, or by "Ctrl+click" to search multiple categories.
>
> I also had the choice of choosing the specific company of employment. Like the category search above, you have the choice to search all employers, multiple employers, or a single employer.
>
> For this research project, I didn't use the job category search or the employer search. I skipped right to the keyword search and typed in "retail analyst."
>
> I then clicked "Get Results." Within all of the United States, "retail analyst" received 179 results. After these results were listed, you are also given the option to search for specifics within results by entering keywords into this figure below. (For my research, I used total results only.)
>
> You may also modify the dates. (For my research, I used all dates by default.)
>
> I continued my research using the steps above. The tables below summarize my results. Each table demonstrates the area searched (all of the United States, Wisconsin, and Minnesota), as well as the three job titles **Retail Analyst, Construction Project Engineer, and Packaging Engineer.** *Note:* For results over 200, FlipDog reported as 200.

2. Review the following paper for tone. By indicating specific phrases and words, determine whether the author has adopted the correct tone. If your instructor requires it, rewrite the report with a different tone.

HOW TO CHANGE THE BACKGROUND COLOR IN PHOTOSHOP

INTRODUCTION

PhotoShop was always something that scared me a little when mentioned, but after exploring it for a couple of hours I got the hang of it. I am working with PhotoShop with other people in my class and they are doing examples of other projects to do in PhotoShop also. (I have links to their sites below.) My topic is how to change the background color and it was very simple to do.

METHOD

The first thing that I explored with ended up to be the right thing so I got a little lucky when doing this, but a little knowledge of Auto Cad 14 helped too. I clicked on the magic wand, which looks just like it sounds, and then clicked onto the color in the background that I wanted to change. This produces a blinking outline around everything that touches that color in the picture.

I then knew that I needed to be able to select a new color so I clicked on "Window" and scrolled down to "Show Color." This brought up a new smaller window that had a palette full of colors. It asked me to select a color and I chose red. After clicking "Okay" absolutely nothing happened and this had me stumped for a while. Here is where my Auto Cad experience helped me out. I went to "Edit" and scrolled down to "Fill" which basically regenerates your picture the same as in Auto Cad. My background is now red.

RESULT

As you can see from the above photos [not shown here] I was able to change the background color to red. In the corners the gray background is still there because that is a slightly different shade and all I would have to do is go through the process again and click on the color and make sure I had the same shade of red. Another thing to notice is that when you look through the bottles near the top you can see a white background yet, which I could also click on and give a pinkish color.

DISCUSSION

This option in PhotoShop can allow you to do virtually whatever you can come up with in your head. I believe that I am going to take some photos of vacations where the sky is really dark and make it a little brighter day. (You won't remember the difference in thirty years anyway.) This feature could also help if you need to eliminate something in the background completely like an indecent sign or person. If you have any needs similar to this then this should be something that you look into.

3. Analyze Example 9.3 on p. 225 to determine the sentence strategies and tone used for the intended audience.

4. Compare strategies in Examples 9.3 and 4.1. Write a brief memo in which you give examples that illustrate how sentence tone creates a definition of an intended audience. Alternative: In groups of three or four, compare Examples 9.3

and 4.1. Rewrite one of them for a different audience. Read your new memo to the class, who will identify the audience and strategies you used.

▶ You Revise

5. Rewrite two paragraphs of Examples 9.1 (p. 223) or 10.4 (pp. 249–250) in order to relate to a different audience. Keep the content the same, but change the sentence strategies and the tone.

6. Rewrite part or all of Example 9.3 to apply to a different audience.

7. Revise the following paragraphs so that the sentences focus on the Simulation Modeling Engineer as the actor in the process of model building.

> The process of abstraction of the system into mathematical-logical relationships with the problem formulation is the Model Building Phase. The assumptions and decision variables are used to mathematically determine the system responses. Also, the desired performance measures and design alternatives are evaluated. The system being modeled is broken down into events. For each event, relevant activities are identified. The basic model of the system is now an abstraction of the "real" system.
>
> The Data Acquisition Phase involves the identification and collection of data. This phase is often the most time consuming and critical. If the data collected are not valid, the simulation will produce results that are not valid. The data to be acquired are determined by the decision variables and assumptions. All organizations collect large quantities of data for day-to-day management and for accounting purposes. To collect data for simulation, the SMEs need cooperation of management in order to gain knowledge and access to the information sources. When the existing data sources are inadequate, a special data-collection exercise is required or the data are estimated.

▶ You Create

8. Write a memo in which you reject an employee's solution to a problem. Give several reasons, including at least one key item that the employee overlooked.

▶ Group

9. In groups of three or four, review the memos you wrote in Exercise 8 for appropriate (or inappropriate) tone. Select the most (or least) effective memo and explain to the class why you chose it.

Writing Assignments

1. You are a respected expert in your field. Your friend, an editor of a popular (not scholarly) magazine, has asked you to write an article describing basic terms

employed in a newly developing area in your field. Write a two- or three-page article, using Examples 9.1 (p. 223), 9.2 (p. 224), or 10.4 (pp. 249–250) as a guide.

2. Write a learning report for the writing assignment you just completed. See Chapter 5, Assignment 7, pages 134–135, for details of the assignment.

Web Exercise

Investigate any website to analyze the style and organizational devices used in the site. Use the principles discussed in this chapter: Are the sentences written differently from what they might be if they were in a print document? What devices are used to make organization obvious? Write an analytical report (see Chapter 12) to alert your classmates or coworkers to changes they should make as they develop their websites.

Works Cited

Bonime, Andrew, and Ken C. Pohlman. *Writing for New Media: The Essential Guide to Writing for Interactive Media, CD-ROMs, and the Web.* New York: Wiley, 1998.

Fielden, John S. "What Do You Mean You Don't Like My Style?" *Harvard Business Review* 60 (1982): 128–138.

Kohl, John R. "Improving Translatability and Readability with Syntactic Cues." *Technical Communication* 46.2 (Second Quarter 1999): 149–166.

LaBerge, Gene L. *Geology of the Lake Superior Region.* Tucson, AZ: Geoscience Press, 1994.

Mulcahy, Patricia. "Writing Reader-Based Instructions: Strategies to Build Coherence." *Technical Writing Teacher* 15.3 (1988): 234–243.

Pielou, E. C. *After the Ice Age: The Return of Life to Glaciated North America.* Chicago: University of Chicago Press, 1991.

Schriver, Karen. *Dynamics in Document Design: Creating Texts for Readers,* New York: Wiley, 1997.

Shaw, John. *John Shaw's Landscape Photography.* New York: Watson-Guptill, 1994.

Slater, Wayne H. "Current Theory and Research on What Constitutes Readable Expository Text." *Technical Writing Teacher* 15.3 (1988): 195–206.

Focus on
Bias in Language

Current theory has made clear just how much language and language labels affect our feelings. Biased language always turns into biased attitudes and actions that perpetuate demeaning attitudes and assumptions. It is not hard to write in an unbiased way if you apply a few basic rules. The American Psychological Association (APA) publication manual suggests that the most basic rule focuses on exclusion. A sentence that makes someone feel excluded from a group needs to be revised. It's rather like hearing yourself discussed while you are in the room. That feeling is often uncomfortable, and you should not write sentences that give that feeling to others. The APA manual (62–76, based heavily on Maggio) offers several guidelines.

Describe People at the Appropriate Level of Specificity

This guideline helps whenever you have to describe people. Technical writing has always encouraged precise description of technical objects. You should apply the same principle to the people who use and are affected by those objects. So, when referring to a group of humans of both sexes, say "men and women," not just "men."

Be Sensitive to Labels

Call people what they want to be called. However, be aware that these preferences change over time. In the 1960s, one segment of the American population preferred to be called "black"; in the 1990s, that preference changed to "African American."

Basically, do not write about people as if they were objects—"the complainers," "the strikers." Try instead to put the person first—"people who complain," "people who are striking." Because this can get cumbersome, you can begin by using a precise description and after that use a shortened form

as long as it is not offensive. The issue of what is offensive is a difficult question. How do you know that "elderly" is offensive but "older" is not? There is no easy answer. Ask members of that group. Listen to the words that national TV news applies to members of the group.

Acknowledge Participation

This guideline asks you to treat people as action initiators, not as the recipients of action. In particular, it suggests using the active voice to talk about people who are involved in large mass activities. So say, "The secretaries completed the survey," not "The secretaries were given the survey."

Avoid Ambiguity in Sex Identity or Sex Role

This guideline deals with the widespread use of masculine words, especially *he,* when referring to all people. This usage has been changing for some time but still causes much discussion and controversy. The basic rule is to be specific. If the referent of the word is male, use *he;* if female, use *she;* if generic, use *he or she* or, more informally, *they.*

Choose Correct Terms to Indicate Sexual Orientation

Currently the preferred terms are *lesbians* and *gay men, straight women* and *straight men.* The terms *homosexual* and *heterosexual* could refer to men or women or just men, so their use is not encouraged.

Use the Preferred Designations of Racial Groups

The preferred designations change and sometimes are not agreed upon even by members of the designated group. Be sensitive to the wishes of the group you are serving. At times, *Hispanic* is not a good choice because individuals might prefer *Latino* or *Latina, Chicano* or *Chicana,* or even a word related to a specific country, like *Mexican.* Similar issues arise when you discuss Americans of African, Asian, and Arabic heritage. If you don't know, ask.

Do Not Use Language That Equates a Person with His or Her Condition

"Disability" refers to an attribute of a person. Say, "person with diabetes" to focus in a neutral way on the attribute; do not say, "diabetic," which equates the person with the condition.

Choose Specific Age Designations

Use *boy* and *girl* for people up to 18; use *man* and *woman* for people over 18. Prefer *older* to *elderly*.

Works Cited

American Psychological Association. *Publication Manual.* 5th ed. Washington, DC: APA, 2001.

Maggio, R. *The Bias-Free Word Finder: A Dictionary of Nondiscriminatory Language.* Boston: Beacon, 1991.

5 Researching

Chapter 5
In a Nutshell

When you conduct research, you are finding the relevant facts about the subject. Two strategies are *asking questions* and *using keywords*.

Ask questions. Start with predictable primary-level *questions:* How much does it cost? What are its parts? What is the basic concept you need to know? The trick, however, is to ask secondary-level questions that help you establish relationships. Secondary questions include cause (Why does it do this? Why does it cost this much?) and comparison-contrast (How is this like that? Why did it act differently this time?).

Use keywords. Type in *keywords,* following search rules, to search all library databases and Web databases. The two basic skills are knowing how to use this database's "search rules" and knowing which *words* to use.

Spend time learning the database "search rules." Typing in one word is easy, but how do you handle combinations—either phrases (municipal waste disposal) or strings (packaging, corrugated, fluting)? All search engines use logical connectors—*and, but, not, or*—in some fashion. "Recycle" *and* "plastic" narrows the results to those that contain both terms; "recycle" *or* "plastic" broadens the results to all those that contain just one of the two terms.

Finding which *words* to use is a matter of educated guesses and observation. "Packaging" is too broad (that is, it will give you too many choices—so many results that you cannot use them), so use "corrugated"; "fluting" (the wavy material in the middle of corrugated cardboard) will yield narrower results.

People research everything from how high above the floor to position a computer screen to how feasible it is to build a manufacturing plant. This chapter discusses the purpose of research, explores the essential activity of questioning, and suggests practical methods of finding information.

The Purpose of Research

The purpose of research is to find out about a particular subject that has significance for you. Your subject can be broad and general, such as recycling plastics, or narrow and specific, such as purchasing a new photocopier for your office. The significance is the importance of the subject to you or your community. Will the new method of recycling plastics make a profit for the company? Will that new photocopier make the office run more smoothly?

Generally, the goal of research is to solve or eliminate a problem (Why does the photocopier break down?) or to answer a question (What differences are there in photocopier technologies?). You can use two strategies: talking to people and searching through printed information. To find out about the photocopier, for instance, you would talk to various users to discover features that they need, and you could read sales material that explained those features and reviews that evaluated performance.

Questioning—The Basic Skill of Researching

Asking questions is fundamental to research. The answers are the facts you need. This section explains how to discover and formulate questions that will "open up" a topic, providing you with the essential information you and your readers need.

How to Discover Questions

To learn about any topic—such as which photocopier to buy for the office—ask questions. Formulate questions that will help you investigate the situation effectively and that will provide a basis for a report. For instance, the question "In what ways does our staff use the photocopier?" will not only produce important data but will also be the basis for a section on "usage patterns" in a report. Several strategies for discovering helpful questions are to ask basic questions, ask questions about significance, consult the right sources, and interact flexibly.

Ask Basic Questions

Basic questions lead you to the essential information about your topic. They include

> ◗ What are the appropriate terms and their definitions?
> ◗ What mechanisms are involved?
> ◗ What materials are involved?
> ◗ What processes are involved?

Ask Questions About Significance

Questions about significance help you "get the big picture" and grasp the context of your topic. They include

> ◗ Who needs it and why?
> ◗ How is it related to other items?
> ◗ How is it related to current systems?
> ◗ What is its end goal?
> ◗ How do parts and processes contribute to the end goal?
> ◗ What controversies exist?
> ◗ What alternatives exist?
> ◗ What are the implications of those alternatives?
> ◗ What costs are involved?

Consult the Right Sources

The right sources are the people or the printed information that has the facts you need.

People who are involved in the situation can answer your basic questions and your questions about significance. They can give you basic facts and identify their needs. The basic facts about photocopying machines can come from engineers, experienced users, or salespeople. Information about needs comes from people who use the product. They expect a photocopier to perform certain functions, and they know the conditions that make performing those functions possible.

Printed information also answers basic questions and questions of significance, often more thoroughly than people can. It can be hard copy (words printed on paper) and on-line copy (words available electronically). Printed information includes everything from sales brochures to encyclopedias to bulletin board discussions. For the photocopier, sales literature would give you prices, features, and specifications; review articles would evaluate performance; and bulletin boards could give you user testimony.

Interact Flexibly

To ask questions productively, be flexible. People have the information you need, and you must elicit as much of it as you can. Sometimes questions produce a useful answer, sometimes not. If you ask, "Which feature of the photocopier is most important to you?" the respondent might say, "Its speed," which is a broad answer. To narrow the answer, try an "echo technique" question, in which you repeat the key term of the answer: "Speed?" On the other hand, if you ask, "Is the ability to collate pages important to you?", the respondent might say, "No, I seldom use that, but I often copy on both sides of a sheet of paper." That answer opens two lines of questioning for you. Why is collating not important? Why is back-to-back copying important? How frequently do you do it? for what type of job?

You can also use questions to decide what material to read. If your question is "How does this model compare to others?" an article that would obviously interest you is "A Comparison of Photocopiers." Carefully formulating your questions makes your reading more efficient. Read actively, searching for particular facts that answer your questions (Spivey and King).

While reading, take notes, constantly reviewing the answers and information you have obtained. You will find patterns in the material or gaps in your knowledge. If three articles present similar evaluations of the photocopiers, you have a pattern on which to base a decision.

How to Formulate Questions

Basically, researchers ask two kinds of questions: closed and open (Stewart and Cash). You can use both types for interviewing and reading.

A closed question generates a specific, often restricted answer. Technically it allows only certain predetermined answers.

Closed question   **How many times a week do you use the copier?**

An open question allows a longer, more involved answer.

Open question   **Why do you use red ink in the copier?**

In general, ask closed questions first to get basic, specific information. Then ask open questions to probe the subtleties of the topic.

Collecting Information from People

You collect information, or find answers to your questions, in a number of ways. You can interview, survey, observe, test, and read. This section explains the first four approaches. Collecting published information, especially in a library, is treated in a later section.

Interviewing

One effective way to acquire information is to conduct an information interview (Stewart and Cash). Your goal is to discover the appropriate facts from a person who knows them. To conduct an effective interview, you must prepare carefully, maintain a professional attitude, probe, and record.

Prepare Carefully

To prepare carefully, inform yourself beforehand about your topic. Read background material, and list several questions you think will produce helpful answers.

If you are going to ask about photocopiers, read about them before you interview anyone so that you will understand the significance of your answers. Listing specific questions will help you focus on the issue and discourage you and the respondent from digressing. To generate the list, brainstorm questions based on the basic and significance questions we have suggested. A specific issue to focus on could be photocopier problems: How exactly does the photocopier malfunction? Has that happened before? How often?

Maintain a Professional Attitude

Schedule an appointment for the interview, explaining why you need to find out what the respondent knows. Make sure she or he knows that the answers you seek are important. Most people are happy to answer questions for people who treat their answers seriously.

Be Willing to Probe

Most people know more than they reveal in their initial answers, so you must be able to get at the material that's left unsaid. Four common probing strategies are as follows:

- ❯ Ask open-ended questions.
- ❯ Use the echo technique.
- ❯ Reformulate.
- ❯ Ask for a process description.

The basic probing strategy is to ask an *open-ended* question and then develop the answer through the echo technique or reformulation. The *echo technique* is repeating significant words. If an interviewee says, "Red really messes up a print run," you respond with "Messes up?" This technique almost always prompts a longer, more specific answer. *Reformulation* means repeating in your own words what the interviewee just said. The standard phrase is "I seem to hear you saying. . . ." If your reformulation is accurate, your interviewee will agree; if it is wrong, he or she will usually point out where. *Asking for a process description* produces many facts because people tend to organize details around narrative. As the interviewee describes, step by step, how he or she uses the machine, you will find many points where you'll need to ask probing questions.

Record the Answers

As you receive answers, write them down in a form you can use later. Put the questions on an 8½ × 11 sheet of paper, leaving enough room to record answers. Record the answers legibly; avoid listing terms and abbreviations. Ask people to repeat if you didn't get the whole answer written down. After a session, review your notes to clarify them so they will be meaningful later and to discover any unclear points about which you must ask more questions.

Surveying

To survey people is to ask them to supply answers, usually written, to your questions. You use a printed survey to receive answers from many people, more than you could possibly interview in the time you have allotted to the project. Surveys help you determine basic facts or conditions and assess the significance or importance of facts. They have three elements: a context-setting introduction, closed or open questions, and a form that enables you to tabulate all the answers easily.

A context-setting introduction explains (1) why you chose this person for your survey; (2) what your goal is in collecting this information; and (3) how you will use the information. The questions may be either closed or open. The answers to closed questions are easier to tabulate, but the answers to open questions can give you more insight. A good general rule is to avoid questions that require the respondent to research past records or to depend heavily on memory.

The form you use is the key to any survey. It must be well designed (Warwick and Lininger). Your goal is to design it so that it is both easy to read (so that people will be willing to respond) and easy to tabulate (so you can tally the answers quickly). For instance, if all the answers appear at the right margin of a page, you can easily transfer them to another page. Here is a sample survey.

SURVEY

Context-setting introduction

In the past two weeks, we have had many complaints about how difficult the new photocopier is to operate. In order to reduce frustration, we plan to develop a brief manual and to hold training sessions. To help us choose the most effective topics, please take a moment to fill in the attached survey. Please return it to Peter Arc, 150 M Nutrition Building, by Friday, January 30. Thanks.

Closed question

How often do you use the copier?

once a week ____

once every 2–3 days ____

once a day ____

several times a day ____

Closed question   Do you use any of these functions?

                                                          Yes   No

    2-sided copying   ____   ____

    overlay copying   ____   ____

    2-page copying   ____   ____

Closed question   Do you know how to do the following?

                                                          Yes   No

    select the proper paper key   ____   ____

    fill an empty paper tray   ____   ____

    get the number of copies you want   ____   ____

    lay the paper on the glass   ____   ____
    with the correct orientation

Open question   Please describe your problems when you use the machine. Use the back of this sheet if you need more space.

Open question   Are you available at any of the following times during the week of 2/20–2/25 for a training session? Give first and second preferences.

    M ____   8–10 ____

    W ____   1–3 ____

    Th ____   3–5 ____

Closed question   How much notice do you need so that you can attend a training session?

    1 day ____

    1 week ____

Observing and Testing

In both observing and testing, you are in effect carrying out a questioning strategy. You are interacting with the machine or process yourself.

Observing

Observing is watching intently. You place yourself in the situation to observe and record your observations. When you observe to collect information, you do so with the same questions in mind as when you interview: What are the basic facts? What is their significance?

To discover more about problems with the office photocopier, you could simply watch people use the machine. You would notice where people stand, where they place their originals, how carefully they follow the instructions, which buttons they push, how they read the signals sent by the control panel, and so on. If you discover that all steps move along easily except reading the control panel, you may have found a possible source of the complaints. By observation—looking in a specific way for facts and their significance—you might find the data you need to solve the office problem.

Testing

To test is to compare items in terms of some criterion or set of criteria. Testing, which is at the heart of many scientific and technical disciplines, is much broader and more complex than this discussion of it. Nevertheless, simple testing is often a useful method of collecting information. Before you begin a test, you must decide what type of information you are seeking. In other words, what questions should the test answer?

In the case of deciding which photocopier to buy, the questions should reflect the users' concerns. They become your criteria, the standards you will use to evaluate the two machines. Typical questions might be

- Which one produces 100 copies faster?
- Which one makes clearer back-to-back copies?
- Which one generates less heat?

After determining suitable questions, you have people use both machines and then record their answers to your questions. To record the answers, you need a recording form much like the one used for surveys.

Collecting Published Information

This section discusses the basic techniques for gathering published information. As with all writing projects, you must plan carefully. You must develop a search strategy, search helpful sources, and record your findings.

Develop a Search Strategy

With its thousands of books and periodicals, the library can be an overwhelming place. The problem is to locate the relatively small number of sources that you actually need. To do so, develop a "search strategy" ("Tracking Information") by determining your audience, generating questions, predicting probable sources, and searching for "keys."

Determine Your Audience

As in any writing situation, determine your audience and their needs. Are you writing for specialists or nonspecialists? Do they already understand the concepts in the report? Will they use your report for reference or background information, or will they act on your findings? Experts expect to see information from standard sources. Thus quoting from a specialized encyclopedia is more credible than using a popular one, and mentioning articles from technical journals is more credible than citing material from the popular press. However, in some areas, particularly computing but also in subjects like photography, monthly magazines are often the best source of technical information. For computers, *Macworld, PC World,* and *SUN Expert* are excellent sources for technical decisions. Nonspecialists may not know standard sources, but they expect you to have consulted them.

Generate Questions

Generate questions about the topic and its subtopics. These questions fall into the same general categories as those for interviews: What are the basic facts? What is their significance? They include

- What is it made of?
- How is it made?
- Who uses it?
- Where is it used?
- What is its history?
- Do experts disagree about any of these questions?
- Who makes it?
- What are its effects?
- How is it regulated?

Such questions help you focus your research, enabling you to select source materials and to categorize information as you collect it.

Predict Probable Sources

All concepts have a growth pattern, from new and unusual to established and respected. Throughout the pattern they are discussed in predictable—but very different—types of sources. New and unusual information is available only from a few people, probably in the form of letters, conversations, e-mail, answers to listserv queries, and personal websites. More established information appears in conference proceedings and technical journal articles. Established information appears in textbooks, encyclopedias, and general periodicals and newspapers ("Tracking Information").

If you understand this growth pattern, you can predict where to look. Two helpful ways to use the pattern are by age and by technical level. Use the following guidelines to help you find relevant material quickly:

- *Consider the age of the information.* If your topic demands information less than a year old, consult periodicals, government documents, annual reviews, and on-line databases. Write letters, call individuals, ask on a listserv, or search the Web (see "Web Searching" below). If your topic requires older, standard information, consult bibliographies, annual reviews, yearbooks, encyclopedias, almanacs, and textbooks.

▶ *Consider the technical level of the information.* If you need information at a high technical level, use technical journals, interviews with professionals, and specialized encyclopedias or handbooks. On the job, also use technical reports from the company's technical information department. If you need general information, use popular magazines and newspapers. Books can provide both technical and general information.

Search for "Keys"

A helpful concept to guide your searching is the "key," an item that writers constantly repeat. Look for keywords and key documents.

▶ *Find keywords.* Keywords are the specific words or phrases that all writers in a particular field use to discuss a topic. For instance, if you start to read about the Internet, you will quickly find the terms *navigation* and *hyperlink* in many sources. Watch for terms like these, and master their definitions. If you need more information on a term, look in specialized encyclopedias, the card catalog, periodical indexes, abstracts, and databases. Keywords can also lead you to other useful terms through cross-references and indexes.

▶ *Watch for key documents.* As you collect articles, review their bibliographies. Some works will be cited repeatedly. These documents—whether articles, books, or technical reports—are *key documents.* If you were searching for information about the World Wide Web, you would quickly discover that three or four books are the "bibles" of the Web. Obviously, you should find and review those books. Key documents contain discussions that experts agree are basic to understanding the topic. To research efficiently, read these documents as soon as you become aware of them.

Search Helpful Sources

To locate ideas and material, you can use Web searching, electronic catalogs, and electronic databases.

Web Searching

The World Wide Web contains stunning amounts of information, so much that, curiously, the problem is to cull out what is usable. The key is to know how to search effectively.

Search the Web as you would any other database. Choose a search engine, a software program (like Yahoo! or AltaVista) whose purpose is to search a database for instances of the word you ask it to look for. Type keywords into the appropriate box on the search engine's "search page." Review the resulting list of sites, or hits, that contain the word you looked for. The list is the problem—you can generate a list that tells you there are a million sites. Which ones do you look at? Unfortunately, there is no standard answer. You will have to undergo a certain amount of trial and error to find which search engine and which method produce the best results for you.

You can help yourself by understanding how the search engines work, by developing strategies for finding sites, and by learning how to use keywords ("Searching"; Seiter).

How Search Engines Work. Search engines are not all the same. Find out how various search engines work. Investigate the following characteristics.

Engines look for words or phrases, but they do not all look in the same place. Various engines search for the word in just one of the following: the URL (e.g., <www.college.edu/communication/textbook.html>), the abstract, the keywords that the website creator sent to the search engine database, the title, and the full text (all the words in the site). (See "Use a search engine," below.)

All search engines have Advanced, or Custom, or Power options—the terms vary—that tell you how to limit the number of hits you receive. (See "Using Keywords and Boolean Logic," p. 123.)

Strategies for Finding Sites. Several strategies can help you locate appropriate sites:

Surf. You can find information by surfing—simply starting at one site and clicking on a hyperlink and then on another until you find something interesting.

Locate hub sites. Because a hub site lists many other sites that have a similar interest, finding one is often a way to shorten your search. Hub sites exist for every topic. You can find them for writing centers, for dietetics, for packaging. The problem is that there is no directory of these. Often you can find them in "Web Columns" that appear in many magazines and newspapers. You can often find them by accident, perhaps by surfing out from the first site listed in a search result.

Use a search engine. Search engines usually search just their database. Common search engines include Lycos, AltaVista, Yahoo!, and Google. Each of these engines searches by a different technique (AltaVista searches full text; Lycos searches abstracts). Each engine reports a list with a relevancy factor, the database's "guess" of which sites will give the best information. Evaluate the effectiveness of this characteristic; it can be very helpful or misleading. Take the time to perform a comparative search—try the same keyword or phrase in three search engines and compare the results.

Use a megasearch engine. Megasearch engines search a number of databases simultaneously. Common megasearch engines are Metacrawler and Sherlock.

Using Keywords and Boolean Logic. Most search engines allow you to join keywords with Boolean connectors—*and, or,* and *not.* The engine then reports results that conform to the restrictions that the connectors cause.

The basic guidelines are

▶ *Choose specific keywords.* Be willing to try synonyms. For example, if *plastic* gives an impossibly long list, try *polymer* or *monomer.*

▶ *Understand how to use Boolean terms.* Read the instructions in the custom search or help sections of the search engine. AltaVista, for instance, has an excellent help section.
  • *And* (sometimes you have to type + and sometimes AND; it depends) asks the engine to report only sites that contain all the terms. Generally because *and* narrows a search, eliminating many sites, it is an excellent strategy.
  • *OR* causes the engine to list any site that contains the term. Entering the words *French and wine* generates a list of just those sites that deal with French wine. *French OR wine* generates a much larger list that includes everything that deals with either France or wine.
  • *NOT* excludes specific terms. Entering *French and wine not burgundy* generates a list that contains information about the other kinds of French wine.

Be willing to experiment with various strings of keywords.

Search the Electronic Catalog

The electronic catalog is your most efficient information-gathering aid. You can easily find information by subject, author, title, keyword, or call number. Many systems allow you to search periodical indexes. All of them allow you to print out an instant bibliography.

Because several major systems and many local variations exist, no textbook can give you all the information you need to search electronically. Take the time to learn how to use your local system. Each system differs, so take the time to learn them. Do several practice searches to find the capabilities of the system and the "paths," or sequences of commands, you must follow to produce useful bibliographies. This section focuses on a few items available in a typical catalog and offers information on using keywords, with electronic catalogs.

The Typical Catalog. The typical catalog presents you with screens for individual items, bibliographies of related items, and categories of searching. An individual item screen appears in Figure 5.1 (p. 124). This screen describes the book in detail: its authors, printing information, call number, and, very importantly, the subject keywords that can be used to locate it. If you use those words as keywords in the database, you will find even more books on this subject.

A bibliography screen appears in Figure 5.2 on page 124. This screen lists the first 8 of 282 books contained in the library under the subject heading "Internet." Note, however, that the books' topics are not all the same; the term *Internet* is too broad to create a focused list. A researcher would have to use Boolean logic

Figure 5.1

General
Information
Screen

Source: Screen shot
from University of
Wisconsin–Stout
Voyager catalog.
Reprinted by
permission.

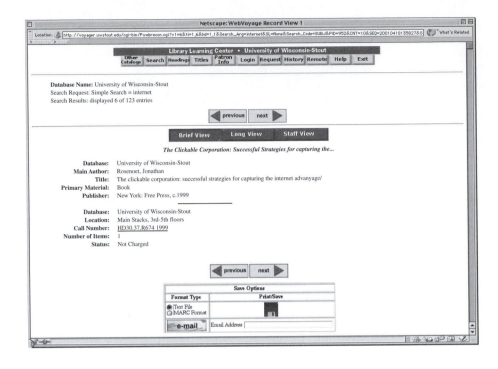

Figure 5.2

Subject
Bibliography
Screen

Source: Screen shot
from University of
Wisconsin–Stout
Voyager catalog.
Reprinted by
permission.

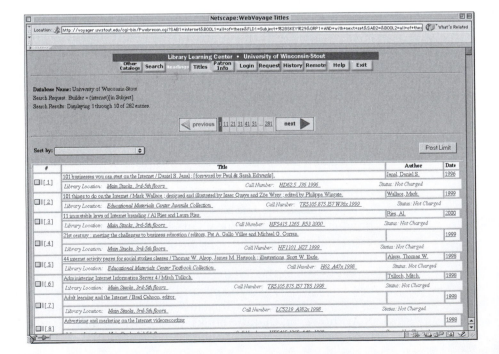

(Internet and corporation) to narrow the 282 items down to a group on the same topic.

Obviously, these systems can generate a working bibliography on any subject almost instantly. You can search by subject, title, keyword, author, date, and call number. Use the method that agrees with what you know. For instance, if you know the author—perhaps she has written a key document, and you want to find what else she has written—use the author method. If you know little about the topic, start with a subject search.

Keywords. By now you can see that the "trick" to using an electronic system is the effective use of *keywords,* any word for which the system will conduct a search. Figure 5.3 shows that, if you want to search for all the items that have *Internet* in their title, enter your keyword ("internet") in the locator box and select "subject" in the *Search in:* pull-down menu. An effective initial strategy is to use the keyword category, which searches for the word anywhere on the individual item screen. If the word appears, for instance, in the item's title, subject heading, or abstract, the system includes the item in the bibliography.

To find keywords, you can use the *Library of Congress Subject Headings* (LCSH) or your own common sense. Because most of the systems are keyed to Library of Congress headings, you can use the two-volume *Library of Congress Subject Headings* to find your headings. All electronic card catalog systems use these

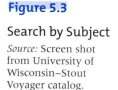

Figure 5.3

Search by Subject

Source: Screen shot from University of Wisconsin–Stout Voyager catalog. Reprinted by permission.

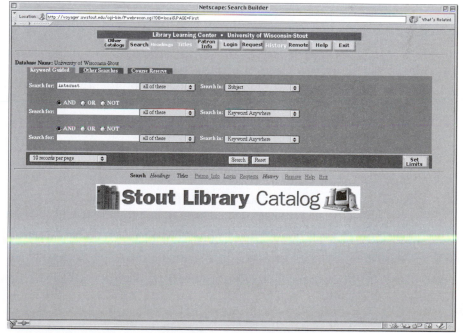

Figure 5.4

Sample Subject
Heading Entry

Source: Reprinted with
permission from the
Academic Index™.
©1998 Information
Access Company.

Internet (Computer network)
 (May Subd Geog)
  [TK5105.875.I57]
  UF   DARPA Internet (Computer network)
  BT   Wide area networks (Computer
       networks)
  RT   World Wide Web (Information
       retrieval system)
  NT   Gopher servers
       WAIS (Information retrieval systems)

words to classify their material. Figure 5.4 is an example of a subject heading listing. The boldfaced word (**Internet**) is used in the catalog. If you enter it into the computer, the system responds with a list of books on that subject. Various symbols can also help you; for example, *BT* means "broader term," and *RT* means "restricted term."

If the LCSH books are not available, brainstorm and use the system's rules. Brainstorm a list of general words that apply to your topic—Internet, Web, hypermedia, cyberspace, URL, domain. This method, however, might turn up hundreds or thousands of titles—too many to use. To limit the size of your bibliography, use the Boolean "operators" such as *and, not,* and *or.* Notice, in Figure 5.3, that you can enter a new keyword in each locator box, choose an appropriate Boolean operator, and choose a different category for every *Search in:* choice. In effect, then, you can mix types of searching, such as author and subject.

Search On-Line Databases

On-line databases are as efficient as the electronic catalog in generating resources for your topic. If you access the database correctly (by using the correct keyword), it produces a list of the relevant articles on a particular subject—essentially a customized bibliography.

Databases are particularly helpful for obtaining current information; sometimes entries are available within a day of when they appear in print, or even before. Most university and corporate libraries provide their patrons with many databases for free. If your library offers databases such as EBSCOhost, use them.

To search a database effectively, you must choose your keywords carefully. Just as in the computerized catalog, if you pick a common term, like *manufacturing* or *packaging* or *retail,* the database might tell you it has found 10,000 items. To narrow the choices, combine descriptors. For example, if you combine *packaging* with such descriptors as *plastic* and *microwavable,* the computer searches for titles that contain those three words and generates a much smaller list of perhaps 10 to 50 items. Figure 5.5 shows a list of 25,000 articles generated for the keyword *Internet.* Obviously, this search would have to be narrowed (by using Refine Search, the button at the top middle). Figure 5.6 shows an item entry, giving all the relevant information. Notice especially the subject line. These words are

Figure 5.5

Generated List of Articles

Source: Screen shot from University of Wisconsin–Stout Voyager catalog. Reprinted by permission.

Figure 5.6

On-Line Bibliographic Entry with Abstract

Source: Screen shot from University of Wisconsin–Stout Voyager catalog. Reprinted by permission.

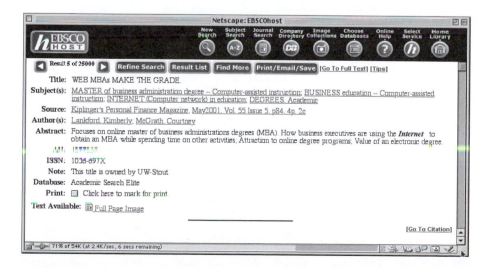

keywords. Type them into the locator to find articles close in topic to the one described. Also notice the abstract, which you can use to decide whether or not to find the full text of the article. (Most systems will let you retrieve full text articles, an invaluable aid as you collect information for your report.)

Databases provide information on almost every topic. EBSCOhost offers many indexes, including Applied Science and Technology, ERIC, Hoover's Company Profiles, and Health Service Plus. Contact your library for a list of the services available to you.

Record Your Findings

As you proceed with your search strategy, record your findings. Construct a bibliography, take notes, consider using visual aids, and decide whether to quote or paraphrase important information.

Make Bibliography Cards

List potential sources of information on separate 3-by-5-inch cards. These bibliography cards should contain the name of the author, the title of the article or book, and facts about the book or article's publication (Figure 5.7). Record this information in the form that you will use in your bibliography. Also record the call number and any special information about the source (for instance, that you used a microfiche version). Such information will help you relocate the source later.

Take Notes

As you read, take notes and put these ideas on cards as well (Figure 5.8). On each card, write the topic, the name of the author, and the page number from

Figure 5.7

Bibliography Card

20

Williams, Thomas R.

"Guidelines for Designing and Evaluating the Display of Information on the Web"

Technical Communication 47.3 (August 2000): 383–396

which you are recording information. Then write down a single idea you got from that source. Each card should contain notes on a single subject and from a single source, no more. This practice greatly simplifies arranging your notes when you finally organize the report.

Make Visual Aids

Visual aids always boost reader comprehension. You either find them in your research and/or create them yourself. If a key source has a visual aid that clarifies your topic, use it, citing it as explained in Appendix B (pp. 571–596). As you read, however, be creative and construct your own visual aids. Use flow charts to show processes, tables to give numerical data, and drawings to explain machines—whatever will help you (and ultimately your readers) grasp the topic.

Quoting and Paraphrasing

It is essential in writing research reports to know how and when to quote and paraphrase. *Quoting* is using another writer's words verbatim. Use a quote when the exact words of the author clearly support an assertion you have made or when they contain a precise statement of information needed for your report. Copy the exact wording of

▶ Definitions
▶ Comments about significance
▶ Important statistics

Paraphrasing means conveying the meaning of the passage in your own words. Learning to paraphrase is tricky. You cannot simply change a few words and then claim that your passage is not "the exact words" of the author. To

Figure 5.8

Note Card

20

Guidelines for Legible Display

Make visuals large enough to be seen and interpreted

Avoid distracting backgrounds

Use white or blue background (383–384)

paraphrase, you must express the message in your own original language. Write paraphrases when you want to

▶ Outline processes or describe machines.
▶ Give illustrative examples.
▶ Explain causes, effects, or significance.

The rest of this section explains some basic rules for quoting and paraphrasing. Complete rules for documenting sources appear in Appendix B.

Consider this excerpt from J. C. R. Licklider's 1968 essay on the benefits of the Internet, written after one of the first "technical meeting[s] held through a computer" (276).

> When people do their informational work "at the console" and "through the network," telecommunication will be as natural an extension of individual work as face-to-face communication is now. The impact of that fact, and of the marked facilitation of the communication process will be very great—both on the individual and on society.
>
> First, life will be happier for the on-line individual because the people with whom one interacts most strongly will be selected more by commonality of interests and goals than by accidents of proximity. Second, communication will be more effective and productive, and therefore more enjoyable. Third, much communication and interactions will be with programs and programmed models, which will be (a) highly responsive, (b) supplementary to one's own capabilities, rather than competitive, and (c) capable of representing progressively more complex ideas without necessarily displaying all the levels of their structure at the same time—and which will therefore be both challenging and rewarding. And fourth, there will be plenty of opportunity for everyone (who can afford a console) to find his calling, for the whole world of information, with all its field and disciplines, will be open to him. . . .
>
> For the society, the impact will be good or bad, depending mainly on the question: Will "to be on line" be a privilege or a right? If only a favored segment of the population gets a chance to enjoy the advantage of "intelligence amplification," the network may exaggerate the discontinuity in the spectrum of intellectual opportunity (Licklider 277).

To quote, place quotation marks before and after the exact words of the author. You generally precede the quotation with a brief introductory phrase.

> According to Licklider, "If only a favored segment of the population gets a chance to enjoy the advantage of 'intelligence amplification,' the network may exaggerate the discontinuity in the spectrum of intellectual opportunity" (277).

Note that the quotes around *intelligence amplification* are single, not double, because they occur within a quotation.

If you want to delete part of a quotation from the middle of a sentence, use ellipsis dots (. . .).

> Licklider notes that "The impact of that fact . . . will be very great—both on the individual and on society" (277).

If you want to insert your own words into a quotation, use brackets.

> Licklider points out that "For the society [certainly the global society], the impact will be good or bad, depending mainly on the question: Will 'to be on line' be a privilege or a right?" (277).

To paraphrase, rewrite the passage using your own words. Be sure to indicate in your text the source of your idea: the author and the page number on which the idea is found in the original.

> After one of the first technical meetings held via a computer, Licklider theorizes that the new possibilities of interaction will have enormous benefits to individuals, who will no longer be limited by physical proximity. The benefits include the possibility to interact with people of like minds from many different locations, to develop complex ideas in greater depth, and to explore one's interests more fully.
>
> While the benefits are clear, the potential information and communication explosion also throws the issue of parity into the spotlight. Those who don't have access to a computer will not be able to participate fully in this intellectual challenge. Current inequalities will be made even more pronounced (277).

Remember when you quote or paraphrase that you have ethical obligations both to the original author and to the report reader.

1. When in doubt about whether an idea is yours or an author's, give credit to the author.

2. Do not quote or paraphrase in a way that misrepresents the original author's meaning.

3. Avoid stringing one quote after another, which makes the passage hard to read.

Worksheet for Research Planning

☐ Name the basic problem that you perceive or a question that you want answered.

☐ Determine your audience. Why are they interested in this topic?

☐ List three questions about the topic that you feel must be answered.

☐ List three or four search words that describe your topic.

(continued)

(continued)

☐ Determine how you find information about this topic. Do you need to read? interview? survey? search library databases? search the Web? perform some combination of these?

☐ List the steps you will follow to find the information. Include a time line on which you estimate how many hours or days you need for each step.

☐ Name people to interview or survey, outline a test, or list potential sources (technical or nontechnical) of printed information.

☐ Select a form on which to record the information you discover. If you use interviews or surveys, create this form carefully so that you can later collate answers easily.

Exercises

▶ You Create

1. Develop a research plan and implement it.

   a. Create a list of five to eight questions about a topic you want to research. For each question, indicate the kind of resource you need (book, recent article, website) and a probable search source (*Applied Science and Technology Index, Engineering Index,* library electronic catalog, Google, AltaVista). Explain your list to a group of two or three. Ask for their evaluation, changing your plan as they suggest.

   b. Based on the list you created in Exercise 1a, create a list of five to ten keywords. In groups of two or three, evaluate the words. Try to delete half of them and replace them with better ones.

   c. Select a database (e.g., Compendex) or a Web search engine (e.g., AltaVista).

   d. Go on to Writing Assignment 3.

2. Conduct a subject search of your library's computerized catalog. Start with a general term (*Internet*) and then, using the system's capabilities, limit the search in at least three different ways (e.g., Internet not e-mail, Internet and manufacturing engineering, Internet or World Wide Web). Print out the bibliography from each search. Write a description of the process you used to derive the bibliographies, and evaluate the effectiveness of your methods.

   Alternate: Write a memo to your classmates on at least two tips that will make their use of the catalog easier.

3. Select a topic of interest. Generate a list of three to five questions about the topic. Read a relevant article in one standard reference source. Based on the article,

answer at least one of your questions and pose at least two more questions about the topic.

4. Use Example 9.1 (p. 223) and the information provided in the excerpts of Appendix B, Exercise 4 (pp. 594–596), to create a brief research report on pixels. Your audience is people who are just beginning to use color in documents.

▶ You Analyze

5. Write a memo in which you analyze and evaluate an index or abstracting service. Use a service from your field of interest, or ask your instructor to assign one. Explain which periodicals and subjects the service lists. Discuss whether it is easy to use. For instance, does it have a cross-referencing system? Can a reader find keywords easily? Explain at what level of knowledge the abstracts are aimed: beginner? expert? The audience for your report is other class members.

6. Write a memo in which you analyze and evaluate a reference book or website in your field of interest. Explain its arrangement, sections, and intended audience. Is it aimed at a lay or a technical audience? Is it introductory or advanced? Can you use it easily? Your audience is other class members.

▶ Group

7. Form into groups of three. One person is the interviewer, one the interviewee, and one the recorder. Your goal is to evaluate an interview. The interviewer asks open and closed questions to discover basic facts about a technological topic that the interviewee knows well. The recorder keeps track of the types of questions, the answers, and the effectiveness of each question in generating a useful answer. Present an oral report that explains and evaluates your process. Did open questions work better? Did the echo technique work?

Writing Assignments

1. Write a short research report explaining a recent innovation in your area of interest. Your goal is to recommend whether your company should become committed to this innovation. Consult at least six recent sources. Use quotations, paraphrases, and one of the citation formats explained in Appendix B (pp. 571–596). Organize your material into sections that give the reader a good sense of the dimensions of the topic. The kinds of information you might present include

   - Problems in the development of the innovation and potential solutions.
   - Issues debated in the topic area.
   - Effects of the innovation on your field or on the industry in general.
   - Methods of implementing the innovation.

   Your instructor might require that you form groups to research and write this report. If so, he or she will give you a more detailed schedule, but you must formulate questions, research sources of information, and write the report. Use the guidelines for group work outlined in Chapter 3.

2. You (or your group, if your instructor so designates) are assigned to purchase a word processing package for your campus computer lab (or your company network). Research three actual programs and recommend one. Investigate your situation carefully. Talk to users, discover the capabilities of the current computers (RAM, etc.), discover their ability to support this new program (training), and investigate cost and site licenses. Read several reviews.

3. Following one of the documentation formats, write a brief research paper in which you complete the process you began in Exercise 1.

   a. In your chosen database or Web search engine, use your search words and combinations of them to generate a bibliography.
   b. Read two to five articles that will answer one of your questions.

   Alternate: Write a brief report in which you explain the questions you asked, the method you used, and the results you achieved.

4. Divide into groups of three or four. Construct a three- or four-item question-naire to give to your classmates. Write an introduction, use open and closed questions, and tabulate the answers. At a later class period, give an oral report on the results. Use easy topics, such as demographic inquiries (size of each class member's native city, year in college, length of employment) or inquiries into their knowledge of some common area in a field chosen by the group (such as using search engines on the Web).

5. Interview four people in a workplace to determine their attitude toward a technology (fax, phone system, PDAs, wireless printing, instant messaging). Present to an administrator a memo recommending a course of action based on the responses. One likely topic is the need for training.

6. Write the report your instructor assigns.

   a. Describe your actions, the number of items and type of information, the value of the entries.
   b. Answer the question in several paragraphs. See Appendix B for listing sources.
   c. Describe your database or search engine. Explain why you selected it; whether it was easy to use; whether it was helpful.
   d. Describe your article. Summarize it and explain how it relates to your topic.
   e. List two questions that you can research further as a result of reading your article.

7. After you have completed your writing assignment, write a learning report, a memo to your instructor. Explain, using details from your work, what new things you have learned or old things confirmed. Use some or all of this list of topics: writing to accommodate an audience, presenting your identity, selecting a strategy, organizing, formatting, creating and using visual aids, using an appropriate style, developing a sense of what is "good enough" for any of the pre-

vious topics. In addition, explain why you are proud of your recent work, and tell what aspect of writing you want to work on for the next assignment.

Web Exercise

Decide on a topic relevant to your career area. Using the Web, find three full-text professional articles that previously appeared in print and three documents that have appeared only on a website. Usually, access through a major university library will achieve the first goal; access to a corporate site will usually achieve the second goal.

Do either of the following, whichever your instructor designates:

a. In an analytical report (see Chapter 12) compare the credibility and the usability of the information in the two types of sources.
b. Write a research paper in which you develop a thesis you have generated as a result of reading the material you collected.

Works Cited

Library of Congress. *Library of Congress Subject Headings*. 20th ed. 4 vols. Washington, DC: Library of Congress, 1997.

Licklider, J. C. R. "The Computer as a Communication Device." *Science and Technology* (April 1968). Rpt. as "The Internet Primeval." *Visions of Technology*. Ed. Richard Rhodes. New York: Simon & Schuster, 1999. 274–282.

Peters, Tom. *Thriving on Chaos: Handbook for a Management Revolution*. New York: Harper, 1987.

"Searching the World Wide Web." 4 Aug. 1997 <www.uwstout.edu/lib/srchwshp .html>.

Seiter, Charles. "Better, Faster Web Searching." *Macworld* (December 1996): 159–162.

Spivey, Nancy Nelson, and James R. King. "Readers as Writers Composing from Sources." *Reading Research Quarterly* (Winter 1989): 7–26.

Stewart, Charles J., and William B. Cash, Jr. *Interviewing Principles and Practices*. 8th ed. Boston: McGraw-Hill, 1997.

"Tracking Information." *INSR* 33. Menomonie: University of Wisconsin–Stout Library Learning Center, January 2000.

Warwick, Donald P., and Charles A. Lininger. *The Sample Survey: Theory and Practice*. New York: McGraw-Hill, 1975.

Focus on
Using Google

Google is a search engine that uses the PageRank system to help you find relevant information. Page-Rank uses your keyword or -words to search the World Wide Web, and then rates a webpage's relevancy, based on the number of links to it from other webpages and on the content of the page in relation to your search. The result is a list of webpages that should be most relevant to your keywords.

However, that is the problem with search engines: They operate via keywords. In order to search effectively, you need to develop efficient keyword strategies. The basic strategies suggested by Google are described next.

The most common strategy, of course, is to enter a single word into the search box. Simply type in any word and the search engine returns relevant pages.

Often, however, effective searching requires using several words or phrases. The way you enter the words into the search box can make a huge difference. For instance, if you type just

packaging global communication

you receive a list of 371,000 webpages (see Figure 5.9). But if you type in

packaging + "global communication"

you receive a list of 4900 webpages (see Figure 5.10), a much more focused list.

The strategies that Google recommends are:

Use the Advanced Search Page (see Figure 5.11). Fill in the blanks with the terms relevant to your search, and pick relevant parameters from the drop-down boxes, such as File Format or Date. This strategy is probably the easiest way to refine your search.

Figure 5.9
Google Search
Source: Reprinted by permission of Google.

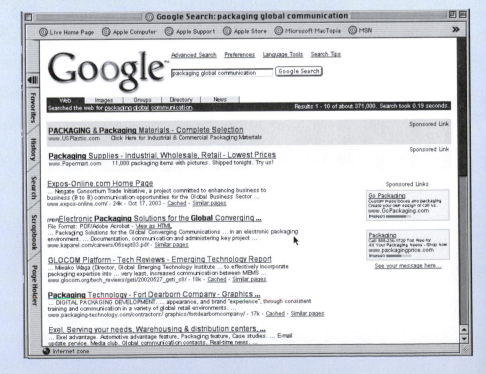

Figure 5.10
A More Focused
Google Search

Source: Reprinted by
permission of Google.

Figure 5.11
Google's Advanced
Search Page

Source: Reprinted by
permission of Google.

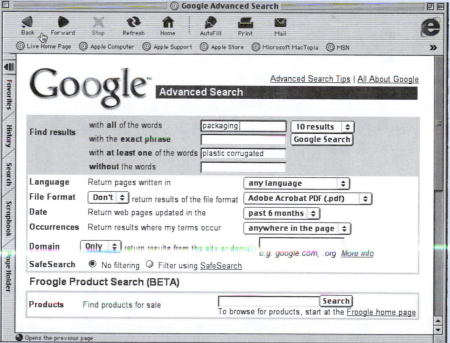

(continued)

(continued)

Use quotation marks. If you put quotation marks around phrases, the search engine looks for that exact phrase. If you don't use quotation marks, the search engine accepts any occurrence of any individual word, with the dramatically larger results shown in Figure 5.9. Along with using the Advanced Search Page, this is probably the easiest strategy to use for refined searching.

Use +. To use the plus sign, skip a space before the sign. Use the sign to ensure that a word or symbol is included in the search. Google ignores common words like *the* and *a* and symbols like roman numeral I. If you wanted results on the first Super Bowl, you could type

Super Bowl + I [or 1].

Use -. The minus search eliminates webpages from the results. If you wish to investigate packaging but do not wish to receive results on plastic packaging, type

packaging -plastic

Use ~. The tilde function allows you to search for synonyms. To use this function, type the tilde (which is to the left of the 1 key) before each word:

~packaging ~corrugated

Use OR. To find pages that contain one or the other of your keywords place the OR between the words:

packaging plastic OR corrugated

A hint about these strategies: Sometimes using the strategy increases rather than decreases the number of hits. For instance, a search using

packaging plastic corrugated

returned 74,400 hits, but

Packaging plastic OR corrugated

returned 7,430,000 hits.

Works Cited

"Advanced Search Made Easy." Google.com. 2003. 19 Oct. 2003 <www.google.com/help/refinesearch .html>.

"Google Searches More Sites More Quickly, Delivering the Most Relevant Results." Google.com. 2003. 19 Oct. 2003 <www.google.com/technology/index .html>.

6

Designing Pages

Chapter Contents

Chapter 6 in a Nutshell
Using Visual Features to Reveal Content
Using Text Features to Convey Meaning
Developing a Style Sheet and Template
Focus on Color

Chapter 6
In a Nutshell

Design is the integration of words and visuals in ways that help readers achieve their goals for using the document. The key idea is to establish a *visual logic*—the same kind of information always looks the same way and appears in the same place (page numbers are italicized in the upper-right corner, for instance).

Visual logic establishes your credibility, because you demonstrate that you know enough about the topic and about communicating to be consistent. Visual logic helps your audience to see the "big picture" of your topic, and as a result they grasp your point more quickly. Both visual and textual features establish visual logic. Two key visual features are *heads* and *chunks.*

Heads tell the content of the next section. Heads should inform and attract attention—use a phrase or ask a question; avoid cryptic, one-word heads.

Heads have levels—one or two are most common. The levels should look different and make their contents helpful for readers.

Chunks are any pieces of text surrounded by white space. Typically, readers find a topic presented in several smaller chunks easier to grasp than one longer chunk.

A key textual feature is *highlighting*—changing the look of the text to draw attention, for instance, by using boldface or italics. In addition, *standardization* and *consistency* are effective ways to orchestrate textual design. Standardization means that each feature, such as boldface, has a purpose. For instance, in instructions for using software, boldfaced words could indicate which menu to access. Consistency means that all items with a similar purpose have a similar design; for instance, all level-one heads in the document have the same look (e.g., Arial 12 point boldfaced).

A *style sheet* and *template* are effective methods to plan design. A style sheet lists the specifications of the design (e.g., "All level-one heads appear in all caps, Arial 12 pt., flush left"). A template is a representative page that indicates the correct look of each item of design.

Technical communicators design their document pages to produce what Paul Tyson, a designer, describes as a document "from which readers can quickly get accurate information" (27). Although *design* is a word with many meanings, in this book *design* means the integration of words and visuals in ways that help readers achieve their goals for using the document (Schriver). This definition implies two major concerns. First, design is about the look of the page—its margins, the placement of the visual aids, the size of the type. The pattern of these items is called a *template*. Second, design is about helping readers relate to the content. This aspect of design is called *visual logic.*

The second concern is actually more important than the first. Karen Schriver, a document design expert, indicates the relationship of the two concerns by saying that the look or design must reveal the structure of the document, and in order to achieve that goal communicators must orchestrate the look to achieve a visual logic which makes structural relationships clear. Tyson explains the relationship this way: "In a well-designed document, the writing and formatting styles expose the logical structure of the content so readers can quickly get what they want from the document" (27).

This chapter will familiarize you with both goals. You will learn how to reveal a document's logical structure by using both text and visual features. This chapter covers using visual features to reveal content, using type features to convey meaning, and developing a style sheet and template.

Using Visual Features to Reveal Content

The visual features that reveal content are white space and chunks; bullets; head systems; and headers, pagination, and rules.

White Space and Chunks

The key visual feature of a document is its white space. *White space* is any place where there is no text or visual aid. White space creates *chunks*—blocks of text—and chunks reveal logical structure to readers, thus helping them grasp the meaning. The rule for creating chunks is very simple: Use white space to make individual units of meaning stand out. You can apply this rule on many levels. The contrast in the two examples here gives you the basic idea.

Here is a memo produced as one chunk, which makes it seem there is only one message. The number of points and the content of the message are not at all clear.

> Hi John, The reports that were presented at the meeting won't be as effective on the company website as the people in the meeting suggested. The tables are too complicated and the actual explanations are unclear and not positioned near enough to refer back and forth easily. These reports will not make it much easier for our intended audience to use our

> data. This group has members who don't really belong, and a few who do belong are missing. I would like to be able to remove two of the marketing people and add the director of library services. Will you call me with your suggestions?

Now, here is the same memo with the content units turned into chunks. Notice that you can see that the content really has three parts—two issues and a request for advice.

> Hi John,
>
> The reports that were presented at the meeting won't be as effective on the company website as the people in the meeting suggested. The tables are too complicated and the actual explanations are unclear and not positioned near enough to refer back and forth easily. These reports will not make it much easier for our intended audience to use our data.
>
> This group has members who don't really belong, and a few who do belong are missing. I would like to be able to remove two of the marketing people and add the director of library services.
>
> Will you call me with your suggestions?

In the second example, the chunks are used to divide a large chunk into units, thus showing sequence—that is, you can tell that the memo has three points, all of which seem to be of equal importance.

However, chunks can also show hierarchy; that is, they can indicate which material is subordinate to other material. Let's take the first example memo and chunk it so that the design clarifies the main idea and the support idea.

> The reports that were presented at the meeting can't go on the website as easily as the people in the meeting suggested.
>
>> The tables are too complicated.
>> The actual explanations are unclear and not positioned near enough to refer back and forth easily.
>
> These reports will not make it easier for our intended audience to use our data.

Notice that indenting the two reasons makes them appear subordinate to the main objection. As a result, the design shows the reader the logical structure of the chunk.

Bullets

Another visual feature that facilitates conveying meaning is introductory symbols—either numbers or *bullets,* dots placed in front of the first word in a unit. Here is the paragraph with bullets added; the resulting list is called a *bulleted list.* Notice that the bullets emphasize the list items, causing the reader to focus on them. For even more emphasis, the author could have used numbers instead, emphasizing that there are two reasons.

Also notice that the second reason has two lines; the second line starts under the first letter of the first line, not under the bullet. This strategy of indenting is called a *hanging indent* and is commonly used in lists to make the parts stand out.

> The reports that were presented at the meeting can't go on the website as easily as the people in the meeting suggested.
>
> - The tables are too complicated.
> - The actual explanations are unclear and not positioned near enough to refer back and forth easily.
>
> These reports will not make it easier for our intended audience to use our data.

Head Systems

A *head* is a word or phrase that indicates the contents of the section that follows. A *head system* is a pattern of heads (called *levels*)to indicate both the content and the relationship (or hierarchy) of the sections in the document. With chunks, heads are a key way to help readers find information and also to see the relationship of the parts of the information.

Figure 6.1 illustrates this idea. Note that each head summarizes the contents of the section below it. Also note that there are two levels. Level 1 condenses the overall topic of the section into a few words and the two level 2 heads show that the topic has two subdivisions. But in particular notice the design of the two levels. Level 2 is indented, and the clear content of the phrases reveals the logical structure of the document—a claim (Good News) and the reasons for the claim (Production Doubles; Sales Increase).

Head systems vary. The goal of each variation is to indicate the hierarchy of the contents of the document—the main sections and the subsections. Head systems are subject to certain norms.

All Caps (each letter is capitalized) Is Superior to "Up and Down" Style (capitals and small letters mixed)

GOOD NEWS FOR WIDGETS
Production Doubles
Sales Increase

Big Is Superior to Little

Good News for Widgets
Production Doubles
Sales Increase

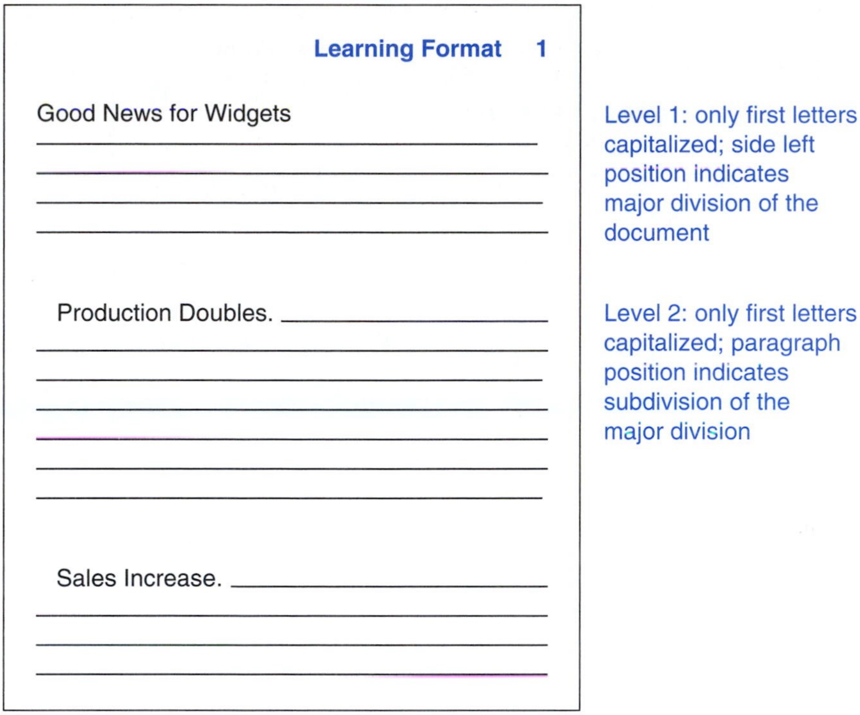

Figure 6.1

Two Levels of
Heads

Learning Format   1

Good News for Widgets
_____
_____
_____
_____

Level 1: only first letters
capitalized; side left
position indicates
major division of the
document

Production Doubles. _____
_____
_____
_____
_____
_____
_____

Level 2: only first letters
capitalized; paragraph
position indicates
subdivision of the
major division

Sales Increase. _____
_____
_____
_____

Dark Is Superior to Light

Good News for Widgets
Production Doubles
Sales Increase

Far Left Is Superior to Indented

Good News for Widgets
Production Doubles
Sales Increase

Head systems also have two basic styles: open and closed. An open system uses only the position and size of the heads to indicate hierarchy. Figure 6.2 (p. 144) illustrates an open system. A closed system uses a number arrangement to indicate hierarchy. Level 1 is preceded by 1, a subsection is 1.1, and a sub-subsection 1.1.1. Figure 6.3 (p. 144) shows a closed system.

Headers or Footers, Pagination, and Rules

Three other features of visual layout are headers or footers, pagination, and rules. *Headers* or *footers* appear in the upper or lower margin of a page. They

Figure 6.2

Open System

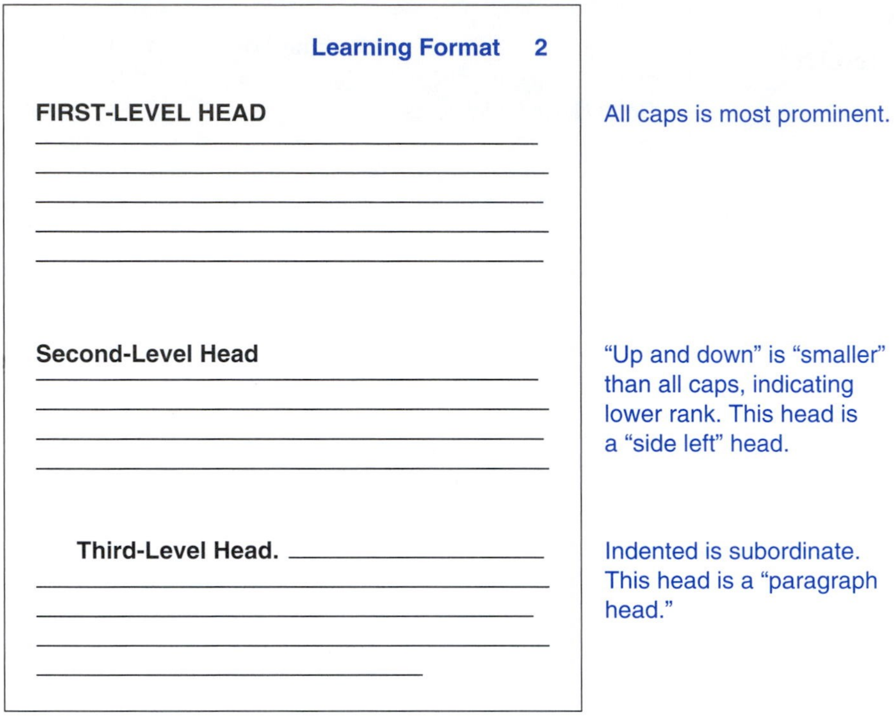

Learning Format   2

FIRST-LEVEL HEAD
_____
_____
_____
_____
_____

All caps is most prominent.

Second-Level Head
_____
_____
_____
_____

"Up and down" is "smaller" than all caps, indicating lower rank. This head is a "side left" head.

    Third-Level Head. _____
_____
_____
_____
_____

Indented is subordinate. This head is a "paragraph head."

Figure 6.3

Closed System

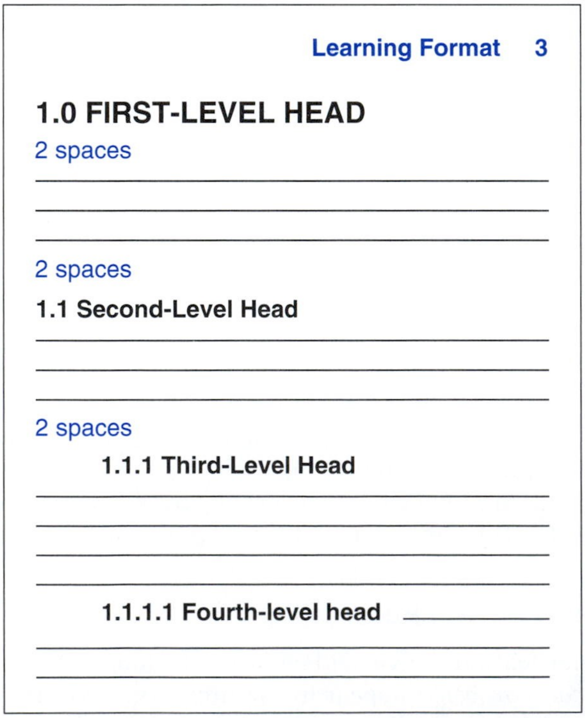

Learning Format   3

1.0 FIRST-LEVEL HEAD

2 spaces
_____
_____
_____

2 spaces
1.1 Second-Level Head
_____
_____
_____

2 spaces
    1.1.1 Third-Level Head
_____
_____
_____
_____

    1.1.1.1 Fourth-level head _____
_____
_____
_____

usually name the section of the document for the reader. *Page numbers* usually appear at the top right or top left (depending on whether the page is a right-hand or a left-hand page), or bottom center of the page. Usually both headers and footers and page numbers are presented in a different type size or font from that of the body text. *Rules,* or lines on the page, act like heads—they divide text into identifiable sections and they can indicate hierarchy. A thinner rule is subordinate to a thicker rule. Figure 6.4 shows headers and footers and page numbers.

Figure 6.4

Basic Page Parts

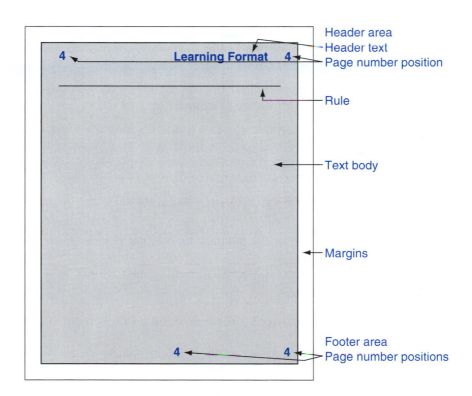

Header area
Header text
Page number position

Rule

Text body

Margins

Footer area
Page number positions

Using Text Features to Convey Meaning

Text features that are used to convey meaning are highlighters, font, font size, leading, columns and line length, and justification. You can use text features to emphasize words or groups of words and to give the text a certain personality.

Highlighters

Highlighters focus the reader's attention on an idea by making a word or phrase stand out from other words.

Types of Highlighters

Common highlighters are

> **Boldface**
> *Italics*
> ALL CAPITALS
> Vertical lists
> Quotation marks

You can see the effect of highlighters by comparing the use of boldface in the following two sentences:

> Your phone comes from the factory set to "700 msec." The talk indicator must be off before programming.

> Your phone comes from the factory set to "700 msec." **The TALK indicator must be off before programming.**

In the first example, the two sentences look the same. Nothing is emphasized. In the second example, the important condition stands out, and the key word (*talk*) stands out even more because it appears in all caps.

Here is a second example of the use of boldface:

> Another must is **Brazos Bend State Park,** one of the best places in Texas to photograph alligators in their natural habitat (Miller).

In this example, the boldface focuses the reader on the important name in the sentence.

Here is another example, from the instructions for a scanner:

> Click **Scan.**

The boldfaced word is the one found on the screen.

Use Highlighters to Help Your Readers

The key to effective usage is to define the way you will use the highlighter. Give it a function. For instance, in the scanner instructions, the highlighting is used to signal that the word in the text is the one to look for on the screen. As soon as you use formatting to indicate a special use or meaning once, you set up a convention that the reader will look for: You have defined a style guide rule for your document. Once you establish a convention, maintain it. In the scanner manual, every time the writer uses boldface, the readers know that they will find that word on their computer screen.

Other Ways to Use Highlighters

Use italics to emphasize a word that you will define:

> "Each element on your form will have a *name* and *value* associated with it. The name identifies the data that is being sent." (Castro 178)

Use quotation marks to introduce a word used ironically or to indicate a special usage.

That was a "normal" sale in their opinion.
The "dense page" issue affects all designers.

Use all caps as a variant of boldface, usually for short phrases or sentences. All caps has the written effect of orally shouting.

Your phone comes from the factory set to "700 msec." THE TALK INDICATOR MUST BE OFF BEFORE PROGRAMMING.

Use vertical lists to emphasize the individual items in the list and to create the expectation in the reader that these are important terms that will be used later in the discussion. Notice below that commas are not used after the items in the list here. The indentation heightens the sense that these words are different from the words in a usual sentence.

Highlighters include boldface, italics, quotation marks, and all caps.

Highlighters include
- boldface
- italics
- quotation marks
- all caps

Font, Font Size, Leading, Columns and Line Length, and Justification

Use text features such as fonts, font size, leading, columns and line length, and justification to affect the reader's ability to relate to the text. Features can seem appropriate or inappropriate, helpful or not helpful.

Font

Font, or typeface, is the style of type. Fonts have personality—some seem frivolous, some interesting, some serious, some workaday.

Consider this sentence in four different typefaces. Alleycat and Sand seem frivolous; Shelley and Alien Ghost are illegible.

The TALK indicator must be off before programming.

The TALK indicator must be off before programming.

The TALK indicator must be off before programming

THE TALK INDICATOR MUST BE OFF BEFORE PROGRAMMING.

Typefaces that routinely appear in reports and letters are Times, Helvetica, and Palatino, all of which appear average or usual, the normal way to deliver information.

This is Times
This is Helvetica
This is Palatino

Fonts belong to one of two major groups: serif and sans serif. The letters in serif faces have extenders at the ends of their straight lines. Sans serif faces do not. Serif faces give a classical, more formal impression, whereas sans serif faces appear more modern and informal. There is some evidence that serif faces are easier to read. However, some tests have indicated the readers prefer serif for longer, continuous text (like this chapter), and prefer sans serif for shorter, more telegraphic text—manuals, for instance. In addition, sans serif fonts are preferable for use in on-line material (Schriver 298, 508).

Many designers suggest that you use the same font for heads and text. Some designers suggest that you use a sans serif font for heads and titles (display text) and a serif font for body text.

Font Size

Font size is the height of the letters. Size is measured in points: 1 point equals $1/72$ inch. Common text sizes are 9, 10, and 12 points. Common heading sizes are 14, 18, and 24 points. Most magazines use 10-point type, but most reports use 12 point.

Font size affects characters in a line; the larger the point size, the fewer characters in a line.

18-point type allows this many characters in this line

9-point type allows many more characters in a line of the same length, causing a different sense of width (Felker).

As Figure 6.5 shows, type size affects the appearance, length, and readability of your document.

Leading

Related to size is *leading,* or the amount of space between lines. Leading is also measured in points and is always greater than the font size. Word processing programs select leading automatically so it is not usually a concern. However, too much or too little leading can cause text to look odd. Notice the effects of leading on the same sentence:

▶ Twelve point text with 12 point leading:

Technical communication is "writing that aims to get work done, to change people by changing the way they do things."

▶ Twelve point text with 18 point leading:

Technical communication is "writing that aims to get work done, to change people by changing the way they do things."

Figure 6.5

Text Features

Learning Format   5

18-Point Helvetica

This paragraph appears in 12-point Palatino, a *serif* font. The paragraph is "ragged right," which means I turned off the "right-justification" command in my word processor. It is set in 12-point type because it is extremely easy to read.

This paragraph appears in 9-point Helvetica, a **sans serif** font. The paragraph is right-justified. The right margin appears as a straight line. Research and practice vary on right-justifying. Research suggests not to.
• *Time* magazine does not justify.
• The *New Yorker* does.
Long paragraphs of Helvetica are not comfortable to read.

One column

Rule

Italics highlight
Serif font

Ragged right margin

Bold highlight
Sans serif font

Right-justified

Vertical list highlight

Columns

Columns are vertical lines of type; a normal typed page is just one wide column. Many word processing programs allow multiple columns (12 or more); in practice, however, reports seldom require more than two columns. In general, use a single column for reports. To achieve a contemporary design, consider using a 2- or 2-½-inch-wide left margin. In other cases, two columns are especially useful for reports and manuals if you plan to include several graphics. For various column widths, see Examples 6.1 and 6.2, pages 156–158.

Column width affects line length, the number of characters that will fit into one line of type. Line length affects readability (Felker; Schriver). If the lines are too long, readers must concentrate hard as their eyes travel across the page and then painstakingly locate the next correct line back at the left margin. If the lines are too short, readers become aware of shifting back and forth more frequently than normal. Short lines also cause too much hyphenation.

Typographers use three rules of thumb to choose a line length and a type size:

Use one and a half alphabets (39 characters) or 8 words of average length per line.
Use 60 to 70 characters per line (common in books).
Use 10 words of average length (about 50 characters).

Unfortunately, no rule exists for all situations. You must experiment with each situation. In general, increase readability by adding more leading to lines that contain more characters (White).

Justification

Justification (see Figure 6.5) means aligning the first or last letters of the lines of a column. Documents are almost always *left-justified,* that is, the first letter of each line starts at the left margin. *Right-justified* means that all the letters that end lines are aligned at the right margin. Research shows that ragged-right text reads more easily than right-justified text (Felker).

Combining Features to Orchestrate the Text for Readers

Given all the possibilities for combining features in order to help readers quickly get accurate information, what are some guidelines to help with that task? The goal of design is twofold: to help readers easily find the information; and to reveal the logical structure of the document. This section will give you several guidelines to help you with your orchestration (Schriver).

Analyze: Identify the Rhetorical Clusters in Your Document

Rhetorical clusters are visual and verbal elements that help the reader interpret the content in a certain way. Every document has many rhetorical clusters: titles, heads, visuals, captions, paragraphs, warnings, numbers, and types of links (in online reports). You must be aware of all of these items and treat them appropriately.

Standardize: Give Each Text or Visual Feature a Purpose

In the scanner example (p. 146), a boldfaced word in the text indicated a word that appeared on the computer screen. Thus, boldfaced words are a rhetorical cluster. They tell the reader how to interpret a word that is treated differently, from the main text. Readers will quickly interpret your purpose and cluster design, counting on it to help them with the contents of the document.

Be Consistent: Treat All Like Items Consistently Throughout the Document

In effect, you repeat the design of any item and that repetition sets up the expectation of the reader. Once the expectation is set up, readers look for the same item to cue them to interpret the content. They know that all cap heads indicate the start of a new section, that boldface indicates a special item, that indented lists are important, that 12-point links go to major sections of a document and 9-point links lead to other sites.

Be Neat: Align Items

Aligning basically means to create a system of margins and start similar features at the same margin. Figure 6.6 shows items haphazardly related and aligned.

Notice that the left edges of all the visuals align with one another, as do the left edges of all the text chunks. In addition, the top edge of each visual is aligned with the top edge of each text chunk. Alignment creates meaningful units.

Figure 6.6

Ineffective Versus Effective Use of Edges

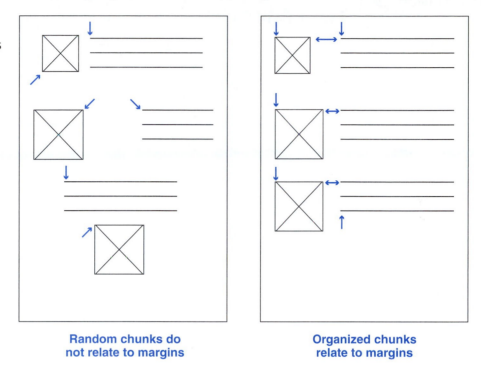

Random chunks do not relate to margins

Organized chunks relate to margins

Learn: Use the Design Tips of Experts

Designers have researched many features to determine what is most effective:

1. Use top-to-bottom orientation to gain emphasis. Typically, readers rank the item at the top as the most important. Put your most important material near the top of the page (Sevilla).

2. Use brightness to gain emphasis. Readers' eyes will travel to the brightest object on the page. If you want to draw their attention, make that item brighter than the others (Sevilla).

3. Use larger-to-smaller orientation (Figure 6.7, p. 152). Readers react to size by looking at larger items first (Sadowski). Put important material (such as main heads) in larger type. *Note:* Boldfacing causes a similar effect. Boldfaced 12-point type seems larger than normal (roman) type.

4. Use left-to-right orientation to lead your readers through the text (Rubens). Place larger heads or key visuals to the left and text to the right in order to draw readers into your message. See Figure 6.8 (p. 152), where the large left margin and heads perform this function.

Figure 6.7

Larger-to-Smaller
Orientation

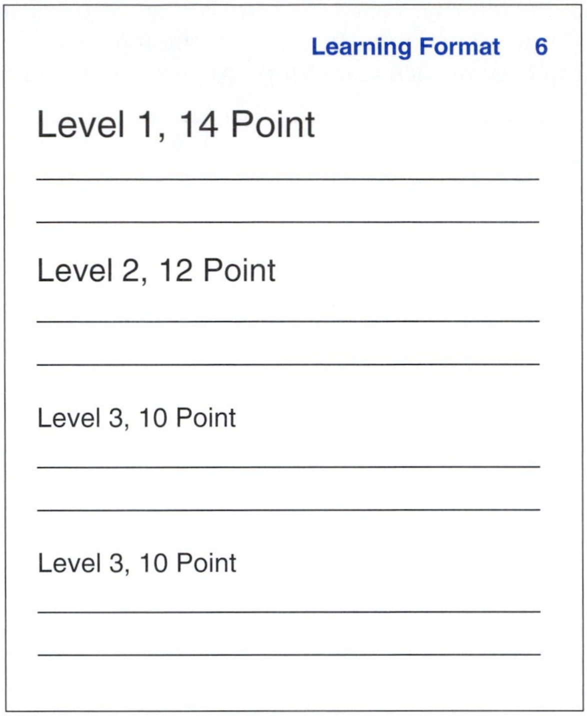

Learning Format   6

Level 1, 14 Point

Level 2, 12 Point

Level 3, 10 Point

Level 3, 10 Point

Figure 6.8

Left-to-Right
Orientation

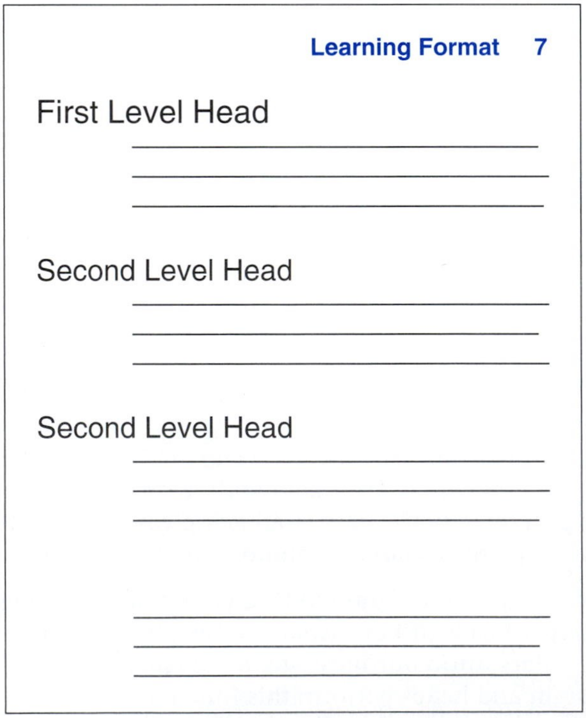

Learning Format   7

First Level Head

Second Level Head

Second Level Head

Modern large
left margin

Figure 6.9

Placement
of Visuals

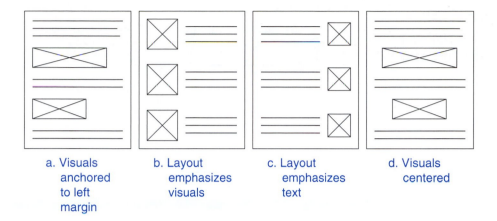

a. Visuals
anchored
to left
margin

b. Layout
emphasizes
visuals

c. Layout
emphasizes
text

d. Visuals
centered

5. Place visuals so that they move the reader's attention from left to right (Rubens; *Xerox*). In a two-column format (Figure 6.9), you can place the visuals to the left and the text to the right or vice versa, depending on which you want to emphasize. Whatever you do, always anchor visuals by having one edge relate to a text margin (Sadowski).

6. In a multiple-page document, "hang" items from the top margin. In other words, keep a consistent distance from the top margin to the top of the first element on the page (whether head, text, or visual) (Cook and Kellogg). See Figure 6.10 (p. 154).

7. Learn to use color effectively. Guidelines for effective use of color can be found in "Focus on Color," pages 166–173.

Developing a Style Sheet and Template

In order to remain consistent, especially if you are working as part of a group, develop a *style sheet,* a list of specifications for each element in your document. You develop this list as part of your planning process. For brief documents you may not need to write it out, but you do need to think it through. Longer documents or group projects require a written or electronic style sheet. For instance, for a two-page memo, the style sheet would be quite short:

▶ Margins: 1-inch margin on all four sides
▶ Line treatment: no right justification
▶ Spacing within text: single-space within paragraphs, double-space between paragraphs
▶ Heads: heads flush left and boldfaced, triple-space above heads and double-space below
▶ Footers: page numbers at bottom center

Figure 6.10

Items Hanging
from Top Margin

Hang items from the same top margin

For a more complicated document, you need to make a much more detailed style sheet. In addition to margins, justification, and paragraph spacing, you need to include specifications for

- A multilevel system of heads
- Page numbers
- Rules for page top and bottom
- Rules to offset visuals
- Captions for visuals
- Headers and footers—for instance, whether the chapter title is placed in the top (header) or bottom (footer) margins
- Lists

Figure 6.11 shows a common way to handle style sheets. Instead of writing out the rules in a list, you make a *template* that both explains and illustrates the rules. (For more on planning style sheets, see the section in Chapter 17, "Format the Pages," pp. 447–449.)

The electronic style sheet is a particularly useful development. Many word processing and desktop publishing programs allow you to define specifications for each style element, such as captions and levels of heads. Suppose you want all level 1 heads to be Helvetica, 18-point, bold, flush left, and you want all figure captions to be Palatino, 8-point, italic. The style feature allows you to enter these commands into the electronic style sheet for the document. You can then direct the program to apply the style to any set of words.

Figure 6.11

Sample Template

CHAPTER HEADER
_____

HEAD LEVEL 1

Introductory level 1 text is not indented. Double-space between head and text. Text at all levels is not right-justified. Triple-space above level 2 heads.

Head Level 2
Introductory level 2 text is not indented. Skip no space between text and head. Subsequent paragraphs are indented 5 spaces, single space text.

   Head Level 3. Third level head is indented 5 spaces. Text is placed at left margin. Double space above level 3 head.

   To make a list:
- use a colon
- use bullets, but no end punctuation
- skip 2 spaces to the right and start the second line under the first letter of the first line

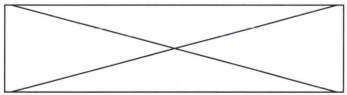

Fig 1. *Italicize caption at left edge*

    Usually you can also make global changes with an electronic style sheet. If you decide to change all level 1 heads to Palatino, 16-point, bold, flush left, you need only make the change in the style sheet, and the program will change all the instances in the document.

Worksheet for a Style Sheet

☐ Select margins.

☐ Decide how many levels of heads you will need.

☐ Select a style for each level.

☐ Select a location and format for your page numbers.

(continued)

(continued)

☐ Determine the number of columns and the amount of space between them.

☐ Choose a font size and leading for the text.

☐ Place appropriate information in the header or footer area.

☐ Establish a method for handling vertical lists.
Determine how far you will indent the first line. Use a bullet, number, letter, or some other character at the beginning of each item. Determine how many spaces will follow the initial character. Determine where the second and subsequent lines will start.

☐ Choose a method for distinguishing visuals from the text.
- Will you enclose them in a box or use a rule above and below?
- Where will you place visuals within the text?
- How will you present captions?

Examples

Examples 6.1 and 6.2 present the same report section in two different formats, each the result of a different style sheet.

Example 6.1

Two-Column Design

DISCUSSION

High-Protein Diets

Introduction

The goal of this search was to determine if the Internet was a valuable source of information regarding high-protein diets. To define my information as usable, it must meet three criteria.

The three criteria are as follows: the information must be no older than 1997, the sites must be found to be credible sites, and it must take no longer than 10 minutes to find information pertinent to the topic on each site.

Findings

I used the Dogpile search engine to find my sites on high-protein diets. The keywords I used were *protein, high protein*, and *fad diets*. These keywords led me to the sites listed below (Table 1).

As seen in Table 1, most of the sites fit the criteria. Using Dogpile to search for nutrition information yielded mixed results of commercial and professional sites. The Internet provided a vast amount of information regarding high-protein diets.

In my results I determined that the information from the website cyberdiet.com was credible even though it was a commercial site. The Tufts University Nutrition Navigator, a well-known, credible website that evaluates nutrition websites, recommended Cyberdiet and gave it a score of 24 out of 25.

Conclusion   The credibility and recency of the information did not all meet the criteria. Because of this, I conclude that the Internet does have valuable information regarding high-protein diets, but that the Web user must use caution and be critical in determining the validity of each website.

Table 1. Standards of High-Protein Diet Search

	Less Than 10 Min.?	Recency (>1997)	Credibility
Cyberdiet.com "High Protein, Low Carbohydrate Diet"	Yes	1999 Yes	Yes
Heartinfo.org "The Reincarnation of the High Protein Diet"	Yes	1997 Yes	Yes-Professional
more.com "Information on High Protein Diets"	Yes	——	No-Commercial
Prevention.com "A Day in the Zone"	Yes	1995 No	No-Consumer
Eatright.org (ADA) "In the News: High-Protein/Low Carbohydrate Diets"	Yes	1998 Yes	Yes-Professional

Example 6.2

One-Column Design

DISCUSSION

High-Protein Diets

Introduction.   The goal of this search was to determine if the Internet was a valuable source of information regarding high-protein diets. To define my information as usable, it must meet three criteria.

The three criteria are as follows: the information must be no older than 1997, the sites must be found to be credible sites, and it must take no longer than 10 minutes to find information pertinent to the topic on each site.

Findings.   I used the Dogpile search engine to find my sites on high-protein diets. The keywords I used were *protein, high protein,* and *fad diets.* These keywords led me to the sites listed below (Table 1).

(continued)

Example 6.2

(continued)

Table 1. Standards of High-Protein Diet Search

	Less Than 10 Min.?	Recency (>1997)	Credibility
Cyberdiet.com "High Protein, Low Carbohydrate Diet"	Yes	1999 Yes	Yes
Heartinfo.org "The Reincarnation of the High Protein Diet"	Yes	1997 Yes	Yes-Professional
more.com "Information on High Protein Diets"	Yes	——	No-Commercial
Prevention.com "A Day in the Zone"	Yes	1995 No	No-Consumer
Eatright.org (ADA) "In the News: High-Protein/Low Carbohydrate Diets"	Yes	1998 Yes	Yes-Professional

As seen in Table 1, most of the sites fit the criteria. Using Dogpile to search for nutrition information yielded mixed results of commercial and professional sites. The Internet provided a vast amount of information regarding high-protein diets.

In my results I determined that the information from the website cyberdiet.com was credible even though it was a commercial site. The Tufts University Nutrition Navigator, a well-known, credible website that evaluates nutrition websites, recommended Cyberdiet and gave it a score of 24 out of 25.

Conclusion.   The credibility and recency of the information did not all meet the criteria. Because of this, I conclude that the Internet does have valuable information regarding high-protein diets, but that the Web user must use caution and be critical in determining the validity of each website.

Exercises

▶ You Create

1. For a nonexpert audience, write a three- to five-paragraph description of a machine or process you know well. Your goal is to give the audience a general familiarity with the topic. Create two versions. In version 1, use only chunked text. In version 2, use at least two levels of heads, a bulleted list, and a visual aid. Alternate: Using the same instructions, create a description for an expert audience.

2. Write a paragraph that briefly describes a room in which you work. Create and hand in at least two versions with different designs (more if your instructor requires). Alter the size of the type, the font, and the treatment of the right margin. For instance, produce one in a 12-point sans serif font, right-justified format, and another in a 10-point serif font, ragged-right format. Label each version clearly with a head. Write a brief memo explaining which one you like the best.

3. Create a style sheet and template for either Example 6.1 or Example 6.2.

4. Create a layout for the following instructions and visual aids. Correct any typographical or formatting errors. Develop a style sheet to submit to your instructor before you redo the text.

SOLDERING PROCESS

Introduction

The following information will show you how to make a correct solder joint between two wires. A solder joint is needed to electrically and physically connect two wires. A correct solder joint is one that is that connects the wire internally as well as externally (Figure 1).

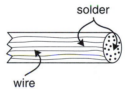

Figure 1
Correct Joint

A incorrect solder joint does not maximize the electrical connection and also will probably break if handled. Due to the wide varieties of wire and soldering irons, this demonstration will be done using a 35-watt soldering iron and 16-gauge insulated wire. Generally, the thicker the wire, the more energy it takes to heat the wire. This means that any wire larger than 16 gauge would require a larger-wattage soldering iron.

Process

Step 1—Preheat Iron
The soldering iron must be plugged in and allowed to heat for at least 5 minutes. This will assure that the iron has reached its heating potential. This means that the iron will properly heat the iron and solder.
Step 2—Prepare Wire
Prepare both wires by cutting the ends to be soldered. Be sure to leave no frayed wire or insulation hanging.
Step 3—Strip Wire
Using a wire stripper, strip off approximately $1/2''$ of insulation from both wires (Figure 2). Visually check to be sure there is no hanging or frayed wire. If there is trim them off.

Figure 2

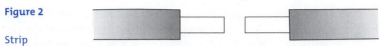

Strip

Step 4—Clean Iron

Using a damp cloth, wipe off the end of the soldering iron. Be sure the tip is clear of any residue. This will assure that the solder will be free of impurities.

Step 5—Tin Iron

Using Rosin Core solder, apply a small amount of solder to the iron tip (Figure 3). Rosin Core solder is a special solder that has cleaning solvents internally in the solder. This makes soldering possible without the need of using other solvents such as Flux. Tinning the iron tip will allow maximum heat transfer between the iron and the wire.

Figure 3
Tinning

Step 6—Twist Wires

Setting wires end-to-end, wrap them around each other. Be sure they are twisted together tight to be sure that they are making a good physical connection. The solder will only join the wires where they are touching (Figure 4).

Figure 4
Wire Twist

Step 7—Apply Heat

Touch the soldering iron tip firmly to the middle of the joined wire. Set the tip at a 45-degree angle to the wire (Figure 5). Heat focuses to a point so the most energy is at the tip of the iron.

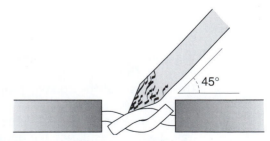

45°

Figure 5
45-Degrees

Step 8—Apply Solder

Apply solder where tip meets the wire (Figure 6). Observe solder to see if it is flowing into the wire. If the solder is globbing to the tip, either the tip is dirty and must be cleaned or the iron is not being firmly pressed to the wire. Watch insulation to be sure it does not melt. Apply solder until the wire appears to be full. Do not apply too much solder for it will begin to glob on the surface.

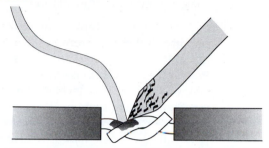

Figure 6
Solder

Step 9—Let Cool
Allow the soldered joint to cool slowly. Rapid cooling will cause the solder to become brittle.

Step 10—Inspect Solder Joint
Visually inspect the solder joint. Check to see if the solder has flowed into the wire. Make sure that it has not globbed around one point in the joint. Also check to be sure that the solder has made a solid physical connection (Figure 7). If the solder joint fails either of these qualifications, you must cut the wire and start over.

Good   Bad

Figure 7   Inspect

Step 11—Tape Joint
Using electrical tape, wrap joint to be sure that no metal is exposed (Figure 8).

Figure 8
Tape

5. Use the following information to create a poster on your desktop system or word processor.

   a. The campus "golden oldies" club will hold a workshop for interested potential members on Wednesday, November 4, from 6:30 to 8:30 p.m. at the main stage in the University Student Center. No prior experience is necessary.

   b. The local Digital Photo Club will hold a benefit auction for the area United Way Food Pantry on Sunday, May 8, at 1:00 p.m. at the local fairgrounds. Admission is $10.00. The event will consist of photos for sale and workshops on buying digital cameras and creating digital photos.

6. Create a poster for an event that interests you.

▶ **You Revise**

7. Use principles of design and visual logic to revise the following paragraph. Also eliminate unnecessary information.

> **Capabilities of System.**   The new system will need to provide every capability that the current system does. I spoke with Dr. Franklin Pierce about this new system. Dr. Pierce has a vast amount of experience with Pascal (the old system is written in it) and in Ada (the new system is to be written in it). He also developed many software systems, including NASA's weather tracking system. After looking at the code for the current system, Dr. Pierce assured me that every feature in the old system can be mapped to a feature written in Ada. He also said that using Ada will allow us *multiple versions* of the program by using a capability of Ada to determine what type of computer is used for the menu. Furthermore, he said that using Ada will allow us to *improve the performance* of some of the capabilities, such as allowing the clock to continually be updated instead of stopping while another function is being performed and then being updated after that function has been completed. Thus, the first criterion could be met.

8. Rearrange the following layout so that it is more pleasing. Write a memo that explains why you made your changes.

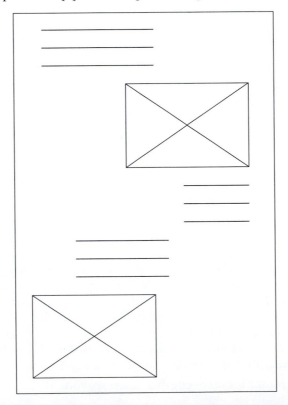

▶ Group

9. In groups of three or four, analyze Examples 6.1 and 6.2, pages 156–158. Decide which one you prefer, and explain your decision in a group memo to the class.

10. In groups of two, create a design for the following text, which recommends that a student center purchase a particular sound board. Your instructor will require each group to use one of the various design strategies discussed in this chapter. Be prepared to present an oral report explaining your format decisions.

> This is a recommendation on whether to purchase a Goober sound board or a Deco sound board for the Technical Services Crew at the Student Center. Over the past few years, we have rented a sound board when bands or other large speaking groups come to the Student Center. The rental costs run from $200 to $500, and now that money has been allocated to buy a sound board, we should seriously consider purchasing one. The use we will get out of the board will make it pay for itself in two or three years. This memo will detail my recommendation as to which board to purchase.
>
> I recommend we purchase the Goober XS-2000 24-channel sound board. Greg Newman, the Tech Crew Chief, and I have compared the Goober XS-2000 to the Deco TXS-260 24-channel board in audio magazines and by talking to people who have used the Goober board and the Deco board. We found the Goober to be an overall better board. We have also experienced using both boards on many occasions and have liked the Goober board better. We compared the boards in terms of these criteria: features, reliability/experience, and cost.
>
> The first criterion is the features. This is the most important factor in deciding which sound board to purchase. We are looking for a sound board that provides 24 channels in and 8 lines out. The lines out are used for sound effects and equalization. The more "outs" you have, the better the sound will be. This is what the Goober XS-2000 has: 24 channels in and 8 lines out. The Deco TXS-260 has 24 channels in and only 6 lines out. We find it necessary to have 8 lines out because we want to offer bands the best-quality sound possible.
>
> Reliability is another important factor. Goober and Deco are both reputable companies that make good sound boards, but when it comes down to operating at maximum efficiency, we feel the Goober board can offer us better reliability. I had the opportunity to speak with Will Hodges of Southern Thunder Sound, a sound rental company from which we have rented sound boards. He said that most bands that come in to rent his equipment rent the Goober boards because they are easier to work with and don't break down as often as the Deco boards. Mr. Hodges's opinion lends further credibility to my recommendation that we purchase the Goober board.
>
> Greg and I have had experience working with both boards. Once when we rented a Deco board, a loud buzzing sound began to be emitted from

the speakers halfway through the concert. We did everything we could, but the buzzing continued throughout the concert. On the occasions when we have worked with a Goober board, it has operated without any problems.

The cost is the last criterion. We have allocated an amount of $3500 to purchase a sound board. The Deco TXS-260 is $2800 with a two-year limited warranty. The Goober XS-2000 is $3600 with a five-year limited warranty. We have dealt with this Goober dealer before, and she said she would throw in a 100-foot 24-channel snake (a $700 value) free and would knock $300 off the total price—a $4300 value for $3300. A $1000 savings. Although the Deco board is less expensive, we should spend the extra $500 for a better board with a longer warranty, greater reliability, and a free 100-foot snake.

Writing Assignments

1. In one page of text (just paragraphs), describe an opportunity for your firm, and ask for permission to explore it. Your instructor will help you select a topic. Your audience is a committee that has resources to take advantage of the opportunity. Use a visual aid if possible. Then use the design guidelines in this chapter to develop the same text into an appropriate document for the committee. In groups of three or four, review all the documents to see where the format best conveys the message. Report to the class your conclusion about which was best and why it works.

2. Write a learning report for the assignment you just completed. See Chapter 5, Writing Assignment 7, pages 134–135, for details of the assignment.

Web Exercises

1. Analyze two webpages in order to explain to a beginner audience how to lay one out—where to place titles, how large to make them, etc. To give your advice a range of examples, use a homepage/index and an information link by following a particular path through several links, e.g., About Us/Our Products/Cameras/FX20MicroZoom/Technical Spec.

2. Analyze the color scheme in a website. Use a major corporation like AT&T or Sun Micro. Write a memo to your classmates that explains how the site uses color to make its message clear.

Works Cited

Castro, Elizabeth. *HTML4 for the World Wide Web*. Berkeley, CA: Peachpit, 1998.
Cook, Marshall, and Blake R. Kellogg. *The Brochure: How to Write and Design It*. Madison, WI: privately printed, 1980.

Felker, Daniel B., Frances Pickering, Veda R. Charrow, V. Melissa Holland, and Janice C. Redish. *Guidelines for Document Designers*. Washington, DC: American Institutes for Research, 1981.

Kostelnick, Charles. "Supra-Textual Design: The Visual Rhetoric of Whole Documents." *Technical Communication Quarterly* 5.1 (1995): 9–33.

Kramer, Robert, and Stephen A. Bernhardt. "Teaching Text Design." *Technical Communication Quarterly* 5.1 (1995): 35–60.

Miller, Brian. "The Uniquely Southern Landscape." *Outdoor Photographer* (June 2003): 66.

Panasonic. *Operating Instructions for Model KX-TC 1800B*. Secaucus, NJ: Panasonic, n.d.

Rubens, Phillip M. "A Reader's View of Text and Graphics: Implications for Transactional Text." *Journal of Technical Writing and Communication* 16 (1986): 73–86.

Sadowski, Mary A. "Elements of Composition." *Technical Communication* 34 (1987): 29–30.

Schriver, Karen A. *Dynamics in Document Design: Creating Texts for Readers*. New York: Wiley, 1997.

Sevilla, Christine. "Page Design: Directing the Reader's Eye." *intercom* (June 2002): 6–9.

Southard, Sherry. "Practical Considerations in Formatting Manuals." *Technical Communication* 35.3 (1988): 173–178.

Tyson, Paul. "Designing Documents." *intercom* (December 2002): 27–29.

White, Jan V. *Graphic Design for the Electronic Age*. New York: Watson-Guptill, 1988.

Xerox Publishing Standards: A Manual of Style and Design. New York: Xerox Press, 1988.

Focus on
Color

Color invites us in; it entices. New technology has allowed color into newspapers, and Web pages explode with color. Cheaper technology that allows individuals to add color to documents has opened a new world for communicators. Now anyone can produce a multiple-color document. Working effectively with color means knowing

- How color relationships cause effects.
- How color can be used in documents.

Effects Produced by Color Relationships

When colors are placed next to each other, their relationships cause many effects. The concepts of the color wheel and value illustrate how color affects visibility. Various hues cause emotional and associational reactions.

The Color Wheel and Visibility   The key concept is the color wheel, which provides a way to see how colors relate to one another. Figure 1 shows the six basic rainbow colors, from which all colors can be made.

The colors that occur across from one another are called *complementary* (e.g., red and green), those that touch are called *adjacent* (e.g., red and orange or violet), and those that are two apart are called *contrasting* (e.g., red and yellow, red and blue, orange and green, orange and violet). The relationships are shown in Figures 2 through 4. Each of the relationships affects visibility.

Maximum Visibility   Figure 2 shows that sets of complementary colors cause high contrast or maximum visibility. This relationship strongly calls attention to itself. Notice, however, that the colors tend to "dance." Most people find them harsh and can view them for only a short time.

Minimum Visibility   Figure 3 shows that sets of adjacent colors cause low contrast or minimal visibility. These relationships tend to have a relaxing, pleasing effect. The colors, however, tend to blend in with each other, making it difficult to distinguish the object from the background.

Pleasing Visibility   Figure 4 shows that sets of contrasting colors cause medium contrast or pleasing visibility. These relationships have strong contrast, but they do not dance. Most people find them bold and vivid.

Value Affects Visibility of Individual Hues   In addition to having relationships with other colors, any color has relationships with itself. So a basic color, like blue, is called a *hue.*

However, you can mix white or black with the hue and so create a value—a *tint* (a hue and white) or a *shade* (a hue and black). Figure 5 shows blue as a hue in the center, but as a tint to the left and a shade to the right. The more white you add, the lighter the tint; the more black, the darker the shade.

Reduce Harsh Contrast by Changing Value   Understanding value allows you to affect the impact of color relationships. Although complementary colors of equal value are quite jarring, as shown in Figure 2, notice how the jarring decreases when the background color changes value either as a tint (Figure 6) or a shade (Figure 7).

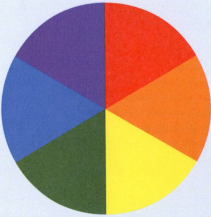

Figure 1
The Basic Color Wheel

Figure 2
Complementary Colors

Figure 3
Adjacent Colors

Figure 4
Contrasting Colors

Figure 5
Tint, Hue, Shade

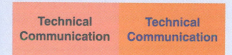

Figure 6
Tinted Background

Figure 7
Shaded Background

Increase Visibility by Using Innate Value Colors have innate value. In other words, the basic hue of some colors is perceived as brighter than the basic hue of others. The brightest basic hue is yellow and the darkest is violet (see Figure 8, p. 168). The brighter hues are more difficult to read (because they do not contrast as well with white, the color of most pages) than the darker ones.

Colors and Emotions As shown in Figure 9 (p. 168) colors are divided into warm (red, orange, yellow) and cool (green, blue, violet). Warm colors appear to most people to be soft, cozy, linked to passions, celebrations, and excitement. Cool colors appear to most people as harder, icier, linked to rational, serious, reliable decorum.

(continued)

(continued)

Figure 8
Innate Values

Colors and Associations   Colors have traditional associations. In the United States, red, for instance, implies stop, hot, desire, passion, energy. Blue implies peace, heavenliness, space, cold, calm. Green implies jealousy, nature, safety. Yellow implies joy, happiness, intellect. Using colors appropriately in context makes your work more effective. Consider:

That pipe is hot and That pipe is hot.

Try to remain calm and Try to remain calm.

Colors have different associations in different cultures. See "Globalization and Visual Aids" in Chapter 7.

How Color Can Be Used in Documents

Color has four important functions in documents. Use color to

- Make text stand out.
- Target information.
- Indicate organization.
- Indicate the point in a visual aid.

Use Color to Make Text Stand Out   As Figures 2 through 7 illustrate, one common use of color is to make text clearly visible, or legible. Legibility is a matter of contrast. If the text contrasts with its background, it will be legible. Two principles are helpful.

- The best contrast is black type on a white background.
- Colored type appears to recede from the reader. Higher values (yellow) appear farther away and less legible than lower values (blue).

Color alone will not cause individual words to stand out in black text. To make colored text stand out from black text, you must also change its size. Figures 10 and 11 show that size and color cause minimal emphasis in text, but Figure 12 shows that colored words made larger do show more emphasis. Figure 13 shows that words will stand out if the color contrasts with the color of the surrounding text.

Technical Communication	Technical Communication	Technical Communication

Figure 9a
Warm Colors

Technical Communication	Technical Communication	Technical Communication

Figure 9b
Cool Colors

It is a well-known
fact that technical
communication is
the best of all
possible subjects
to study.

Figure 10
Minimal Emphasis via Size

It is a well-known
fact that technical
communication is
the best of all
possible subjects
to study.

Figure 11
Minimal Emphasis via Color

It is a well-known
fact that technical
communication is
the best of all
possible subjects
to study.

Figure 12
Emphasis via Size and Color

It is a well-known
fact that technical
communication is
the best of all
possible subjects
to study.

Figure 13
Emphasis via Contrast

Use Color to Target Information   Color focuses attention. It does this so strongly that color creates "information targets." In other words, people see color before they see anything else. As a consequence, you should follow these guidelines:

- Use color to draw attention to "independent focus" text—types of text that readers must focus on independently of other types.
- Make each type of text look different from the other types. Use different value (tint or shade, not hue) and different areas or shape to cause the difference.

  Common examples of independent focus texts are

- Warnings.
- Hints.
- Cross references.
- Material the reader should type.
- Sidebars.

Figure 14 (p. 170) shows a page that has each of these items. Notice that the difference is achieved by differences in value, shape, and location.

Use Color to Indicate Organization   Color creates a visual logic. Readers quickly realize that color indicates a function. How strongly a color indicates that function is increased when that color is combined with shape and area.

You can use color to indicate different levels in your document's hierarchy. Common functions that color can indicate are

- Marginal material.
- Running information—in headers and footers, or in the head system.

Figure 15 (p. 171) shows a page that has each of these items. Notice that the difference is achieved by differences in value, shape, and location.

(continued)

(continued)

Figure 14
Examples of
Independent-Focus
Text

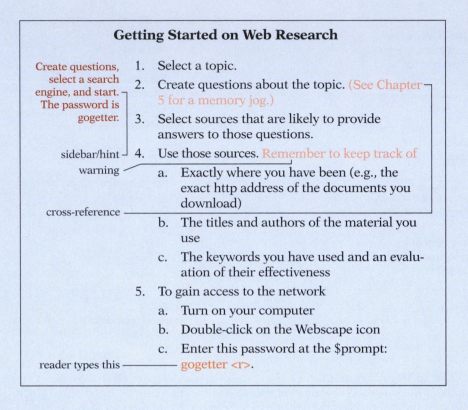

Getting Started on Web Research

Create questions, select a search engine, and start. The password is gogetter.

1. Select a topic.
2. Create questions about the topic. (See Chapter 5 for a memory jog.)
3. Select sources that are likely to provide answers to those questions.

sidebar/hint

warning

4. Use those sources. Remember to keep track of
   a. Exactly where you have been (e.g., the exact http address of the documents you download)

cross-reference

   b. The titles and authors of the material you use
   c. The keywords you have used and an evaluation of their effectiveness
5. To gain access to the network
   a. Turn on your computer
   b. Double-click on the Webscape icon
   c. Enter this password at the $prompt:

reader types this ——————— gogetter <r>.

Use Color to Indicate the Point of a Visual Aid Use color in visual aids to draw readers' attention to specific items(see Figure 16, p. 172). For instance,

- To highlight a single line in a table.
- To highlight the data line in a line graph.
- To focus attention on a particular bar or set of bars in a bar graph.
- To differentiate callouts and leaders from the actual visual.

Be aware that many of the graphics programs provide you with color but that the color is not chosen to make this visual aid make more sense.

Summary Guidelines for Using Color Follow these basic guidelines in your handling of color in your documents:

- Be consistent. For each type of item, use the same hue, value, shape, and location.
- Correctly use contrast. Follow the color wheel to select combinations of colors that create high visibility. Remember that black and white create the highest contrast. Colored text must change size to contrast with surrounding black text.
- Correctly use feeling and association. Use warm colors for action items, especially warnings. Use cool colors for reflection items.

Figure 15
Examples of Color
Function

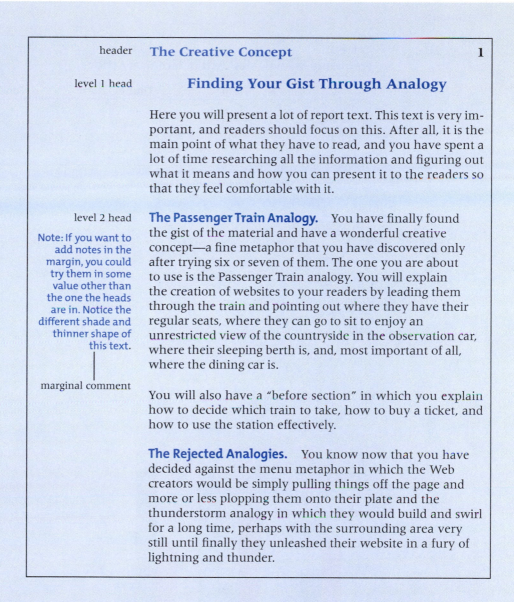

header   The Creative Concept   1

level 1 head

Finding Your Gist Through Analogy

Here you will present a lot of report text. This text is very important, and readers should focus on this. After all, it is the main point of what they have to read, and you have spent a lot of time researching all the information and figuring out what it means and how you can present it to the readers so that they feel comfortable with it.

level 2 head

Note: If you want to add notes in the margin, you could try them in some value other than the one the heads are in. Notice the different shade and thinner shape of this text.

marginal comment

The Passenger Train Analogy. You have finally found the gist of the material and have a wonderful creative concept—a fine metaphor that you have discovered only after trying six or seven of them. The one you are about to use is the Passenger Train analogy. You will explain the creation of websites to your readers by leading them through the train and pointing out where they have their regular seats, where they can go to sit to enjoy an unrestricted view of the countryside in the observation car, where their sleeping berth is, and, most important of all, where the dining car is.

You will also have a "before section" in which you explain how to decide which train to take, how to buy a ticket, and how to use the station effectively.

The Rejected Analogies. You know now that you have decided against the menu metaphor in which the Web creators would be simply pulling things off the page and more or less plopping them onto their plate and the thunderstorm analogy in which they would build and swirl for a long time, perhaps with the surrounding area very still until finally they unleashed their website in a fury of lightning and thunder.

- Generally use only one hue with varying tints and shades. Use two or more colors after you have practiced with color and have had your creations critiqued by readers and users.
- Help color-blind readers by using different brightnesses of the same color. Remember that location on the page will help color-blind readers also. (Even if they cannot "read" the red, they will know it is a marginal comment because of where it is located.)

(continued)

(continued)

Figure 16
Use of Color in Visual
Aids

Table 1
Frequency of Analogies Used in Technical Articles

	Train	Thunderstorm	Food Menu
Packaging	30	40	30
Hospitality	20	15	65
Automotive	10	75	15

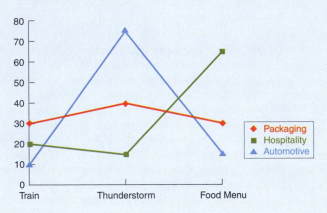

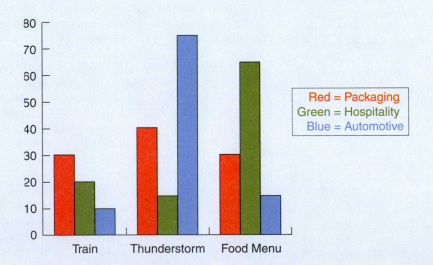

Works Consulted

The Basics of Color Design. Cupertino, CA: Apple, 1992.

"Color Wheel Pro—See Color Theory in Action." 10 Apr. 2004 <www.colorwheel-pro.com/color-meaning .html>.

Horton, William. "Overcoming Chromophobia: A Guide to the Confident and Appropriate Use of Color." *IEEE Transactions of Professional Communication* 34.3 (1991): 160–171.

Jones, Scott L. "A Guide to Using Color Effectively in Business Communication." *Business Communication Quarterly* 60.2 (1997): 76–88.

Keyes, Elizabeth. "Typography, Color, and Information Structure." *Technical Communication* 40.4 (1993): 638–654.

Mazur, Beth. "Coming to Grips with WWW Color." *intercom* 44.2 (1997): 4–6.

White, Jan. *Color for the Electronic Age.* New York: Watson-Guptill, 1990.

Chapter

7

Using Visual Aids

Chapter 7
In a Nutshell

Visual aids help readers by summarizing data and by showing patterns.

Common types of visual aids.

Tables have rows and columns of figures. Place the items to compare (corporate sales regions) down the left side and the ways to compare them (by monthly sales) along the top. The data fill in all the appropriate spaces.

Line graphs illustrate a trend. Place the items to compare along the bottom axis (days of the week) and the pattern to illustrate (the price of the stock) along the left axis. The line shows the fluctuation.

Bar graphs illustrate a moment in time. Place the items to compare along the bottom axis (four cities) and the terms to compare (population figures) along the left axis. The bars show the difference immediately.

Pie charts represent parts of a whole. They work best when they compare magnitude.

Illustrations, either photos or drawings, show a sequence or a pattern—the correct orientation for inserting a disk into a computer for example.

Guidelines for using visual aids.

▶ Develop a *visual logic*—place visuals in the same position on the page, make them about the same size, treat captions and rules (black lines) the same way. Be consistent.

▶ Create neat visuals to enhance your clarity and credibility.

▶ Tell readers what to notice and explain pertinent aspects, such as the source or significance of the data.

▶ Present visual aids below or next to the appropriate text.

Visual aids are an essential part of technical writing. Graphics programs for personal computers allow writers to create and refine visual aids within reports. Many of the bar graphs, line graphs, and pie charts shown in this chapter are computer generated. This chapter presents an overview of visual aids. It explains general information on using, creating, and discussing visual aids, as well as specific information on all the common types of visual aids.

The Uses of Visual Aids

Visual aids have a simple purpose. According to noted theorist Edward Tufte, visual aids "reveal data" (13). This key concept controls all other considerations in using visual aids. You will communicate effectively if your visual aids "draw the reader's attention to the sense and substance of the data, not to something else" (91). Technical writers use visual aids for four purposes:

▶ To summarize data
▶ To give readers an opportunity to explore data
▶ To provide a different entry point into the discussion
▶ To engage reader expectations

To summarize data means to present information in concise form. Figure 7.1, a graph of a stock's price for one week, presents the day-end price of the stock for each day. A reader can tell at a glance how the stock fared on any day that week.

Figure 7.1

Graph That
Summarizes
and Engages

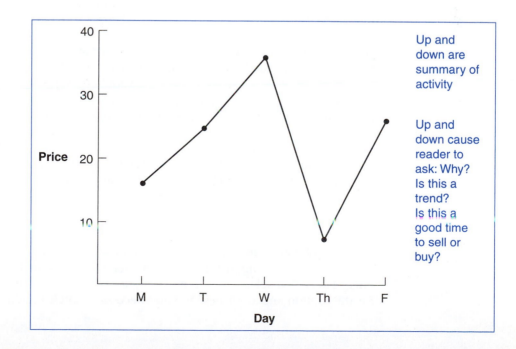

To give readers an opportunity to explore data means to allow them to investigate on their own. Readers can focus on any aspects that are relevant to their needs. For instance, they might focus on the fact that the stock rose at the beginning and again at the end of the week, or that the one-day rebound on Friday equaled the two-day climb on Monday and Tuesday.

To provide a different entry point into the discussion means to orient readers to the topic even before they begin to read the text. Studying the graph of a stock's price could introduce the reader to the concept of price fluctuation or could provide a framework of dollar ranges and fluctuation patterns.

To engage reader expectations is to cause readers to develop questions about the topic. Simply glancing at the line that traces the stock's fluctuation in price would immediately raise questions about causes, market trends, and even the timeliness of buying or selling.

Creating and Discussing Visual Aids

How to Create Visual Aids

The best way to create a visual aid is to follow the basic communication process: plan, draft, and finish.

Plan the Visual Aid Carefully

Make the visual aid an opportunity both to present your data and to engage your readers. Your overall goal is to help your readers as they research the information they need and help them to make sense of it once they find it (Schriver). Consider your audience's knowledge and what they will do with the information. Are they experts who need to make a decision? Consider your goal in presenting the information. Is it to summarize data? to offer an opportunity to explore data? to provide a visual entry point to the discussion? to engage reader expectations? Your visual aid should have just one main point. Take into account the fact that graphs have an emotional impact. For example, in graphs about income, lines that slope up to the right cause pleased reactions, whereas those that slope down cause anxiety. Consider also any constraints on your process. How much time do you need to make a clear visual aid? Consider the layout of the visual aid. In tables, which items should be in rows and which in columns? In graphs, which items should appear on the horizontal axis and which on the vertical?

Draft the Visual Aid Just as You Draft Text

Revise until you produce the version that presents the data most effectively. For graphs, select wording, tick marks, and data line characteristics (solid, broken, dots). For tables, select column and row heads, enter the data, create a format for the caption and any rules, and add necessary notes. For charts and illustrations, select symbols, overall dimensions, a font for words, and a width for rules. Reread the visual aid to find and change unclear elements, in either content or form.

Finish by Making the Visual Aid Pleasant to View

Treat all items consistently. Reduce clutter as much as possible by eliminating unnecessary lines and words.

How to Discuss Visual Aids

Carefully guide the reader's attention to the aspect of the visual aid you want to discuss. Your goal is to enrich the reader's understanding of the topic (Schriver). For instance, you could choose to explain elementary, intermediate, or overall information (Killingsworth and Gilbertson). *Elementary* information is one fact: On Wednesday, the stock's price rose. *Intermediate* information is a trend in one category: As the week progressed, the stock price fluctuated. *Overall* information is a trend that relates several categories: After the price dropped, investors rushed to buy at a "low."

In addition, you must explain background, methodology, and significance. *Background* includes who ordered or conducted the study of the stock, their reasons for doing so, and the problem they wanted to investigate. *Methodology* is how the data were collected. *Significance* is the impact of the data for some other concern—for instance, investor confidence. See pages 180–181 and 183–185 for examples of discussions of visual aids.

How to Reference Visual Aids

Refer to the visual aid by number. If it is several pages away, include the page number in your reference. You can make the references textual or parenthetical.

Textual Reference

A *textual* reference is simply a statement in the text itself, often a subordinate clause, that calls attention to the visual aid.

> . . . as seen in Table 1 (p. 10).
>
> If you look at Figure 4, . . .
>
> The data in Table 1 show . . .

Parenthetical Reference

A *parenthetical* reference names the visual aid in parentheses in the sentence. A complete reference is used more in reports, and an abbreviated reference in sets of instructions. In reports, use *see* and spell out *Table*. Although *Figure* or *Fig.* are both used, *Figure* is preferable in formal writing.

> The profits for the second quarter (see Figure 1) are . . .
> A cost analysis reveals that we must reconsider our plans for purchasing new printers (see Table 1).

In instructions, you do not need to use *see,* and you may refer to figures as *Fig.*

> Insert the disk into slot A (Fig. 1).
> Set the CPM readout (Fig. 2) before you go on to the next step.

Do not capitalize *see* unless the parenthetical reference stands alone as a separate sentence. In that case, also place the period inside the parentheses.

> All of these data were described above. (See Tables 1 and 2.)

Guidelines for Effective Visual Aids

The following five guidelines (Felker et al.; MacDonald-Ross; Schriver) will help you develop effective visual aids. Later sections of the chapter explain which types of visual aids to use and when to use them.

1. Develop visual aids as you plan a document. Because they are so effective, you should put their power to work as early as possible in your project. Many authors construct visuals first and then start to write.

2. As you draft, make sure each visual aid conveys only one point. If you include too many data, readers cannot grasp the meaning easily. (Note, however, that tables often make several points successfully.)

3. Position visual aids within the draft at logical and convenient places, generally as close to their mention in the text as possible.

4. Revise to reduce clutter. Eliminate all words, lines, and design features (such as the needless inclusion of three dimensions) that do not convey data.

5. Construct high-quality visual aids, using clear lines, words, numbers, and organization. Research shows that the quality of the finished visual aid is the most important factor in its effectiveness (Felker et al.; MacDonald-Ross).

Using Tables

A table is a collection of information expressed in numbers or words and presented in columns and rows. It shows the data that result from the interaction of an independent and a dependent variable. An *independent variable* is the topic itself. The *dependent variable* is the type of information you discover about the topic (White, *Graphic*). In a table of weather conditions, the independent variable, or topic, is the months. The dependent variables are the factors that describe weather in any month: average temperature, average precipitation, and whatever else you might want to compare. The data—and the point of the table—are the facts that appear for each month.

Parts and Guidelines

Tables have conventional parts: a caption that contains the number and title, rules, column heads, data, and notes, as shown in Figure 7.2. The following guidelines will help you use these parts correctly (based in part on *Publication*).

1. Number tables consecutively throughout a report with arabic numerals in the order of their appearance. Put the number and title above the table. Use the "double-number" method (e.g., "Table 6.3") only in long reports that contain chapters.

2. Use the table title to identify the main point of the table. Write brief but informative titles. Do not place punctuation after the title.

3. Use horizontal rules to separate parts of the table. Place a rule above and below the column heads and below the last row of data. Seldom use vertical rules to separate columns; use white space instead. If the report is more informal, use fewer or no rules.

4. Use a *spanner* head to characterize the column headings below it. Spanners eliminate repetition in column heads.

5. Arrange the data into columns and rows. Put the topics you want to compare (the independent variables) down the left side of the table in the *stub* column. Put the factors of comparison (the dependent variables) across the top in the column headings. Remember that columns are easier to compare than rows.

Figure 7.2

Elements
of a Table

TABLE 1								Number
Winter Weather Conditions in Minnesota								Title
	Average Temp (°F)		Record Temp (°F)		Average Snow (in.)[a]	Record Snow (in.)		Spanner heads
Month	High	Low	High	Low		High	Low	Column heads
January	20	3	59	241	9.0	46	0.0	Data
February	25	9	64	240	7.7	26	Tr	
March	38	23	83	232	9.6	40	.02	

Rule
Notes

[a]Snowfall data began in 1859.

Note: February includes calculations based on 28 and 29 days.

Source: Based on 1997 Minnesota Weatherguide Calendar. Eds. Bruce Watson and Jim Gilbert. Minneapolis: WCCO, 1996.

Source

6. Place explanatory comments below the bottom rule. Introduce these comments with the word *Note*. Use specific notes to clarify portions of a table. Indicate them by raised (superscript) lowercase letters within the table and at the beginning of each note.

7. Cite the *source* of the data unless the data were obviously collected specifically for the paper. List the sources you used, whether primary or secondary.

When to Use a Table

Because tables present the results of research in complete detail, they generally contain a large amount of information. For this reason, professional and expert audiences grasp tables more quickly than do nonexperts. When your audience knows the topic well, use tables to do the following (Felker et al.):

▶ To present all the numerical data so that the audience can see the context of the relationships you point out
▶ To compare many numbers or features (and eliminate the need for lengthy prose explanations)

In the text, you should add any explanation that the audience needs to understand the data in the table.

A Sample Table and Text

Table 1 in the following example appeared in a recommendation report. Although the data are clear, the numbers do not give enough background to explain the differences in cost. The explanatory text provides this background; it tells what features come with the base price and explains the figures in the two middle rows.

Purchase Cost

Background
Explanation

The Deutz Z291 has a base price of $150,000 (see Table 1). Deutz also offers a 5% discount ($7500) if we purchase before June 1. At this price the Deutz Z291 comes fully assembled, ready for use. The price does not include two tractor trucks and trailers that would be required to transport the conveying system and generators to run the equipment. However, we have no need to purchase these trucks and trailers because we have two trucks and trailers available. The final cost of the Deutz Z291 is $142,500.

Background
Explanation

The Pioneer 7000A has a base price of $120,000 (see Table 1). The Pioneer 7000A comes fully assembled and ready for use. The price includes mounting it on a flat bed truck. When the Pioneer 7000A is mounted, it requires no additional trucks or trailers for transportation. The Pioneer 7000A also comes with 50 feet of pumping hose. However, we feel that some of our projects will require 100 feet of hose. The additional pumping hose costs $5000, boosting the total cost of the Pioneer 7000A to $125,000.

TABLE 1

A Comparison of Purchase Costs

Cost	Deutz Z291	Pioneer 7000A
Base price	$150,000	$120,000
Discount	7500	0
Additional cost	0	5000
Final cost	142,500	125,000

Using Line Graphs

A *line graph* shows the relationship of two variables by a line connecting points inside an X (horizontal) and a Y (vertical) axis. These graphs usually show trends over time, such as profits or losses from year to year. The line connects the points, and its ups and downs illustrate the changes—often dramatically. On the horizontal axis, plot the independent variable, the topic whose effects you are recording, such as months of a year. On the vertical axis, record the values of the dependent variable, the factor that changes when the independent variable changes, such as sales. The line represents the record of change—the fluctuation in sales (Figure 7.3).

Figure 7.3

Dependent and Independent Variables

FIGURE 1
Five-Month Sales Data

Parts and Guidelines

Line graphs have conventional parts: a caption that contains the number and title (with a source note when necessary), axis rules, tick marks and tick identifiers, axis labels, a data line, and a legend. These parts are illustrated in Figure 7.4, page 182. The following guidelines will help you treat these parts correctly.

Figure 7.4

Elements of
a Line Graph

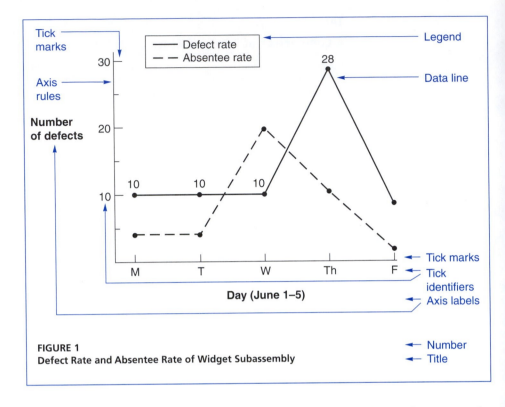

FIGURE 1
Defect Rate and Absentee Rate of Widget Subassembly

1. Number figures consecutively throughout the report, using arabic numerals. Use double-numbering (e.g., "Figure 6.6") only if you have numbered chapters.

2. Use a brief, clear title to specify the content of the graph. Do not punctuate after the title.

3. Put the caption below the figure. (Many computer programs automatically place the caption above the figure. This method is acceptable, especially for informal reports. Choose one placement or the other and use it consistently throughout the document.)

4. Place the independent variable on the horizontal axis; place the dependent variable on the vertical axis.

5. Space tick marks equally along the axis rule. Use varying thicknesses to indicate subordination. (Note that in Figure 7.4 the 15 and 25 tick marks are smaller than the 20 and 30 marks.)

6. Provide clear axis labels. In general, spell out words and write out numbers from left to right. Use abbreviations only if they are common.

7. Present a data line with definite marks that indicate the intersection of the two axes (such as the dot in Figure 7.4 where Monday meets 10). Add explanatory numbers or words inside the graph.

8. If the graph has two or more lines, make them visually distinct and identify them with labels or in a legend, or key.

When to Use a Line Graph

Line graphs depict trends or relationships. They clarify data that would be difficult to grasp quickly in a table. Research shows that expert readers grasp line graphs more easily than nonexperts (Felker et al.). Use a line graph

- To show that a trend exists (see Figure 7.3)
- To show that a relationship exists, say, of pollutant penetration to filter size
- To give an overview or a general conclusion, rather than fine points
- To initiate or supplement a discussion of cause or significance (Figure 7.3 alerts readers to ask why April is unusual)

Add explanatory text that helps the audience grasp the implications of the graph.

The following line graphs are taken from a report (Makki and Durance) on one enzyme's ability to prevent spoilage in beer. The enzyme is lysozyme, a natural food preservative found in egg whites. The cause of the spoilage is two bacteria, *L. brevis* and *P. damnosus*. Figure 4, page 184, reports on two tests on *L. brevis*; Figure 5, page 185, reports on one test on *P. damnosus*.

The paragraph describing the figures uses several strategies that are effective in relating graphs to readers. Notice that as you read the text and review the graphs, you can grasp the point even if you know nothing about microbiology. The basic strategies are to

- Provide a topic sentence that announces the main point of the paper: "Viability . . . is shown."
- Indicate what to notice: "*L. brevis* initially decreased" but "by Day 9 . . . was almost identical."
- Phrase conclusions: "Although the levels of lysozyme did not prevent growth of *L. brevis,* they did delay growth."
- Relate to research context: "Similar conclusions were made by Bottazzi et al. (1978). . . ."

Viability of *L. brevis* and *P. damnosus* in beer containing lysozyme is shown in Figures 4 and 5. *L. brevis* initially decreased in number in bottles containing 10 and 50 ppm lysozyme (Figure 4a). But by Day 9, the survival in bottles with 10 ppm lysozyme was almost identical to that observed in bottles with no lysozyme, with counts reaching 10^6 CFU/ml. The counts in the bottles with 50 ppm lysozyme remained at the Day 3 level (10^3 CFU/ml) throughout the rest of the experiment. Although 10 ppm lysozyme did not have much inhibitory effect on *L. brevis*, 50 ppm seemed to exert some degree of inhibition. The results of a second trial (Figure 4b) suggested that even 50 ppm lysozyme cannot always inhibit *L. brevis*.

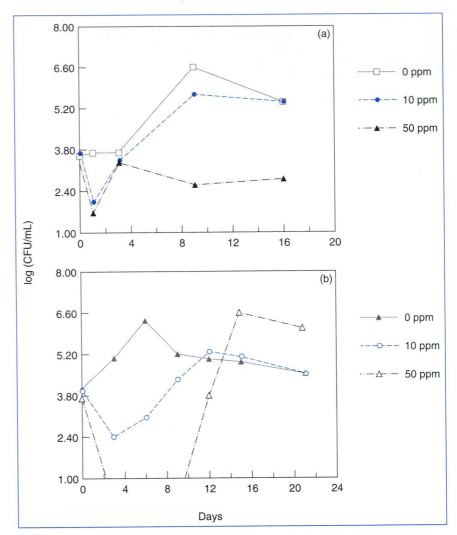

Figure 4
Survival of *Lactobacillus brevis* B-12 in unpasteurized beer in the presence of lysozyme:
(a) first trial; (b) second trial.

Source: F. Makki and T. D. Durance, "Thermal Inactivation of Lysozyme as Influenced by pH,
Sucrose and Sodium Chloride and Inactivation and Preservative Effect in Beer" from *Food
Research International* 29. 7 (October 1996): 635–645. Reprinted by permission of Elsevier.

After a sharp initial drop in the survival curve, the CFU/ml counts in the
bottles with 50 ppm lysozyme recovered after Day 12, reaching even
higher levels than in control bottles in the following days. Although the
levels of lysozyme did not prevent growth of *L. brevis,* they did delay
growth. Similar conclusions were made by Bottazzi et al. (1978) who
studied the effect of lysozyme of thermophilic lactic acid bacteria in

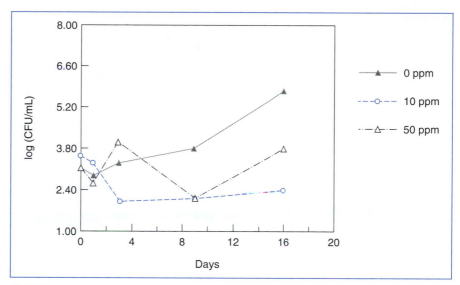

Figure 5
Survival of *Pediococcus damnosus* B-130 in unpasteurized beer in the presence of lysozyme.

Source: F. Makki and T. D. Durance, "Thermal Inactivation of Lysozyme as Influenced by pH, Sucrose and Sodium Chloride and Inactivation and Preservative Effect in Beer" from *Food Research International* 29. 7 (October 1996): 635–645. Reprinted by permission of Elsevier.

milk. They found that *L. helviticus* in milk was inhibited by greater than 50 ppm lysozyme. As for the results obtained with *P. damnosus* (Figure 5), there was little difference between the survival trend observed in the bottles with 10 ppm and 50 ppm lysozyme although both treatment levels reduced the growth of *P. damnosus* in beer. A more complete investigation of the effects of lysozyme on the beer spoilage bacteria used in this study would require the use of a wider range of lysozyme concentrations and a higher upper limit. (643–644)

Using Bar Graphs

A *bar graph* uses rectangles to indicate the relative size of several variables. Bar graphs contrast variables or show magnitude. They can be either horizontal or vertical. Horizontal bar graphs compare similar units, such as the populations of three cities. Vertical bar graphs (often called *column graphs*) are better for showing discrete values over time, such as profits or production at certain intervals.

In bar graphs, the independent variable is named along the base line (Figure 7.5, p. 186). The dependent variable runs parallel to the bars. The bars show the data. In a graph comparing the defect rates of three manufacturing lines, the lines are the independent variable and are named along the base line. The defect rate is the dependent variable, labeled above the line parallel to the bars. The bars represent the data on defects.

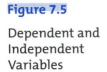

Figure 7.5

Dependent and
Independent
Variables

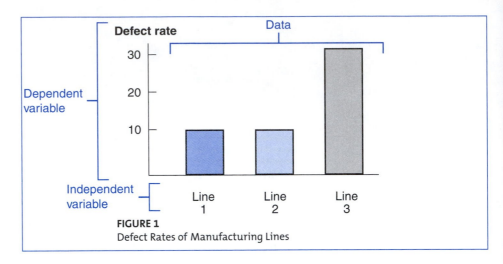

FIGURE 1
Defect Rates of Manufacturing Lines

Parts and Guidelines

Bar graphs have conventional parts: a caption that contains the number and title
(with source line when necessary), axis rules, tick marks and tick identifiers,
axis labels, the bars, and the legend (Figure 7.6). The following guidelines (based
on Tufte) will help you treat these parts effectively. (These guidelines are for vertical bar graphs; rearrange them for horizontal bar graphs.)

1. As with other visual aids, provide an arabic numeral and a title that names
   the contents of the graph. This caption material appears below the figure in
   formal reports but is often found above in informal reports.

2. Place the names of the items you are comparing (the independent variable)
   on the horizontal axis; place the units of comparison (the dependent variable) on the vertical axis.

3. Space tick marks equally along the axis rule. Use varying thicknesses to indicate subordination.

4. Provide clear axis labels. Spell out words and write out numbers.

5. Make the spaces between the bars one-half the width of the bars. (You may
   have to override the default in your computer program.)

6. Use a legend or callouts to identify the meaning of the bars' markings.

7. Avoid elaborate cross hatching and striping, which create a hard-to-read
   "op art" effect.

8. Use explanatory phrases at the end of bars or next to them.

9. Subdivide bars to show additional comparisons (Figure 7.7).

Figure 7.6

Elements of
a Bar Graph

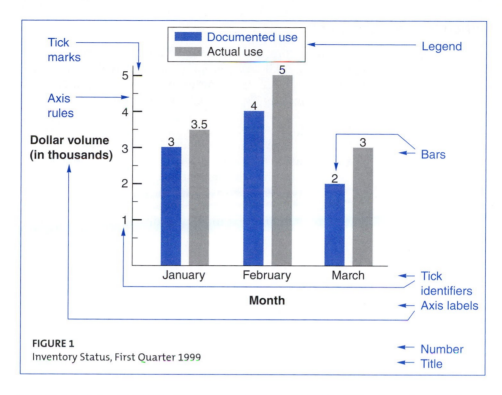

FIGURE 1
Inventory Status, First Quarter 1999

Figure 7.7

Column Graph
with Shading

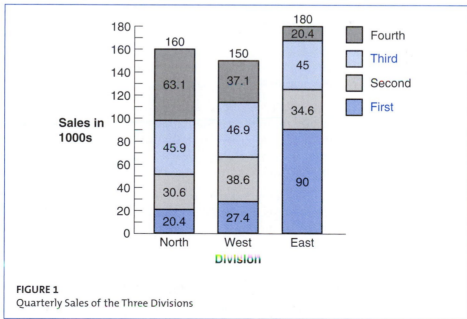

FIGURE 1
Quarterly Sales of the Three Divisions

When to Use a Bar Graph

Bar graphs compare the relative sizes of discrete items, usually at the same point in time. Like line graphs, they clarify data that would be difficult to extract from a table or a lengthy prose paragraph. Nonexpert readers find bar graphs easier to grasp than tables. Use a bar graph

- To compare sizes
- To give an overview or a general conclusion
- To initiate or supplement a discussion of cause or significance (For example, the bar graph of the defect rates of three widget lines [see Figure 7.5, p. 186] prompts a reader to ask why the rate was high on line 3.)

Use accompanying text to help readers grasp the implications of the graph.

Using Pie Charts

A *pie chart* uses segments of a circle to indicate percentages of a total. The whole circle represents 100 percent, the segments of the circle represent each item's percentage of the total, and the callouts identify the segments in the graph. Data words or symbols provide detailed information.

Parts and Guidelines

Like other graphs, pie charts have conventional elements: a caption, the circle, the segments ("pie slices"), and callouts (Figure 7.8). The following guidelines (based in part on *Publication*) will help you treat these parts effectively.

1. The caption may appear above or below the chart. Use arabic numerals and a title that names the contents of the graph. Informal charts often have only a title.

2. Start at "12 o'clock," and run the segments in sequence, clockwise, from largest to smallest. (Sometimes you cannot satisfy all these requirements. If you are comparing the grades in a chemistry class, the A segment is usually smaller than the C segment. However, you would logically start the A segment at 12 o'clock.)

3. Identify segments with callouts or legends. *Callouts* are phrases that name each segment (see Figure 7.8). Use callouts if space permits, and arrange them around the circumference of the circle. A *legend* is a small sample of each segment's markings plus a brief identifying phrase, as shown in Figure 7.9. Cluster all the legend items in one area of the visual.

4. In general, place percentage figures inside the segments.

5. For emphasis, shade important segments, or present only a few important segments rather than the whole circle.

Figure 7.8

Elements of
a Pie Chart

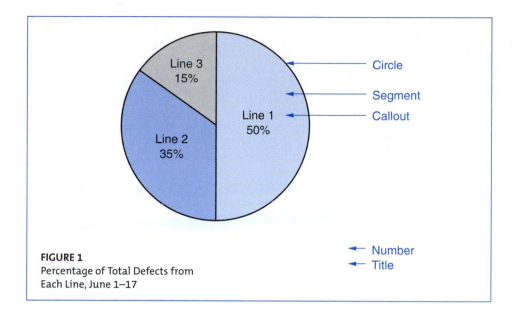

FIGURE 1
Percentage of Total Defects from
Each Line, June 1–17

Figure 7.9

Pie Chart
with Legend

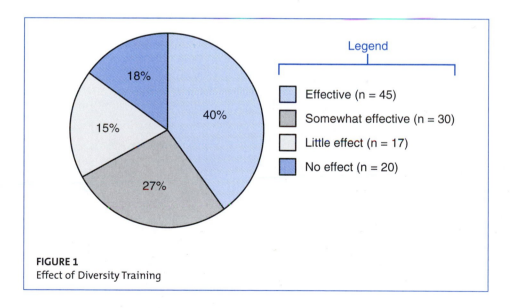

FIGURE 1
Effect of Diversity Training

6. In general, divide pie charts into no more than five segments. Readers have
   difficulty differentiating the sizes of small segments. Also, a chart with many
   segments, and thus many callouts, looks chaotic.

 **E**thics and Visual Effects

Deciding on the visual components of a document is complex, entailing everything from font and page properties to the use of images and their download capabilities for Web use. Clarity and order are paramount to making your point clear to your audience. Hiding or blurring the facts by obscuring the truth with lots of visuals is not only unfair to your audience, it is unethical as well. Make your visual effects user-centered. Carefully considering visual effects means connecting with your audience and understanding their needs, as well as creating a document that is aesthetically pleasing, functional, and ethically sound.

Nancy Allen has suggested several ethical guidelines for writers to consider when incorporating visual material:

Selection: Is all the information included that viewers need to make a decision?

Emphasis: What subject is focused upon? Is that the subject one dealt with in the document?

Framing: What items are being related or grouped by the way the graphic is framed? What purpose is served by relating these items and presenting them in this manner?

Accuracy: Is factual information accurate? Does the overall impression support or distort accurate perception? (102)

Jan White in *Using Charts and Graphs* illustrates ways in which perception can be distorted, thus causing an unethical presentation of data.

1. Changing the width of the units on the X axis alters the viewer's emotional perception of the data. The graphs in Figure 7.10 plot exactly the same data. Using the middle graph could make the situation seem more urgent than it is.

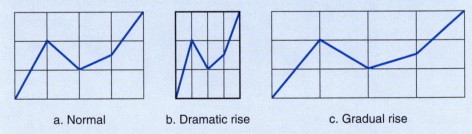

a. Normal   b. Dramatic rise   c. Gradual rise

Figure 7.10
Three Graphs That Plot the Same Data
Source: Jan White, *Using Charts and Graphs* (New York: Bowker, 1984). Reprinted by permission of R. R. Bowker.

2. The nearer a highlighted feature appears, the more impact it has on one's consciousness. The pie charts in Figure 7.11 report the same data. Note that the wedge on the left appears largest and that the wedge on the right is forced into the viewer's consciousness. The middle wedge appears unimportant because it is far away.

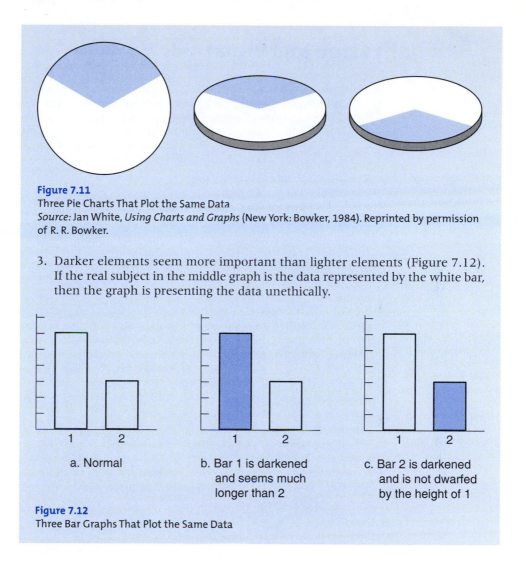

Figure 7.11
Three Pie Charts That Plot the Same Data
Source: Jan White, *Using Charts and Graphs* (New York: Bowker, 1984). Reprinted by permission of R. R. Bowker.

3. Darker elements seem more important than lighter elements (Figure 7.12). If the real subject in the middle graph is the data represented by the white bar, then the graph is presenting the data unethically.

a. Normal

b. Bar 1 is darkened
and seems much
longer than 2

c. Bar 2 is darkened
and is not dwarfed
by the height of 1

Figure 7.12
Three Bar Graphs That Plot the Same Data

When to Use a Pie Chart

Pie charts work best when they are used to compare magnitudes that differ widely (see Figure 7.9, p. 189). Because pie chart segments are not very precise, most people cannot distinguish between close values, such as 17 percent and 20 percent. Nonexpert audiences find pie charts easier to use than tables (Felker). Use pie charts

▶ To compare components to one another
▶ To compare components to the whole
▶ To show gross differences, not fine distinctions

Globalization and Visual Aids

Visual aids can make your documents easier for foreign audiences to understand and to translate. In countries where literacy rates are low, pictures, images, and icons are often the best way to convey the intended message (Twin Cities). However, using visual aids requires careful research on the culture for which you are writing. If possible, consult an expert on that culture, particularly if the culture is non-Western. In order to be effective, visual aids must be chosen with a sensitivity to cultural norms.

Certain symbols are broadly understood by Americans, and so we assume that they are accepted internationally. However, these shorthand devices can be meaningless, inappropriate, or even offensive in other cultures. For example, the computer mailbox icon symbolizing e-mail is confusing to the rest of the world, where mailboxes generally look quite different. An envelope is a better way to signify mail or e-mail (Thrush). Pictures of body parts or hand gestures are best avoided, as they often have very different, sometimes even obscene, meanings.

Color is an interesting dimension to add to visual aids, but requires considerable cultural sensitivity. What seems a normal use of color in one culture is not normal at all in another. For instance, in the United States to "see red" is an indication of anger, but in Russia red is related to the idea of beauty. In countries such as China and India, red is a color of celebration. In the United States, to "feel blue" is to be sad, but in France blue is connected with fear. In the United States, when one is jealous, one is "green-eyed," but in China one has the "red-eye disease" (Borton). While white is a color of purity in the United States, in many Asian countries it symbolizes death. Purple is often a polarizing and unsafe color to use. One company had to completely redo its European theme park because the purple design struck many viewers as morbid (Peterson).

For further reference check out this website: 3M Worldwide at <www.3m.com/meetingnetwork/presentations/pmag_think_global.html> gives in-depth advice on presenting information to an international audience as well as links to articles on giving effective presentations and using visual aids.

Works Cited

Borton, Meg. "When Colors Take on Different Cultural Hues." *International Herald Tribune: The IHT Online.* 28 Sept. 2002. 7 Feb. 2004 <www.iht.com/articles/72080.html>.

Peterson, Dina. "Internationalization: Understanding Cultural Issues in the Context of Web Site Globalization." Godfrey. 6 Feb. 2004 <http://216.239.41.104/search?q=cache:Tyt12KEJOakJ:www.godfrey.com/pdfs/international_whitey.pdf_+%22germany%22+%22colors%22+%22symbols%22&hl=en&ie=UTF-8>.

Thrush, Emily. "Writing for an International Audience." Suite101.com. 6 Feb. 2004 <www.suite101.com/article.cfm/5381/32233>.

Twin Cities Chapter of the Society for Technical Communication. Newsletter. 31 Jan. 2004 <www.stctc.org/newsletter/tt2003_02_c_2.htm>.

Using Charts

Chart is the catchall name for many kinds of visual aids. Charts represent the organization of something: either something dynamic, such as a process, or something static, such as a corporation. They include such varied types as trouble-shooting tables, schematics of electrical systems, diagrams of the sequences of an operation, flow charts, decision charts, and layouts. Use the same techniques to title and number these as you use for graphs. (Some software companies, including Microsoft, use *chart* to mean *graph,* and you will find these terms used interchangeably.)

Troubleshooting Tables

Troubleshooting tables in manuals identify a problem and give its probable cause and cure. Place the problem at the left and the appropriate action at the right. You can also add a column for causes (Figure 7.13). Almost all manuals include these tables.

Figure 7.13

Troubleshooting Table

Source: Jill Adkins, *Rotary Piston Filler: Eight Head* (Menomonie, WI: MRM/Elgin, 1985). Reprinted by permission of MRM/Elgin.

TABLE 1

Troubleshooting Table

You Notice	This May Mean	Caused by	You Should
Containers do not center with nozzle	Infeed Starwheel is out of time	Misadjustment or loose mounting bolts	Readjust & retighten
	Machine speed is too fast	Speed not reset during change-over	Reset speed (see "Speed Adjust," p. 5)
	Wrong change parts being used		Install correct change parts (see "Change Parts Data Sheet," p. 27)
Machine vibrates	Lack of maintenance	Tight roller chains (chains that drive the conveyor) or the two chains on the end of the Spiral Screw Feed	Loosen chains
		Roller chains running dry	Lubricate chains
		Cross beam bearing dry	Lubricate (see Figure 5–148)

Flow Charts

Flow charts show a time sequence or a decision sequence. Arrows indicate the direction of the action, and symbols represent steps or particular points in the action (Figure 7.14). In many cases, especially in computer programming, the symbols have special shapes for certain activities. For instance, a rectangle signals an action to perform and an oval signals the first or last action. Use a flow chart to help readers grasp a process.

Figure 7.14

Flow Chart

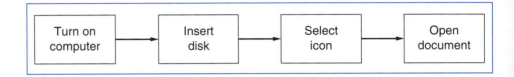

Decision Charts

A *decision chart* (or *decision tree*) is a flow chart that uses graphics to explain whether or not to perform a certain action in a certain situation. At each point, the reader must decide yes or no and then follow the appropriate path until the final goal is reached (Figure 7.15).

Figure 7.15

Decision Chart

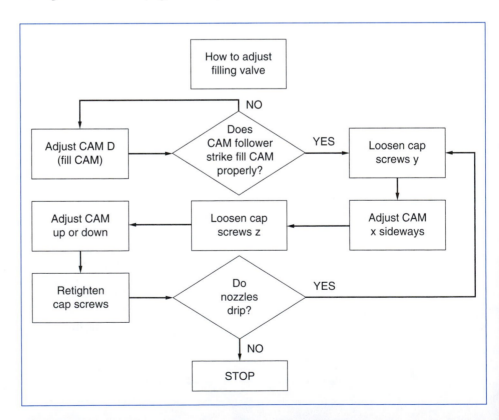

Gantt Charts

A *Gantt chart* (named after its inventor) represents the schedule of a project. Along the horizontal axis are units of time; along the vertical axis are sub-processes of the total project (Figure 7.16). The lines indicate the starting and stopping points of each subprocess.

Figure 7.16

Gantt Chart

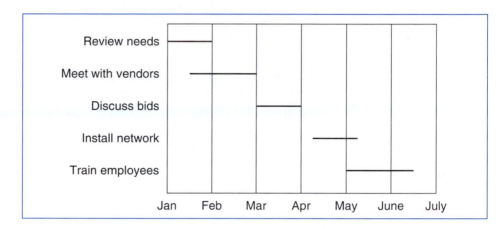

Layouts

A *layout* is a map of an area seen from the top. As Figure 7.17 shows, layouts can easily show before and after arrangements. Draw simple lines, and use callouts and arrows precisely. Callouts may appear inside the figure.

Figure 7.17

Sample Layout

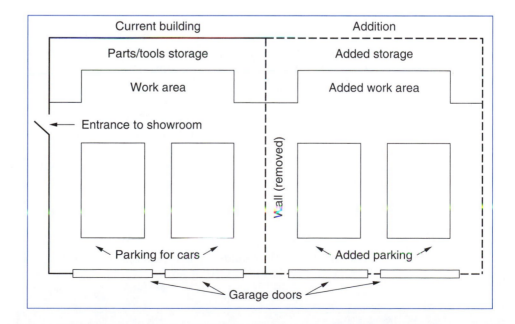

Using Illustrations

Illustrations, usually photographs or drawings of objects, are often used in sets of instructions and manuals.

Guidelines

There are two basic guidelines for using illustrations.

1. Use high-quality illustrations. Make sure they are clear, large enough to be effective, and set off by plenty of white space.

2. Keep the illustrations as simple as possible. Show only items essential to your discussion.

   Use an illustration (Felker)

   ▶ To help explain points in the text
   ▶ To help readers remember a topic
   ▶ To avoid lengthy discussions (A picture of a complex part is generally more helpful than a lengthy description.)
   ▶ To "give the reader permission" (A visual of a computer screen duplicates what is obviously visible before the user, but gives the user permission to believe his or her perception. It reassures the reader.)

Photographs

A good photograph offers several advantages: It is memorable and easy to refer to; it duplicates the item discussed (so audiences can be sure they are looking at what is being discussed); and it shows the relationships among various parts. The disadvantages are that it reduces a three-dimensional reality to two dimensions and that it shows everything, thus emphasizing nothing. Use photographs to provide a general introduction or to orient a reader to the object (Killingsworth and Gilbertson). In manuals, for instance, writers often present a photograph of the object on the first page. Figure 7.18 shows a photograph that clearly indicates not only the part but its relationship to other parts on the machine.

Drawings

Drawings, whether made by computer or by hand, can clearly represent an item and its relationship to other items. Use drawings to eliminate unneces-

Figure 7.18

Photograph
Identifies Part

Source: Rotary Piston Files by Jill Adkins. Reprinted by permission of MRM/Elgin.

sary details so that your reader can focus on what is important. Two commonly used types of drawings are the exploded view and the detail drawing.

Exploded View

As the term implies, an *exploded view* shows the parts disconnected but arranged in the order in which they fit together (Figure 7.19). Use exploded drawings to show the internal parts of a small and intricate object or to explain how it is assembled. Manuals and sets of instructions often use exploded drawings with named or numbered parts.

Figure 7.19

Exploded View

Source: John R. Mancosky, *Manual Controls: Independent Study Workbook.* Used by permission of Microswitch and John R. Mancosky.

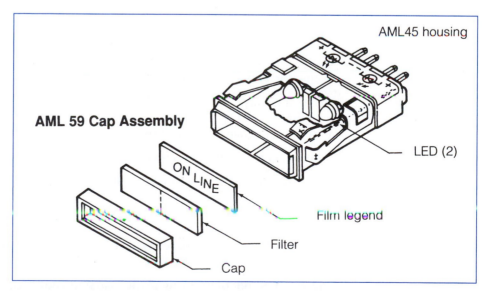

Detail Drawings

Detail drawings are renditions of particular parts or assemblies. Drawings have two common uses in manuals and sets of instructions.

▶ They function much as an uncluttered, well-focused, cropped photograph, showing just the items that the writer wishes.
▶ They show cross-sections; that is, they can cut the entire assembled object in half, both exterior and interior. (In technical terms, the object is cut at right angles to its axis.) A cross-sectional view shows the size and the relationship of all the parts. Two views of the same object—front and side views, for example—are often juxtaposed to give the reader an additional perspective on the object (Figure 7.20).

Figure 7.20

Detail Drawing

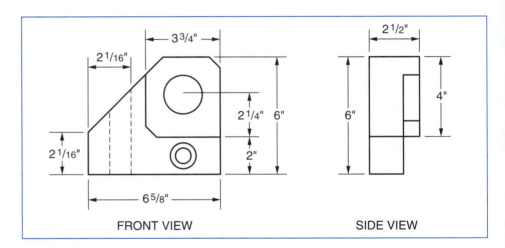

FRONT VIEW   SIDE VIEW

Worksheet for Visual Aids

☐ Name the audience for the visual aid.

☐ What should your visual aid do?
  Summarize data? present an opportunity to explore data? provide a visual entry point? engage expectations?

☐ Choose a format: Where will you place the caption—above or below?

☐ Decide on a way to treat this visual's conventional parts. Follow the guideline lists in this chapter.

☐ How large will the visual aid be? Do you have room for it in your document?

☐ How much time do you have? Will you be able to construct a high-quality visual in that time?

☐ Select a method for referring to the visual in your text.

> ☐ Create a draft of the visual aid. Review it to determine whether you have treated its parts consistently.
>
> ☐ Eliminate unnecessary clutter.

Exercises

▶ You Create

1. Collect data from the entire class on an easy-to-research topic, such as type of major, year in school, years with the company (e.g., 1, 2, 3–5, 6–10, 10+), or population of hometown or birthplace (e.g., 100K+, 50–100K, 5–50K). Collect data from the entire class. Then create two visual aids: a bar graph and a pie chart. Make them as neat and complete as possible, including captions and call-outs. Write a brief memo that states a conclusion you can read by looking at the visual or that tells an audience what to notice in the visual.

   Alternates: (1) Create the visuals in groups of three to four. (2) Create the visuals on a computer. (3) Write a memo explaining the process you used to create the visuals. (4) Write an IMRD (see Chapter 12) to explain your project.

2. Use one of the visuals created in Exercise 1 as the basis for a brief story for the student newspaper or company newsletter, explaining the interesting diversity of your technical writing class.

3. Use a line graph to portray a trend (e.g., sales, absentees, accepted suggestions, defect rate) to a supervisor. Write a brief note in which you name the trend, indicate its significance, and offer your idea of its cause.

4. Use the following data to create a table (Jaehn). Monthly from July to December, the following machines had the following percentages of rejection for mechanical quality defects: #41 — 4.0, 3.1, 3.9, 4.3, 3.5, 3.2; #42 — 3.0, 3.3, 2.4, 3.2, 3.7, 3.1; #43 — 3.4, 3.7, 4.1, 4.5, 4.4, 4.8. Include averages for each month and for each machine. Additional: Write a memo to a supervisor explaining the importance of the trend you see in the data.

5. If you have access to a computer graphics program, make three different graphs of the data in Exercise 4. In a brief paragraph, explain the type of reader and the situation for which each graph would be appropriate.

6. Divide the class into three sections. Have individuals in each section convert the numbers in either of the following paragraphs into visual aids. Section 1 should make a line graph, section 2 a bar graph, and section 3 a pie chart. Have one person from each section put that group's visual on the board. Discuss their effectiveness. Here are the figures:

   Respondents to a survey were asked whether they would pay more for a tamper-evident package: 8.2% said they would pay up to $.15 more; 25.8%

were unwilling to pay more; 51.6% would pay $.05 more; and 14.4% would pay $.10 more.

Industrial designers report that they use various tools to reduce the time they spend on design. Those tools are faster computers (57%), 3D solid modeling (32%), 2D CAD software (26%), Computerized FEA (24%), Rapid Prototyping (15%), other (3%). (Based on Colucci)

7. Convert the following paragraph into a table. Then rewrite the paragraph for your manager, proposing that the company start a recycling program. Refer to specific parts of the table in your paragraph. Alternate: Recommend that the company not start a recycling program. Refer to specific parts of the table.

The company can recycle 100 pounds of aluminum per week. The current rate for aluminum is 25¢ per pound. This rate would earn the company $25.00 per week and $1200.00 per year. The company can also recycle 200 pounds of paper per week. The current rate for paper is 5¢ per pound. Recycling paper would earn the company $10.00 per week and $520.00 per year. The total earnings from recycling is $1820.00 per year.

8. Read over the next four paragraphs from a report. Then construct a visual aid that supports the writer's conclusion.

The pipe cutter must be small enough so it can be transported in our truck. The Grip-Tite model takes up 2 cubic feet, and it comes with a stand so that it can be folded out when in use and then folded up when not in use. The Mentzer model takes up 5 cubic feet and doesn't come with a stand.

The pipe cutter must be able to run off 110-volt electricity. Because 220-volt electricity is not easily accessible on the job site, 110-volt must be used. The Grip-Tite model is capable of running off 220-volt, 110-volt, and DC current. The Mentzer cutter is capable only of running off 220-volt.

The cutter should be able to switch threaders to accommodate $3/8''$ pipe up to 3" pipe. The Grip-Tite model has two threaders. One can be adjusted from $1/4''$ up to 2", and the other can be adjusted from 2" up to 3". The Mentzer model has only one threader that threads $3/4''$ pipe.

The cost of the pipe cutter should not exceed $2000. The Grip-Tite model costs $1750, whereas the Mentzer model costs $2250.

▶ You Revise

9. Redo this table and the paragraph that explains it.

I have estimated construction costs at a price of $50.00 per square foot as quoted by Lamb and Associates Construction Company (see Table 3). This will include construction of perimeter walls, carpeting, decor, and lighting. The cost of this will be $37,500.

TABLE 3

Renovation Costs

Construction:			$37,500.00
carpeting	15,000		
decor	8,000		
lighting	6,000		
perimeter walls	8,500		
Loose Fixtures:			
#5021 (7 ball Chrome)	15 @ $ 5.95 each	$ 89.25	
#4596 (4 straight arm)	4 @ $78.00 each	312.00	
#4597 (2 straight 2 slant)	3 @ $81.00 each	243.00	
			$644.25
			$38,144.25
Personnel:			
2 full-time stock men 2/8 hour days @ $4.50/hr.			$144.00
Total Renovation Costs:			$38,288.25

▶ You Analyze

10. Discuss the differences between these two table designs. Which one is more effective? Why? Alternate: Create your own table for these data.

TABLE 1

Comparison of Three Trucks

Brand	Purchase Price	3-Year Maintenance Cost	Warranty	Warranty
Big Guy	$11,999	$ 700	2 years	24,000 miles
Friend	13,200	1000	3 years	30,000 miles
Haul	13,700	850	5 years	50,000 miles

TABLE 1

Comparison of Three Trucks

Brand	Purchase Price in Dollars	3-Year Maintenance Cost	Warranty Years	Warranty Miles
Big Guy	$11,999	$ 700	2	24K
Friend	13,200	1000	3	30K
Haul	13,700	850	5	50K

▶ Group

11. Divide into groups of three or four, by major if possible. Select a process you are familiar with from your major or from your campus life. Possibilities include constructing a balance sheet, leveling a tripod, focusing a microscope, constructing an isometric projection, threading a film projector, finding a periodical in the library, and making a business plan. As a group, construct a flow chart of the process. For the next class meeting, each person should write a paragraph explaining the process by referring to the chart. Compare paragraphs within your group; then discuss the results with the class.

Writing Assignments

1. Write a report to your manager to alert him or her to a problem you have discovered. Include a visual aid whose data represent the problem (such as a line graph that shows a "suspect" defect rate).

2. Divide into groups of three. From some external source, acquire data on some general topic such as population, budget, production, or volume of sales. Sources could be your Chamber of Commerce, a government agency, an office in your college, or your corporation's human resources office. As a group, decide on a significant trend that the data show. Write a memo to the appropriate authority, informing him or her of this trend and suggesting its significance. Use a visual aid to convey your point.

3. As a class, agree on a situation in which you will present information to alert someone to a trend or problem. Then divide into several groups. Each group should select one of the four reasons for using a visual aid (p. 175). Write a memo and create the visual aid that illustrates the reason your group chose it (e.g., to provide a different entry point). Make copies for each of the other groups. When you are finished, circulate your reports and discuss them.

4. Use a flow chart and an accompanying memo to explain a problem with a process—for example, a bottleneck or a step where a document must go to two places simultaneously. Your instructor may ask you to go on to Writing Assignment 5.

5. Use Writing Assignment 4 as a basis to suggest a solution to a problem. Create a new flow chart that illustrates how your solution solves the problem. Discuss the solution in a memo to your manager requesting permission to put it into effect.

6. Write a learning report for the writing assignment you just completed. See Chapter 5, Writing Assignment 7, pages 134–135, for details of the assignment.

Web Exercise

Write a memo or an IMRD (see Chapter 12) to your classmates in which you analyze the use of visual aids in websites. Your goal is to give hints on how to incorporate visuals effectively into their Web documents. Do the visual aids exemplify one of the four uses of visual aids explained in this chapter? Explain why the visual aid is effective in the Web document.

Works Cited

Adkins, Jill. *Rotary Piston Filler: Eight Head.* Menomonie, WI: MRM/Elgin, 1985.

Allen, Nancy. "Ethics and Visual Rhetoric: Seeing's Not Believing Anymore." *Technical Communication Quarterly* 5 (1996): 87–105.

Colucci, D. "How to Design in Warp Speed." *Design News* (1996): 64–76.

Felker, Daniel B., Francis Pickering, Veda R. Charrow, V. Melissa Holland, and James C. Redish. *Guidelines for Document Designers.* Washington, DC: American Institutes for Research, 1981. This volume was used extensively in the preparation of this chapter.

Jaehn, Alfred H. "How to Effectively Communicate with Data Tables." *Tappi Journal* 70 (1987): 183–184.

Killingsworth, M. Jimmie, and Michael Gilbertson. "How Can Text and Graphics Be Integrated Effectively?" *Solving Problems in Technical Writing.* Ed. Lynn Beene and Peter White. New York: Oxford University Press, 1988. 130–149.

MacDonald-Ross, M. "Graphics in Texts." *Review of Research in Education.* Vol. 5. Ed. L. S. Shulman. Itasca, IL: F. E. Peacock, 1978.

Makki, F., and T. D. Durance. "Thermal Inactivation of Lysozyme as Influenced by pH, Sucrose and Sodium Chloride and Inactivation and Preservative Effect in Beer." *Food Research International* 29.7 (October 1996): 635–645.

Mancosky, John R. *Manual Controls: Independent Study Workbook.* Freeport, IL: Microswitch, 1991.

Publication Manual of the American Psychological Association. 4th ed. Washington, DC: APA, 1994.

Schriver, Karen A. *Dynamics in Document Design: Creating Texts for Readers.* New York: Wiley, 1997.

Tufte, Edward R. *The Visual Display of Quantitative Information.* Cheshire, CT: Graphics Press, 1983.

White, Jan. *Graphic Design for the Electronic Age.* New York: Watson-Guptill, 1988.
. *Using Charts and Graphs.* New York. Bowker, 1984.

Summarizing

Chapter 8
In a Nutshell

Summaries tell readers the main points of an article or a report. Readers may use summaries to decide whether to read the entire article or report, to get the gist of the article or report without reading it, or to preview the material before reading it.

Your goal is to write the main point of the article in one sentence. Then briefly list the topics or divisions of the article, or explain each of the support points in the same order as they appear in the article, or create a minipaper in which you rearrange the way the parts are presented in order to give a helpful sense of the point of the paper. Be brief (usually one paragraph to one page) and give your readers enough detail so that they can carry out their goal (to get the gist).

In a world awash in information, the ability to construct and present concise, short versions of long documents is not only helpful but essential. *Summarizing,* or *abstracting*—the terms are nearly synonymous—is fundamental to technical writing. You will often summarize your own documents, as well as those written by others. This chapter explains those skills.

Summarizing

This section defines summaries and abstracts, explains the various audiences that use them, and presents the skills you need in order to write them.

Definitions of Summaries and Abstracts

Both summaries and abstracts are short restatements of another document (Vaughan). A *summary* restates major findings, conclusions, and support data found in a document. Summaries, which accompany many types of reports, are aimed at readers within an organization, typically executives (a common term for this type of writing is *executive summary*). They appear at the beginning of the report, before the body. An *abstract* is generally a short version of a journal article. Abstracts appear in two places: with the article in the periodical and as an independent unit provided by abstracting services for professionals in the field. Abstracts are either indicative or informative. *Indicative abstracts* list the document's topics; *informative abstracts* present short versions of the document's qualitative and quantitative information. An abstract usually mentions the document's purpose, scope, methodologies, results, and conclusion (ANSI).

Audiences for Summaries and Abstracts

Summaries and abstracts serve similar functions, allowing readers to discover the gist of the report or article without reading the entire document, to determine whether the report or article is relevant to their needs, or to get an overview before focusing on the details.

Readers use summaries to review a short version of a longer document when they don't have the time or the need to read the longer document. Readers use abstracts to keep up with current developments in the field and to review literature relevant to a research project. After reading the abstract, the reader can decide whether to read the entire article.

Planning Summaries

To plan for an abstract or a summary, you need to understand basic summarizing strategies, two methods of organizing a summary or an abstract, and certain details of form.

Use Basic Summarizing Strategies

To summarize effectively, perform two separate activities:

1. Read to find the main terms and concepts.

2. Decide how much detail to include.

In reading to find the main idea, look for various elements:

- ▶ What are the main divisions of the document?
- ▶ What are the key statements?
- ▶ Which sentence expresses the overall purpose of the document?
- ▶ Which sentences tell the main ideas of each paragraph?
- ▶ What details support the main ideas?
- ▶ What are the key terms? Which words are repeated or emphasized?

Consider the underlining in the article on pages 207–209. Note how the summarizer has underlined key sentences and terms. With practice, you will become confident of your ability to find the major divisions, main points, and main support in a document.

To decide how much detail to include, consider your audience's needs. The general rule is to be as complete as your readers require. If they need just a description of the contents, name only the main sections. (See "Descriptive Abstract," p. 210.) If they need to understand the underlying ideas, provide details (see "Summary Using Proportional Reduction" or "Summary Emphasizing Major Idea," pp. 210–211).

Choose an Organization

The two main strategies for organizing a summary are proportional reduction, and main point followed by support.

Proportional reduction refers to the idea that each part in the summary should be proportionally equal to the corresponding part in the original. Suppose the original has four sections, three of which are the same length and one of which is much longer. Your summary of this piece should have the same proportions: three shorter sections of about the same length and a fourth, longer section. You can make the overall summary shorter or longer, depending on how much detail you report for each section, but still maintain the same proportions.

Main point followed by support means you should write a clear topic sentence that repeats the central thesis of the document. This topic sentence could be the purpose of the report or its main findings, conclusions, or recommendations (see p. 211). This method is generally harder to write, but it is often more effective for readers because you can slant the summary to meet the reader's needs (ANSI).

Use the Usual Form

Summaries generally have the following characteristics:

- Length of 250 words to 1 page (Abstracts that will be reprinted must stay brief, about 250–300 words. Summaries can be longer; a 200-page report might need 2 to 5 pages for a clear, inclusive summary.)
- Verbs in the active voice and present tense
- A clear reference to the document (Abstracts always include a complete bibliographic entry. Generally, report summaries contain the title of the report either in the first sentence or in the title.)
- No terms, abbreviations, or symbols unfamiliar to the reader (Do not define terms in a summary unless definition was the main point of the document. Notice the unexplained technical terms in "Summary Emphasizing Major Idea," p. 211.)
- No evaluative comments such as "In findings related tangentially at best to the facts she presents . . ." (Report the contents of the document without bias.)
- Main points first (The first sentence usually gives the purpose of the report or the main findings; support follows.)

Writing Summaries

Read the following article. Then review the abstract and the two types of summaries on pages 210–211. Because the article is rather difficult for nonexperts, the summary can increase your ability to comprehend the article. Many harried managers might prefer the summary to the article.

Corporate Express Goes Direct: The B2B Office Products Vendor Is Slashing Costs by Integrating Tightly with Customer's Procurement Systems

by Gary H. Anthes

When you have consolidated computer systems from 500 corporate acquisitions, you get pretty good at IT integration. Indeed, nothing distinguishes IT at Corporate Express Inc. as the degree to which it seamlessly links major supply systems—its own and those of customers and suppliers. The $5.5 billion, Broomfield, Colo.–based vendor of office supplies for the business market has, in essence, become an extension of the procurement systems of its largest customers.

"Integration is one of our core competencies," says CEO Lisa Peters, who has seen her IT shop grow from 12 people to more than 300 in the last nine years.

Corporate Express has for years taken orders for furniture, paper, computer supplies, and other office products by telephone, fax, and electronic data interchange. In 1997, it began offering customers Internet-based procurement via a simple CD-ROM catalog. Now, more than half of its 75,000 daily orders arrive electronically, most as XML transactions through a richly featured Web portal called E-Way.

The smallest of Corporate Express' 30,000 customers, which typically lack their own automated procurement systems, log onto E-Way and conduct purchasing transactions much as a consumer might at Amazon.com. These buyers have also placed many of their unique procurement rules into E-Way, so that, for example, E-Way checks budgets, buyer authorizations, and other controls for customers.

But 750 of the company's largest customers—which account for some 80% of the sales volume—have a more direct connection to E-Way. Corporate Express has integrated E-Way into the processing fabric of their internal procurement systems. That involved integrating with some 40 different commercial packages from companies ranging from Ariba Inc. to Commerce One Operations Inc. and eScout LLC. These customers start to build orders locally, but then bridge to E-Way by leveraging the integration features offered by each vendor's product. For example, Corporate Express uses PunchOut for integration with Ariba, Roundtrip for Commerce One, ConnectScout for eScout, and so on.

Although customers can maintain their own versions of the Corporate Express catalog, more often the catalogs are maintained by and at Corporate Express. Every catalog is tailored to its user's format, terminology, and buying practices.

"E-Way knows all the customer rules—for example, that they don't buy desks from us, so desks will be blocked out," says Wayne Aiello, vice president of e-business services. "E-Way actually becomes the customer's system, so every customer has to be examined and treated differently."

Corporate Express is doing about 10 new customer integrations per month. They take 10 to 20 days—with requirement definition, coding, and testing taking equal amounts of time—but the first few projects took 10 times that long.

"The hardest thing three to four years ago was that nobody in the industry had done it," Aiello says. "XML was the buzzword, but it was really new. Platforms like Ariba and Commerce One were beginning to get a lot of hype, but there weren't people who had actually implemented them."

The company buys off-the-shelf software when it can, but much of E-Way and its other supply systems were developed in-house. Commercial packages often aren't scalable or flexible enough to accommodate the unique needs of customers, the company says. For example, Corporate Express developed its own search engine tailored to the characteristics of an office supply catalog.

Unocal Corp. in El Segundo, Calif., used to buy from Corporate Express by telephone and fax, but now it has integrated its Oracle Corp. procurement system with E-Way. Michael Comeau, e-procurement tools manager at Unocal, says he likes E-Way's ability to send all the invoices to a central point for payment, its buying controls, and its order-tracking ability. "Maverick spending is reduced tremendously," he says.

Meanwhile, Trisha Smallwood, manager of mail center operations at The Kroger Co. in Cincinnati, says she likes E-Way because it saves her five minutes on every order. No more filling out an order form, faxing it, and waiting for the confirmation. The process is made even simpler, she says, by E-Way's ability to maintain a list of items that Kroger orders frequently as well as the special prices and terms that the grocery chain has negotiated.

Corporate Express used software from webMethods Inc. in Fairfax, Va., to integrate its major logistics and financial applications. The webMethods tool provides flexibility in deciding where data and logic are best placed, says Brett McInnis, vice president for e-business technologies at Corporate Express.

For example, warehouse inventory balances change constantly, so E-Way makes a synchronous call through webMethods to Corporate Express' custom-developed InVision ERP system to retrieve balances when they are requested by an online customer. But pricing data is more static, so pricing algorithms and data about customers and products that are in PL/SQL procedures in InVision are replicated to E-Way using SharePlex from Quest Software Inc. in Irvine, Calif.

"It basically sniffs the Oracle logs and replicates the data in real time," McInnis says. "We wrap that Oracle code in a JavaBean, so we call the exact same algorithm that InVision calls to get the customer's price."

A performance advantage comes from being able to price an item just when it's needed. Previously, every combination of item price and customer was stored in a table that soon grew to an unwieldy 1 billion rows, he says. Moreover, changes to price, customer, and item data can be made in just one place and replicated elsewhere as needed. And use of an integration tool like webMethods effectively allows software to be used in multiple places without rewriting it, McInnis adds.

Optimum placement of data and logic are critical when dealing with a high volume of low-margin transactions, McInnis says. "Performance and scalability are always huge issues for us," he says. "Our average order size is [small], so it takes a lot of transactions for $5 million a day on our website."

While grabbing data using webMethods is "theoretically easy," Aiello says, optimizing performance is not. "At any point, we have thousands of people on the site placing orders. Making something available to 3,000 users simultaneously is a challenge."

Descriptive Abstract

Bibliographic information

Anthes, Gary H. (2003) Corporate Express goes direct. *Computerworld* 37(35): 17–18.

List of article's contents

This article explains how Corporate Express, an IT company, links supply systems from 500 clients with their own and their suppliers to provide quality supply procurement. Using XML and a software called E-Way, they have learned to integrate the best parts of each member's system, thereby creating a supply and delivery technique that is unequaled in the industry, causing Corporate Express to grow exponentially. This article provides insight into how they have accomplished something no one has and describes how data everywhere have been blended into hardware and software realms to create a first-class IT system.

Summary Using Proportional Reduction

Anthes, Gary H. (2003) Corporate Express goes direct. *Computerworld* 37(35): 17–18.

Purpose

Corporate Express Inc. has seamlessly linked major supply systems—its own and those of customers and suppliers to create a $5.5 billion business. The Broomfield, Colo.–based vendor of office supplies for the business market has, through connecting the supply systems of its customers and suppliers to its own system, distinguished itself as a leader in the procurement business.

Part 1

Moving from the era of paper ordering, telephone, and faxes, Corporate Express has moved to a CD-ROM catalog allowing for Internet-based procurement. Daily orders arrive electronically, generally as XML transactions through a Web portal called E-Way. Smaller companies that do not have the computer resources for procurement can use E-Way. E-Way can check budgets, buyer authorizations, and other controls for customers. It works in many ways like shopping on Amazon.com.

Part 2

Some of Corporate Expresses' larger customers can use E-Way more directly. Working together, Corporate Express and a specific company have integrated E-Way into the processing fabric of their internal procurement systems.

Part 3

Corporate Express continues to grow and has become much quicker at integrating a company into its system; the turnaround is approximately 20 days. Using software from webMethods Inc. allows integration of the logistics and financial aspects of the system. Pricing algorithms and data about customers and products, along with Oracle logs and code are wrapped in JavaBean, allowing the company to get the customer's price. This method rendered unwieldy tables obsolete and allowed for real-time pricing.

Part 4

It also allows all information to be stored in one location and accessed more easily and efficiently when needed. Software does not need to be updated as often either. Because the individual transaction is quite small as far as profit margin, this efficiency is critical, especially when thousands of users use the system simultaneously.

Summary Emphasizing Major Idea

Anthes, Gary H. "Corporate Express Goes Direct" *Computerworld* Volume 37, Number 35. (September 1, 2003) p. 17–18

Purpose

Corporate Express Inc. has seamlessly linked major supply systems—its own and those of customers and suppliers to create a $5.5 billion business. Through connecting the supply systems of its customers and suppliers to its own, the Broomfield, Colo.–based vendor of office supplies for the business market has distinguished itself as a leader in the procurement business.

Current situation

Originally taking orders by telephone, fax, and electronic data interchange, in 1997 the company began offering customers Internet-based procurement via a simple CD-ROM catalog. Today 50% or more of its orders arrive electronically, most as XML transactions through a Web portal called E-Way. Moving to such a method has demanded some significant changes in the way the company processes orders and the equipment needed to accomplish this task.

Methods used in current situation

This is accomplished by a two-way flow between customers and E-Way as XML transactions by way of two Internet service providers. Orders from small and midsize customers go to the E-Way Web server and pass through the webMethods Integration Platform server directly into InVision. Larger clients with their own e-procurement systems have been integrated with Integration Platform from webMethods. Integration Platform allows for processing the order and sending it to its own warehouse or to a wholesaler for pickup and delivery the next day.

Effects of current situation on clients and company

Much of E-Way and its other supply systems were developed in-house so that the company could accommodate the unique needs of customers. For each customer, unique integration enhances efficiency because changes to price, customer, and item data can be made in just one place and replicated elsewhere as needed. Efficient control of data is critical when dealing with a high volume of low-margin transactions. Corporate Express's average order size is small, but the effectiveness of the integration system allows the company to transport $5 million a day.

Worksheet for Summarizing

☐ Read over the article or report you intend to summarize.

☐ Determine your audience for this summary.
How much do they know about the topic?

(continued)

(continued)

□ How will your audience use this summary? (Experts want names of people and companies, and they understand common jargon in the field; nonexperts don't.)
Do they want an overview? Are they trying to "keep up"? Will they use it instead of reading the entire document? Will they make a decision on the basis of its contents?

□ Mark all keywords and phrases.

□ Arrange all the words and phrases into groups or clusters.

□ Write a sentence stating the main point you want to convey.

□ Write the body using the groups or clusters you have created.

Examples

Both of the following professional abstracts clearly indicate the contents and main points of their respective articles. Readers can determine the main point of each article, even if they do not have the knowledge background to evaluate the specifics.

Example 8.1

Professional Abstract

From Dale L. Sullivan and Michael S. Martin. Habit formation and story telling: A theory for guiding ethical action. *Technical Communication Quarterly* 10.3 (Summer 2001): 251–272.
This article proposes retrospective narrative justifications combined with classical concepts of habit formation as a theory for ethics appropriate for practicing technical communicators. To explicate the theory, the article draws on Alasdair MacIntyre's ethical theory, which involves habit formation and narrative theory; on *apologia* and account-giving theory; and on traditional ethical stances, such as the teleological and deontological doctrines. Special attention is given to the ends-means relationship and the tension between individual and corporate identity in technical communication environments.

Example 8.2

Professional Abstract

From Keech, Gregory W., Phillip Whiting, and D. Grant Allen. Effect of paper machine additives on the health of activated sludge. *TAPPI Journal* 83.3 (March 2000): 86–90.
This study examines batch respirometry as a screening tool to identify problematic papermaking additives that could disrupt the biological treatment of mill effluent. The method rapidly evaluates the toxicity of paper additives by tracking oxygen consumption of the respiring microorganisms in the

activated sludge. Batch screening tests of 20 paper additives indicated that three paper dyes, a cleaner/solvent, and a microbiocide were the most toxic to the respiring biomass, while polymeric additives had no significant impact. A four-month pilot study with an orange dye confirmed the validity of the rapid respirometric method coupled with microscopic examination. The results also show that biological treatment systems can recover from the impact of harmful additives.

Exercises

▶ You Create

1. Create a new major idea summary to emphasize the processes for ordering at Corporate Express (see pp. 207–209).

2. Create an abstract for the report in Example 12.1 (pp. 306–309).

3. Create a one-paragraph summary for Example 9.2 (p. 224). Make the summary a proportional reduction of the original, keeping the points in the same order as the original.

4. Create a learning report for the first three exercises you have just completed. See Chapter 5, Writing Assignment 7, pages 134–135 for details on how to do this.

▶ You Analyze

5. Evaluate this summary to decide the type and effectiveness. Change the summary if needed to make it more effective.

> In E-Way's system, customer orders, acknowledgments, invoices, and other transactions are generally XML transactions flowing two ways. Small and midsize customers go to the E-Way Web server directly into In-Vision. Larger companies have been integrated with E-Way and interact with Integration Platform from webMethods.
>
> Corporate Express maintains catalogs for its clients. These catalogs are integrated into the client's ordering system in such a way that each customer is treated differently.
>
> Corporate Express created the system by using XML. CE mostly uses systems developed in-house. Unocal uses the system, so does Kroger. The Oracle is wrapped in a JavaBean that allows E-Way to integrate ordering and pricing. The system allows CE to handle thousands of users simultaneously without having to constantly rewrite it.

6. Evaluate Example 8.1 for type and strength. Change if necessary. If you do not change it, please explain why you think it is fine as it is.

7. Evaluate Example 8.2 for type and strength. Change if necessary. If you do not change it, please explain why you think it is fine as it is.

8. Select a website from your chosen field of study that contains an article related to that field. Evaluate that site and write a descriptive abstract of that article for a nontechnical audience.

Web Exercise

Select an innovative technology in your field (biodegradable plastics, digital television, micromachines). Find a website that contains an article or a report on the technology. Write a descriptive abstract and a proportional reduction abstract of the document for a nontechnical audience.

Works Cited

American National Standards Institute (ANSI). *American National Standard for Writing Abstracts* (Z39.14-1979). New York: ANSI, 1979.

Anthes, Gary H. "Corporate Express Goes Direct." *Computerworld* 37.35 (September 1, 2003): 17–18.

Keech, Gregory W., Phillip Whiting, and D. Grant Allen. "Effect of Paper Machine Additives on the Health of Activated Sludge." *TAPPI Journal* 83.3 (March 2000): 86–90.

Sullivan, Dale L., and Michael S. Martin. "Habit Formation and Story Telling: A Theory for Guiding Ethical Action." *Technical Communication Quarterly* 10.3 (Summer 2001): 251–272.

Vaughan, David K. "Abstracts and Summaries: Some Clarifying Distinctions." *Technical Writing Teacher* 18.2 (1991): 132–141.

Chapter 9
In a Nutshell

Definitions orient readers. Definitions help readers place new concepts in context. Definitions explain new terms and concepts to readers. The traditional way to define is to put the term in a class and then explain how it is different from other members of the class:

"A camera is a device (*the class*) for taking photographs (*the difference*)."

Often, however, writers use an extended definition because the reader needs to understand the concept, not just the term. In an extended definition, use one or more strategies to make the term familiar to your audience; for example, compare or contrast, use a common example, explain cause and effect, or add a visual aid.

The following paragraph explains cause (within 300 light-years), uses a common example (spherical ball), and a comparison (mothballs):

"In other places in the sky, thousands or hundreds of thousands of stars of a common origin may be located within 300 light-years or so, forming a huge spherical ball. These groupings are called globular clusters. In the northern sky, the globular cluster M13 in the constellation Hercules is the easiest to see. A globular cluster may look like a hazy mothball to the naked eye or when viewed through a small telescope; larger telescopes are necessary to see individual stars in this type of cluster . . ." (Menzel and Pasachoff 116).

Providing definitions—giving the precise meanings of terms—is an important strategy in presenting new concepts. Definitions help readers relate new material to ideas they already hold. They take readers from the familiar to the new. This chapter explains formal definitions, informal definitions, and extended definitions. It also provides advice on planning definitions.

Creating Formal Definitions

A *formal definition* is one sentence that contains three parts: the term that needs defining, the class to which the item belongs, and the differentiation of that item from all other members of its class. Here are some examples (Ziemian):

> **Long bones,** such as those in the legs, arms, fingers, and toes are long, strong, slightly curved shafts with large ends that are able to absorb shock.

> The kneecap, wrist, and ankle bones are **short bones** that are shaped like irregular cubes and are made of spongy tissue covered by compact tissue.

> **Flat bones** like the skull, ribs, sternum, scapula, and hips help protect organs and act as anchor points for muscles.

> The vertebrae and facial bones are **irregular bones** which help support and protect the body (Ziemian 2).

Classify the Term

To define a term, you first place it in a class, the large group to which the term belongs. A class can be either broad or narrow. For instance, a pen can be classed as a "thing" or as a "writing instrument." A carburetor can be a "part" or a "mixing chamber." The narrower the class, the more meaning conveyed, and the less that needs to be said in the differentiation. The class, however, must be broad enough to be included in the reader's knowledge base. For the definition to be effective, readers must be able to relate it to something they know.

Differentiate the Term

To *differentiate* the term, explain those characteristics that belong only to it and not to the other members of the class. If the differentiation applies to more than one member of the class, the definition is imprecise. For instance, if a writer says, "*Evaporation* is the process of water disappearing from a certain area," the definition is too broad; water can disappear for many reasons, not just from evaporation. The differentiation must explain the characteristics of evaporation that make it unlike any other process: the change of a substance from a liquid to a vapor.

Here are four common methods for differentiating a term:

- Name its essential properties: the characteristic features possessed by all individuals of this type.
- Explain what it does.
- If the term is an object, describe what it is made of and what it looks like.
- If the term is a process, explain how to make or do it.

In the following examples, note that the classification is often deleted. In many cases, it would be a broad statement, such as ". . . is a machine."

Name the Essential Properties

A set of instructions for a computer to follow is called a **program.** The collection of programs used by a computer is referred to as the **software** for that computer. The actual physical machines that make up a computer installation are referred to as **hardware** (Savitch 2).

Explain What It Does

A sequence of precise instructions which leads to a solution is called an **algorithm.** Some approximately equivalent words are *recipe, method, directions, procedure,* and *routine* (Savitch 14).

Describe What It Looks Like

An **input device** is any device that allows a person to communicate information to a computer. Your primary input devices are likely to be a keyboard and a mouse. Among other things, a mouse is used to point to and choose one of a list of alternatives displayed on the screen. The device is called a *mouse* because the cord connecting it to the terminal makes it look somewhat like a mouse with a long tail (Savitch 4).

Describe What It Is Made Of

A **network** consists of a number of computers connected, so they may share resources, such as printers, and may share information. A network may contain a number of workstations and one or more mainframes, as well as shared devices such as printers (Savitch 3).

Explain How to Make or Do It

The result of the **problem-solving phase** is an algorithm, expressed in English, for solving the problem. To produce a program in a programming language such as C++, the algorithm is translated into the programming language (Savitch 16).

Avoid Circular Definitions

Do not use circular definitions, which repeat the word being defined or a term derived from it. You will not help a reader understand *capacitance* if you use the word *capacitor* in the differentiation. Noncrucial words, such as *writing* in the term *technical writing,* may, of course, be repeated.

Creating Informal Definitions

For specialized or technical terms that your readers will not know, you can provide an informal definition. Two common informal definitions are operational definitions and synonyms.

Operational Definitions

An *operational definition* gives the meaning of an abstract word for one particular time and place. Scientists and managers use operational definitions to give measurable meanings to abstractions. The operational definition "creates a test for discriminating in one particular circumstance" (Fahnestock and Secor 84). For instance, to determine whether a marketing program is a success, managers need to define success. If their operational definition of success is "to increase sales by 10 percent" and if the increase occurs, the program is successful. In this sense, the operational definition is an agreed-upon criterion. If everyone agrees, the definition facilitates the discussion and evaluation of a topic.

Synonyms

A *synonym* is a word that means the same as another word. It is effective as a definition only when it is better known than the term being defined. People are more familiar with *cardboard* than with *f-flute corrugated,* the technical term that has the same meaning. If your audience knows less about the topic than you do, use common words to clarify technical terms.

When using synonyms, put the common word or the technical term in parentheses or set it off with dashes, as in the following examples. Writers often highlight the term that they are defining (Brown; Hewitt).

Parentheses

Any quantity that requires both magnitude (the amount of, or how much) and direction (which way) for a complete description is a **vector quantity** (Hewitt 37).

The monosaccharides consist of one molecule and include glucose ("blood sugar" or "Dextrose"), fructose ("fruit sugar"), and galactose—which has no nickname (Brown 14-4).

Dash

There is only one *cardiac muscle*—the heart (Ziemian 8).

End of the sentence | The muscle that initiates the action is the agonist or *prime mover* (Ziewian 8).

Developing Extended Definitions

Extended definitions are expanded explanations of the term being defined. After reading a formal definition, a less knowledgeable reader often needs more explanation to understand the term completely. Seven methods for extending definitions follow.

Explain the Derivation

To explain the *derivation* of a term is to explain its origin. One way is to show how it is a combination of other words. *Technology,* for example, derives from the Greek words *techne,* meaning "an art or a skill," and *logia,* meaning "a science or study." Thus the literal meaning of the word *technology* is the study of an art or skill (Ramanathan). Another is to spell out acronyms. *ASCII* is an acronym for *A*merican *S*tandard *C*ode for *I*nformation *I*nterchange.

Explicate Terms

In this context, to *explicate* means to define difficult words contained in the formal definition. Many readers would need definitions of terms such as *algorithm* in the formal definitions on page 216. When explicating, you can often provide an informal definition rather than another formal one. Note that "poor nutrition" defines "malnutrition" in the following example.

> *Malnutrition* means "poor" nutrition and applies to both ends of the nutrient intake range. You can be malnourished because you eat inadequate amounts of nutrients or because you eat too much of them. Vitamin A toxicity is an example of malnutrition, just as scurvy is (Brown 1-13).

Use an Example

An example gives readers something concrete to help them understand a term. In the following paragraph, a formal definition of *food additives* is amplified by an extended example:

Definition | Chemical substances that are intentionally (or unintentionally) added to foods or that affect the characteristics of foods are considered *food additives.* . . . Without intentional food additives, bread would mold or dry out within a few days, salt would absorb moisture and clump up in the shaker, marshmallows would quickly harden, ice cream would crystallize, salad dressings would separate, and diet sodas would contain only water (Brown 5-3).

Use an Analogy

An *analogy* points out a similarity between otherwise dissimilar things. If something is unknown to readers, it helps if you compare it to something they do know. Here is an analogy that compares train wheels to tapered cups.

Comparison of train wheels to tapered cups

Why does a moving railroad train stay on the tracks? Most people assume the wheel flanges keep the wheels from rolling off. But if you look at these flanges you'll notice they may be rusty. They seldom touch the track, except when they follow slots that switch the train from one set of tracks to another. So how do the wheels of a train stay on the tracks? They stay on the track because their rims are slightly tapered.

If you roll a tapered cup across a surface, it makes a curved path. The larger diameter end rolls a greater distance per revolution and has a greater linear speed than the smaller end. If you fasten a pair of cups together at their wide ends (simply taping them together) and roll the pair along a pair of parallel tracks, the cups will remain on the track and center themselves whenever they roll off center. This occurs because when the pair rolls to the left of center, say, the wider part of the left cup rides on the left track while the narrow part of the right cup rides on the right track. This steers the pair toward the center. If it "overshoots" toward the right, the process repeats, this time toward the left, as the wheels tend to center themselves. Likewise for a railroad train, when passengers feel the train swaying as these corrective actions occur (Hewitt 51).

Compare and Contrast

A *comparison-contrast* definition shows both the similarities of and the differences between similar objects or processes. An example is comparing water flowing through a pipe to electricity flowing through a wire. Like other methods of extending a definition, the comparison-contrast method takes advantage of something the readers know to explain something they do not know. Comparing and contrasting a *semiconductor* with a *conductor* of electricity works only if the reader knows what a conductor is.

Here is an extended definition that compares and contrasts speed and velocity.

First term example

Second term example

Indication of difference in meaning

Loosely speaking, we can use the words *speed* and *velocity* interchangeably. Strictly speaking, however, there is a distinction between the two. When we say that something travels at 60 kilometers per hour, we are specifying its speed. But if we say that something travels at 60 kilometers per hour to the north, we are specifying its velocity. A race-car driver is concerned primarily with his speed—how fast he is moving; an airplane pilot is concerned with her velocity—how fast and in what direction she is moving. When we describe speed and the *direction* of motion, we are specifying **velocity** (Hewitt 25).

Explain Cause and Effect

Some concepts are so elusive that they must be defined in terms of their causes and effects. In the following example, the writer describes the causes and effects of acceleration in order to extend the formal definition.

> We can change the velocity of something by changing its speed, by changing its direction, or by changing both its speed *and* its direction. We define the rate of change of velocity as **acceleration:**
>
> Acceleration = change of velocity / time interval

Effect to explain

> We are all familiar with acceleration in an automobile. In driving, we call it "pickup" or "getaway"; we experience it when we tend to lurch toward the rear of the car. The key idea that defines acceleration is *change.*

Example illustrates cause

> Suppose we are driving and in 1 second we steadily increase our velocity from 30 kilometers per hour to 35 kilometers per hour, and then to 40 kilometers per hour in the next second to 45 in the next second and so on. We change our velocity by 5 kilometers per hour each second. This change of velocity is what we mean by acceleration.

Second example shows cause from another point of view

> The term *acceleration* applies to decreases as well as to increases in velocity. We say the brakes of a car, for example, produce large retarding accelerations; that is, there is a large decrease per second in the velocity of the car. We often call this *deceleration.* We experience deceleration when we tend to lurch toward the front of the car (Hewitt 26).

Analyze the Term

To *analyze* is to divide a term into its parts. Analysis helps readers understand by allowing them to grasp the definition bit by bit. For example, *elemental times,* a term used to analyze the work of a machine operator, is easier to understand when its main parts are discussed individually.

Preview

> The elemental times are the objective for taking the time study in the first place. The elemental time slots on the time-study form contain the following times: overall time, average time, normal time, and standard time.

Definition of part 1
Definition of part 2

> *Overall time* is the whole time from start to finish; it includes total times of all elements (the individual pieces studied). *Average time* is the time it takes to produce or assemble each piece into a product. When the overall time is divided by the number of pieces produced, the average time per piece is obtained.

Definition of part 3

> *Normal time* is the average time per piece multiplied by a leveling factor. This factor is just a conversion of the performance rating found by using a time-study conversion chart. This normal time represents the time required by a qualified worker, working at the normal performance level, to perform the given task.

Definition of part 4

> *Standard time,* which is the time allowed to do a job, is obtained by adjusting the normal times according to some allowance factor.

Planning Your Definition

To plan your definition, consider your audience's level of knowledge and the amount of detail they need in this situation. Either the audience does not know the term at all, or they are not sure which of several possible meanings you are using. Acronyms, for instance, must often be explained to less knowledgeable readers. Common terms such as *planning phase* may need to be defined so that readers will know what the phrase means in a particular context.

The amount of detail that you provide depends on the reader's needs. At the most basic level, defining unfamiliar words provides enough information for the reader to continue reading intelligently. Thus the definition of *file server* as a *network rest area* provides a lay reader with a general framework to grasp the basic relationship involved. However, for a reader who requires a technical orientation, you need to use a more precise formal definition.

Definitions also provide background that enables people to act in a situation. For instance, suppose a group must decide whether to continue a program. If the decision is based on the program's success, the group must first define success before it can act.

Worksheet for Defining Terms

☐ **Name the audience for the definition.**

☐ **What do they know about the concepts on which this definition is based?**

☐ **What is your goal for your readers?**
In other words, how will they use your definition?

☐ **Select a method—one-sentence, informal, or extended.**
For one-sentence definitions, select a class and differentiation that the reader can grasp. Are both items narrow enough to be useful?

For informal definitions, select synonyms or clear operational wording that the reader knows.

For extended definitions, select a method or methods that enable the reader to grasp the new information. Is the new information based on material you can reasonably expect that the reader already knows?

☐ **Will a visual aid help convey the concept?**

Examples

Examples 9.1, 9.2, and 9.3 illustrate many of the methods of defining that this chapter explains.

Example 9.1

Model
Illustrating
Definitions

Source: Excerpted
from "Photo Input:
What's the Best Way
for You to Bring
Film-Based Images
into the Computer?"
from *PC Photo* 1.2
(July/August 1997):
66–69 by permission
of *PC Photo Magazine.*

PHOTO INPUT: WHAT'S THE BEST WAY FOR YOU
TO BRING FILM-BASED IMAGES INTO THE COMPUTER?

One thing that can be quite confusing is comparing resolution of scanners versus what's on a Photo CD or other disk-based digitized image. The problem is best understood if you think of disk-based images as already scanned images. Since they're already scanned, they have a finite size and number of "pixels" (or dots) making up the image that can be counted.

Scanners' numbers don't work that way since a scanner has to scan an image. The image can be infinitely variable in size and so will have a variable number of total pixels in the image area. Each scanner has a "dpi" or dots per inch scan that will recognize a certain number of pixels in a defined area (per square inch). As this defined area expands to include the total image area, it'll also include more total pixels, although the dpi stays the same.

Just remember that *dpi* refers strictly to an area 1×1-inch square—as you tell the scanner to include more 1×1 areas, you'll also be adding to the total number of pixels (since each 1×1 area has the same dpi). A small photo will have the same dpi as a large photo, but will have fewer pixels because the image area only includes so many 1×1-inch areas.

On a disk such as a Photo CD, the image has already been scanned to a certain size, so while there was a dpi for the scanner, it now has no relevance since the overall size has been selected. (You can change that size in image-processing software, which will change the dpi because you're changing area with the same overall number of pixels.) The total size of the image is what's important, so it's given as dimensions in pixels.

This isn't dpi, since once you start using the image, the dots or pixels per inch will change as you change the size of the image (i.e., an image with large height and width dimensions will spread out the finite pixels available for a low "dpi," while a smaller image will cram them together for a higher "dpi").

- **dpi**—dots per inch; used with a scanner to look at the resolution of the unit
- **pixels**—total number of dots per entire image; whether the image is shrunk or enlarged, the number doesn't change; when the image is shrunk, these finite number of pixels come closer together, giving a higher dpi; when the image is enlarged, these pixels spread apart, giving a lower dpi

Example 9.2

Model
Illustrating
Definitions

Source: Excerpt
pp. 49–50 from
Conceptual Physics,
8/e, by Paul G.
Hewitt. Copyright
© 1998 by Paul G.
Hewitt. Reprinted
by permission of
Pearson Education,
Inc.

CIRCULAR MOTION

Which moves faster on a merry-go-round—a horse near the outside rail or a horse near the inside rail? Ask different people this question, and you'll get different answers. That's because it's easy to get linear speed confused with rotational speed.

Linear speed is what we have been calling simply *speed*—the distance in meters or kilometers moved per unit of time. A point on the outside of a merry-go-round or turntable moves a greater distance in one complete rotation than a point on the inside. The linear speed is greater on the outside of a rotating object than inside and closer to the axis. The speed of something moving along a circular path can be called **tangential speed** because the direction of motion is always tangent to the circle. For circular motion we can use the terms *linear speed* and *tangential speed* interchangeably.

Rotational speed (sometimes called angular speed) refers to the number of rotations or revolutions per unit of time. All parts of the rigid merry-go-round and turntable turn about the axis of rotation *in the same amount of time.* All parts share the same rate of rotation, or *number of rotations or revolutions per unit of time.* It is common to express rotational rates in revolutions per minute (RPM). Phonograph records that were common not so long ago, for example, rotate at 33-⅓ RPM. A ladybug sitting anywhere on the surface of the record revolves at 33-⅓ RPM.

Tangential speed and rotational speed are related. Have you ever ridden on a giant, rotating round platform in an amusement park? The faster it turns, the faster is your tangential speed. This makes sense; the greater the RPMs, the faster your speed in meters per second. More exactly, if you are at a given distance from the center, there is a direct proportion between tangential speed and rotational speed. If, for example, you double the RPMs, you double your tangential speed; triple the RPMs and you triple your tangential speed. We say that tangential speed is *directly proportional* to rotational speed (at a fixed radial distance).

Tangential speed, unlike rotational speed, depends on the distance from the axis. At the very center of the rotating platform, you have no speed at all; you merely rotate. But as you approach the edge of the platform you find yourself moving faster and faster. Tangential speed is directly proportional to distance from the axis (for a given rotational speed). Move out twice as far from the rotational axis at the center and you move twice as fast. Move out three times as far and you have three times as much tangential speed. If you find yourself in any rotating system whatever, your tangential speed depends on how far you are from the axis of rotation. When a row of people locked arm in arm at the skating rink make a turn, the motion of "tail-end Charlie" is evidence of this greater speed.

So tangential speed is directly proportional to both rotational speed and radial distance (Hewitt 49–50).

Example 9.3

Model
Illustrating
Definitions

Source: From *Mayo Clinic Women's Health Source,* September 2000, with permission of Mayo Foundation for Medical Education and Research, Rochester, MN 55905.

BASIC GENETICS

You need to know the basics before you can begin to understand the complexity of gene therapy. Here's a guide.

- *Human cell.* Each of the 100 trillion cells in the human body (except blood cells) contains the entire human genome—all the genetic information needed to build a human being. This information is encoded in 6 billion base pairs, subunits of DNA. Egg and sperm cells each have half this amount of DNA.
- *DNA.* Short for deoxyribonucleic acid, the principal carrier of a cell's genetic information. It's organized into two chains that form a double helix. Units that make up these chains include four "bases"—adenine (A), thymine (T), guanine (G), and cytosine (C). These bases form interlocking pairs that can fit together in only one way: A pairs with T; G pairs with C. The organization of these base pairs provides the informational code of the DNA molecule.
- *RNA.* Short for ribonucleic acid, a single-stranded chain made by DNA that helps guide protein synthesis.
- *Cell nucleus.* Six feet of DNA are packed into 23 pairs of chromosomes. One chromosome in each pair comes from each parent inside this central structure of the cell.
- *Chromosomes.* Rod-like structures containing DNA and protein located in the cell nucleus. There are 46 chromosomes in each human cell.
- *Genes.* Genes are the biological units of heredity. They're made up of segments of DNA. Each gene acts as a blueprint for making a specific enzyme or other protein.
- *Protein.* Composed of amino acids, they're the body's workhorses—essential components of all organs and chemical activities within the body. Their function depends on their shapes, determined by the 30,000 to 50,000 genes in the cell nucleus.

Exercises

▶ **You Create**

1. Write one-sentence definitions of four technical terms for a nonexpert reader. Alternate: Define the same four terms for an expert reader.

2. Write an extended definition of a term drawn from your major. Use one of the following methods for extending your definition: analogy, comparison-contrast,

or cause and effect. Use a visual aid, if possible. If necessary, use one of the following terms:

mutual fund	serving size
cyberspace	rejection rate
hypernet	a type of testing common in your field
flat-rate pricing	
anorexia	search engine
detail drawing	keyword

3. Provide a visual aid that clarifies one or several of the items in Example 9.2.

▶ You Analyze

4. Analyze, and rewrite if necessary, these definitions:

   A bibliography is a list of books.

   A manager runs the show.

   The Web is a bunch of computers hooked together.

   Software is what you use to run hardware.

5. Analyze either Example 9.1 or 9.2 to determine the types of definitions used and their intended audience.

▶ You Revise

6. In Example 9.2, rewrite for a different audience the section that explains the three types of speed.

▶ Group

7. In groups of two or three, write one-sentence formal definitions of three different objects or concepts. From that group, pick one definition and decide on any audience. For the next class, each member will write an extended definition about one page long. At the second class, in groups of three or four, compare the definitions. Then read and discuss your choices with the class.

Writing Assignments

1. You have been asked to provide basic background to a committee that will make a purchase decision on an object or a process. Suppose, for instance, that your office wants to install a local area network. Define the basic concepts (such as *server*) that they need to know. Plan by filling out the worksheet (p. 222); then draft the document. Be sure to credit any sources that you use.

2. Write a learning report for the writing assignment you just completed. See Chapter 5, Writing Assignment 7, pages 134–135, for details of the assignment.

Web Exercise

Write an extended definition of an effective website. Choose one of these categories—commercial, entertainment, professional society, university, personal, technical information.

Works Cited

Brown, Judith E. *Nutrition Now.* Minneapolis/St. Paul: West, 1995.

Fahnestock, Jeanne, and Marie Secor. *A Rhetoric of Argument.* New York: Random House, 1982.

Hewitt, Paul G. *Conceptual Physics.* Reading, MA: Addison-Wesley, 1998.

Menzel, Donald M., and Jay M. Pasachoff. *A Field Guide to the Stars.* Boston: Houghton Mifflin, 1983.

"Photo Input: What's the Best Way for You to Bring Film-Based Images into the Computer?" *PC Photo* 1.2 July/Aug. 1997: 66–69.

Ramanathan, K. "The Polytrophic Components of Manufacturing Technology." *Technological Forecasting and Social Change* 46.3 (1994): 221–258.

Savitch, Walter. *Problem Solving with C++.* Menlo Park, CA: Addison-Wesley, 1996.

Ziemian, Joe. *Human Anatomy Coloring Book.* New York: Dover, 1982.

Describing

Chapter 10
In a Nutshell

Description orients readers to objects and processes. Your goal is to make the readers feel in control of the subject.

▶ Show readers how this subject fits into a larger context that is important to them.

▶ Tell them what you are going to say; write this kind of document in a top-down manner.

▶ Choose clear heads; write in manageable chunks; define terms sensibly; use visual aids that help-fully communicate.

▶ Choose a tone that enables readers to see you as a guide.

Mechanism descriptions. In the *introduction*

▶ Define the mechanism and tell its purpose.

▶ Provide an overall description.

▶ List the main parts.

For each *main part*

▶ Define the mechanism and describe it in terms of size, shape, material, location, color.

▶ List and describe any subparts.

Process descriptions. In the *introduction* you need to

▶ Tell the goal and significance of the process.

▶ Explain principles of operation.

▶ List the major sequences.

For each sequence or step

▶ Tell the end goal of the process.

▶ Describe the action in terms of qualities and quantities.

Description is widely used in technical writing. Many reports require that you describe something—a machine, process, or system. Sometimes you will describe in intricate detail, other times in broad outline. This chapter shows you how to describe a mechanism, an operation, and a process focused on a person in action.

Planning the Mechanism Description

The goal of a mechanism description is to make the readers confident that they have all the information they need about the mechanism. Obviously, you can't describe every part in minute detail, so you select various key parts and their functions. When you plan a description of a mechanism, consider the audience, select an organizational principle, choose visual aids, and follow or adopt the usual form for writing descriptions.

Consider the Audience

To make the audience feel confident, consider their knowledge level and why they need the information. Basically, the principle is to give them the physical details that they need to act. The details you choose and the amount of definition you provide reflect your understanding of their knowledge and need. Here is a brief, simple description for an audience that must make a decision about a topic easily understood by most people. The author can safely assume that length and width terms need no definition.

> The truck box size is an important factor because we frequently transport 4 ft by 8 ft sheets of wood. The box size of the Hauler at the floor is 3.5 ft by 6 ft. The box size of the X-200 at the floor is 4 ft by 8 ft. This factor means we should purchase the X-200.

Here, however, is another brief description, also the basis for a decision, directed at an audience that has a specialized technical knowledge. The writer here assumes that the audience understands terms like "dpi resolution." If the writer assumed the audience did not understand the term, he or she would have to enlarge the discussion to define all the terms.

> The ABC scanner has 50–1200 dpi resolution, 24-bit color scanning, and an optional transparency adaptor. The XYZ scanner has 400–1600 dpi resolution, 24-bit color, and an optional transparency adaptor. The higher resolution capabilities make XYZ the preferable purchase to fill our needs.

Select an Organizational Principle

You can choose from several organizational principles. For instance, you can describe an object from

▶ Left to right (or right to left).
▶ Top to bottom (or bottom to top).
▶ Outside to inside (or inside to outside).
▶ Most important to least important (or least important to most important).

Base this decision on your audience's need. For a general introduction, a simple sequence like top to bottom is best. For future action, say, to decide whether or not to accept a recommendation, use most to least important.

An easy way to check the effectiveness of your principle of organization is to look for *backtracking.* Your description should move steadily forward, starting with basic definitions or concepts that the audience needs to understand later statements. If your description is full of sections in which you have to stop and backtrack to define terms or concepts, your sequence is probably inappropriate.

Choose Visual Aids

Use visual aids to enhance your description of a mechanism. As the figure of a paper micrometer (p. 231) demonstrates, overviews show all the parts in relationship. Details focus readers on specific aspects. Often a visual aid of a detail can dramatically shorten a text discussion. Consider this brief discussion of a problem with broken piping joints. It would be much longer and difficult to comprehend if it were all text.

Figure 1a shows a typical cross-section view of a copper-to-copper tube joint soldered together. Notice how the solder covers up the entire opening between the two tubes. Figure 1b shows the break in the solder due to a change in temperature.

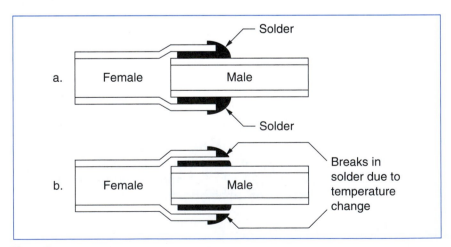

Figure 1
Copper-to-Copper Tube Joints

Often your visual aid focuses the text. In effect, your words describe the visual aid.

Follow the Usual Form for Descriptions

Generally, descriptions do not stand alone but are part of a larger document. However, they still have an introduction and body sections; conclusions are optional. Use conclusions only if you need to point out significance. Make the introduction brief, stating either your goal for the reader or the purpose of the mechanism. To describe a part, point out whatever is necessary about relevant physical details—size, shape, material, weight, relationship to other parts, or method of connection to other parts. If necessary, use analogies and statements of significance to help your reader understand the part.

Writing the Mechanism Description

A stand-alone mechanism description has a brief introduction, a description of each part, and an optional conclusion.

Introduction

The introduction gives the reader a framework for understanding the mechanism. In the introduction, define the mechanism, state its purpose, present an overall description, and preview the main parts.

Definition and purpose
Overall description
Main parts

A paper micrometer is a small measuring instrument used to measure the thickness of a piece of paper. The micrometer, roughly twice as large as a regular stapler (see Figure 1), has four main parts: the frame, the dial, the hand lever, and the piston.

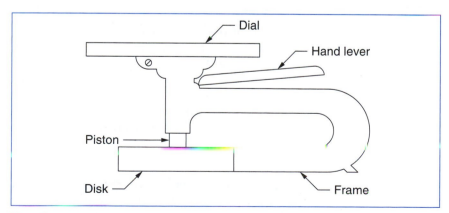

Figure 1
Paper Micrometer

TIP

Mechanism Description

A quick way to plan a mechanism description is to use this outline:

I. Introduction
   A. Definition and Purpose
   B. Overall Description (size, weight, shape, material)
   C. Main Parts

II. Description
   A. Main Part A (definition followed by detailed description of size, shape, material, location, method of attachment)
   B. Main Part B (definition followed by overall description, and then identification of subparts)
      1. Subpart X (definition followed by detailed description of size, shape, material, location, method of attachment)
      2. Subpart Y (same as for X)
   C. Other Main Parts
   D. Etc.

Body: Description of Mechanism

The *body* of the description contains the details. Identify each main part with a heading and then describe it. In a complex description like the following one, begin the paragraph with a definition, then add details. Use coordinate structure (see pp. 71–72) for each section. If you put size first in one section, do so in all of them.

THE FRAME

Definitions

Color and analogy
Size and analogy

Weight

The frame of the paper micrometer is a cast piece of steel that provides a surface to which all the other parts are attached. The frame, painted gray, looks like the letter *C* with a large flat disk on the bottom and a round calibrated dial on top. The disk is 4½ inches in diameter and resembles a flat hockey puck. The frame is 5⅛ inches high and 7½ inches long. Excluding the bottom disk, the frame is approximately 1¼ inches wide. The micrometer weighs 8 pounds.

THE DIAL

Definition and analogy

Analogy
Size

The dial shows the thickness of the paper. The dial looks like a watch dial except that it has only one moving hand. The frame around the dial is made of chrome-plated metal. A piece of glass protects the face of the dial in the same way that the glass crystal on a watch protects the face and hands. The dial, 6 inches in diameter and ⅞ inch thick, is calibrated in

Appearance | .001-inch marks, and the face of the dial is numbered every .010 inch. The hand is made from a thin, stiff metal rod, pointed on the end.

THE HAND LEVER

Analogy and definition

Relationship to other parts

Effect

The hand lever, shaped like a handle on a pair of pliers, raises and lowers the piston. It is made of chrome-plated steel and attaches to the frame near the base of the dial. The hand lever is 4 inches long, $\frac{1}{2}$ inch wide, and $\frac{1}{4}$ inch thick. When the hand lever is depressed, the piston moves up, and the hand on the dial rotates. When the hand lever is released and a piece of paper is positioned under the piston, the dial shows the thickness of the paper.

THE PISTON

Definition

Function

Size

Relationship to other parts

The piston moves up and down when the operator depresses and releases the hand lever. This action causes the paper's thickness to register on the dial. The piston is $\frac{3}{8}$ inch in diameter, flat on the bottom, and made of metal without a finish. The piston slides in a hole in the frame. The piston can measure the thickness of paper up to .300 inch.

Other Patterns for Mechanism Descriptions

Two other patterns are useful for describing mechanisms: the function method and the generalized method.

The Function Method

One common way to describe a machine is to name its main parts and then give only a brief discussion of the function of each part. The *function method* is used extensively in manuals. The following paragraph is an example of a function paragraph:

FUNCTION BUTTONS

List of subparts

Function and size of subpart 1

Function and size of subpart 2

Function and size of subpart 3

Function and size of subpart 4

The four function buttons, located under the liquid crystal display, work in conjunction with the function switches. The four switches are hertz (Hz), decibels (dB), continuity (c), and relative (REL).

The hertz function allows you to measure the frequency of the input signal. Press the button a second time to disable. The decibel function allows you to measure the intensity of the input signal which is valuable for measuring audio signals. It functions the same way as the hertz button.

The continuity function allows you to turn on a visible bar on the display, turn on an audible continuity signal, or disable both of them. The relative function enables you to store a value as a reference value. For example, say you have a value of 1.00 volt stored; every signal that you measure with this value will have 1.00 volt subtracted from it.

The Generalized Method

The *generalized method* does not focus on a part-by-part description; instead, the writer conveys many facts about the machine. This method of describing is commonly found in technical journals and reports. With the generalized method, writers use the following outline (Jordan):

1. General detail

2. Physical description

3. Details of function

4. Other details

General detail consists of a definition and a basic statement of the operational principle. *Physical description* explains such items as shape, size, appearance, and characteristics (weight, hardness, chemical properties, methods of assembly or construction). *Details of function* explain these features of the mechanism:

- How it works, or its operational principle
- Its applications
- How well and how efficiently it works
- Special constraints, such as conditions in the environment
- How it is controlled
- How long it performs before it needs service

Other details include information about background, information about marketing, and general information, such as who makes it.

Here is a sample general description.

> The QMS ColorScript Laser 1000 breaks new ground: it's the first color laser printer to sell for under $10,000 (even if it is just a dollar under), and it's the first color laser printer for the desktop (if your desktop can hold a large picnic cooler that weighs 106 pounds). The ColorScript Laser 1000 uses an 8-pages-per-minute, 300-dots-per-inch print engine whose single paper tray can hold 250 sheets. (A second tray is optional.) The printer comes with 12MB of memory (expandable to 32MB), a 60MB internal hard drive that stores downloadable fonts and incoming print jobs, and 65 fonts. Built-in network ports include LocalTalk, Centronics parallel, and serial; Ethernet and Token Ring network interfaces are optional.
>
> The ColorScript Laser 1000 is about four times as complex to set up as a monochrome laser: you need to install four developer cartridges, four toner cartridges (cyan, magenta, yellow, and black), a cleaning pad, a bottle of oil (it keeps the toner from sticking to the printer's heat rollers), a photosensitive belt, and a hopper that holds waste toner. These consumables need to be replaced at varying intervals: every 3000 pages for the oil, for instance, and once a year or so for the belt.

As with a monochrome laser printer, the time between toner feedings varies—pages with a lot of color or black use more toner than do pages containing very little. This contrasts with thermal-wax and dye-sublimation machines, which use the same amount of ink ribbon for each page regardless of a page's content. This, and the ColorScript Laser 1000's ability to use conventional photocopier bond paper, makes the machine's cost-per-page significantly lower than that of thermal-wax and dye-sub technologies—roughly 5 to 10 cents per color page, compared with about 60 cents for thermal-wax, and several dollars for dye-sub.

How does the output look? Very good. The colors aren't as vivid as those produced by competing technologies, but they're perfectly adequate for business documents—bar charts, colored headlines, and transparencies. The printer's text quality is almost as sharp as that produced by a monochrome laser printer, and generally sharper than thermal-wax and dye-sub output. The print is also more durable than ink-jet or thermal-wax output: it doesn't smear if it gets wet (ink-jet output sometimes does), and it's nearly indestructible: unlike the wax applied by thermal-wax machines and the ink from solid-ink printers such as Tektronix's 300i, the ColorScript Laser 1000's color toner doesn't scratch off when you paper-clip it, or flake when you fold it. The ColorScript Laser 1000 also fared well on Macworld Lab's test track, although its overall performance was half that of Tektronix's Phaser 220i, the fastest thermal-wax machine available. The printer was particularly slow in the Photoshop test. Further incentive to look elsewhere for prepress work is that the printer doesn't do as good a job with scanned images as do thermal-wax machines (such as Tektronix's Phaser 200 series) that provide enhanced halftoning options. Also, color laser technology doesn't allow for as much consistency ("QMS" 75).

Planning the Process Description

Technical writers often describe processes such as methods of testing or evaluating, methods of installing, flow of material through a plant, the schedule for implementing a proposal, and the method for calculating depreciation. Manuals and reports contain many examples of process descriptions.

As with a mechanism description, the writer must consider the audience, select an organizational principle, choose visual aids, and follow the usual form for writing descriptions.

Consider the Audience

Your goal is to make your audience confident that they have all the information they need about the process. The knowledge level of audiences and their potential use of the document will vary. Their knowledge level can range from advanced to beginner. Uses will vary; often they use the description to make a decision. For instance, a plant engineer might propose a change in material flow in a plant because a certain step is inefficient, causing a bottleneck. To get the change approved, he or she would have to describe the old and new processes to a manager, who would use that description to decide whether to implement the new process.

Process descriptions also explain theory, thus answering the audience's need for a background understanding. The writer can describe how a sequence of actions has a cause-effect relationship, thus allowing the reader to understand where the trouble might be in a machine or what the significance of an action might be. Here is a brief process description that allows a reader to analyze his or her own leaking faucet:

> When the hot- and cold-water handles of the stem faucets are turned on, the rotating stems ride upward on their threads. As they rise, the stems draw the washers away from the brass rings, called faucet seats, at the tops of the water supply lines, allowing water to flow. When both hot and cold water flow through faucets, they mix in the faucet body and run from the spout as warm water. When the handles are turned to the off position, the stems ride downward on their threads. The washers press against the faucet seats, shutting off the flow of water (*How* 71).

Process descriptions can also present methodology, the steps a person took to complete a project or solve a problem. These statements often reassure the reader by showing that the writer has done the project the "right" way. Here is a brief methodology statement.

> I started the project by sending out for price quotes on the 21 modular containers. I chose three vendors from the Phoenix area: Box Company, Johnson Packaging, and Packages R Us. I chose the first two based on their past services for the JCN-Tucson plant. I chose Packages R Us because they are the national vendor for the Modular program for the entire JCN Corporation. From these vendors I asked for two price quotes: one based on a just-in-time (JIT) inventory system, and the other based on the existing inventory system.

Select an Organizational Principle

The organizational principle for processes is *chronological:* Start with the first action or step, and continue in order until the last. Also consider whether you need to use cause-effect in the arrangement. Many processes have obvious sequences of steps, but others require careful examination to determine the most logical sequence. If you were describing the fashion cycle, you could easily determine its four parts (introduction, rise, peak, and decline). If you had to describe the complex flow of material through a plant, however, you would want to base your sequence of steps on your audience's knowledge level and intended use of the description. You might treat "receiving" as just one step, or you might break it into "unloading," "sampling," and "accepting." Your decision depends on how much your audience needs to know.

Choose Visual Aids

Choose a visual aid that orients your reader to the process, either to see the entire process at a glance or to see the working of one step. If your subject is a machine in operation, visuals of the machine in different positions will clarify the process. If you are describing a process that involves people, a flow chart can quickly clarify a sequence.

Follow the Usual Form for Writing Descriptions

The process description takes the same form as the mechanism description: a brief introduction, which gives an overview, and the body, which treats each step in detail, usually one step to a section. Make the introduction brief, either a statement of your goal or the purpose of the process. Use conclusions only if you need to point out significance. In each paragraph, first define the step (often in terms of its goal or end product), and then describe it. Use coordinate structure (see pp. 71–72) for each section. If you follow a definition of an end goal with a brief description of the machine and then the action in one section, do so in all sections.

Define a step's end goal or purpose, and then describe the actions that occur during that step. Point out qualities like "fast" and quantities like "60 times." Add statements of significance if you need to.

Writing the Process Description
. . . .

The outline in the Tip on page 239 shows the usual form for a description of a process that does not involve a person. Other examples appear in the Examples section (pp. 245–253). Include an introduction, a description of each step, and an optional conclusion.

Introduction

The introduction provides a context for the reader. Define the process, explain its principles of operation (if necessary), and preview the major sequences. The following introduction performs all three tasks:

Date: March 29, 2005
To: April Bilasky, Second Shift Floor Supervisor
From: Brad Tanck, First Shift Floor Supervisor
Subject: Processes involved in producing wooden puzzle in automated
           manufacturing cell

Background

The following information pertains to your request for a description of the process of continuous production of wooden puzzles by an automated manufacturing cell. This memo explains that process. The wooden

Common examples

Preview of major
sequences

puzzle consists of 9 hex-shaped parts arranged in a spherical configuration when fully assembled. The automated manufacturing consists of five processes: hex-shaping the stock, cutting the stock to size, grooving the parts, deburring the parts, and finishing and packaging the parts (see Figure 1).

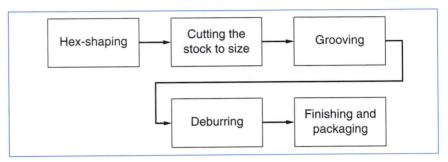

Figure 1
Process for Producing Wooden Puzzle Parts

Body: Description of the Operation

In the body of the paper, write one paragraph for each step of the process. Each paragraph should begin with a general statement about the end goal or main activity. Then the remainder explains in more detail the action necessary to achieve that goal. In the following example, each paragraph starts with an overview, and all the paragraphs are constructed in the same pattern. The flow chart gives readers an overview before they begin to read.

HEX-SHAPING THE STOCK

End goal

Action
Action

The hex-shaping process entails running 30-inch lengths of ¾-inch-square maple stock through a set of shapers to cut the stock into a hex shape. The maple stock lengths are hand-placed in a gravity hopper, where they drop down onto a moving conveyor system. From here, the conveyor transports the stock until it reaches a set of push rollers, one on each side of the stock. Friction from the push rollers forces the stock through a set of diamond-shaped shaper bits, which are offset at 60-degree angles to one another in order to shape the square stock into perfect hex stock. From here, the hex stock rides on a conveyor belt to the next station.

CUTTING THE STOCK TO SIZE

End goal

Action

The purpose of cutting the stock to size is to get 9 parts, all 3 inches in length. When the hex stock reaches the cut-to-size station, it trips a limit switch. This switch causes a stop to eject and prevents the stock from further advancing. At this point, a pneumatic circular saw is activated, which cuts a 3-inch part off the hex stock.

Process Description

To create a process description quickly, follow this outline:

I. Introduction
   A. Define process
   B. Explain principles of operation or give common examples
   C. Preview main steps in the process

II. Describe process
   A. Main Step One
      1. Define the step's goal
      2. Add necessary background material
      3. Present details of action
   B. Main steps
   C. Etc.

Action

When this process is completed, the stop is retracted and the stock moves onward, pushing the cut-to-size part down a chute to the next station and signaling the stop in preparation for cutting another part from the stock.

GROOVING THE PARTS

End goal

Background

In order for the puzzle parts to fit together, several different grooves must be cut into the parts at various angles. Of the 9 parts that are processed, 3 parts have 2 grooves in them, 3 parts have 1 groove in them, and 3 parts have 3 grooves in them.

Action

Background

Action

After the part arrives at the grooving station, it is grasped by a set of pinch rollers. The pinch rollers feed the part into a hex-shaped aluminum collet mounted on an x-y table. This collet is specially designed with 2 slots cut out of the top of it, both at 70-degree angles, and one slot cut out of its bottom at a 120-degree angle. The table moves the part and collet through a set of 2 routers, one on top and one on the bottom, which cut the required grooves in the part through the slots in the collet.

Action

When this grooving process is completed, the pressure on the collet is released. The ram of a pneumatic cylinder pushes the part onto an out-feed chute, where it travels to the next station.

DEBURRING THE PARTS

End goal

Deburring the parts includes removing any burrs and shavings from the part and cleaning up its surface before it is finished.

Action

When the part arrives at the deburring station, it lands in a gravity hopper at the head of the station. The orientation of the part is changed

Action

at this time. As the part comes through the hopper, it is handled side for side instead of end for end.

After the part drops down onto the stationary feed table, pressure from a small pneumatic cylinder secures it to the back edge of the moving feed table. At this time, a larger secondary pneumatic cylinder is actuated. This

Action

cylinder pushes the attached moving feed table with the part through a set of abrasive wheels, one above the part and one below it. These wheels take the burrs off the part and clean it up for finishing.

Action

At the conclusion of its stroke, the larger cylinder pauses and the smaller cylinder retracts, releasing the part to the next station. The larger cylinder then extends in order to deburr the next part.

FINISHING AND PACKAGING THE PARTS

End goal

The finishing and packaging process includes spraying a lacquer finish on the part, drying the finish, and packaging the puzzle parts in a sealed

Action

plastic carton. As the part drops from the deburring station, it lands on a moving conveyor system. The conveyor system's ¾-inch pins protruding

Background

from the belt keep the part from coming in contact with the belt itself. The conveyor system transports the part through a spray booth, where a lac-

Action

quer finish is applied. From here the conveyor keeps moving through a drying booth, where high-intensity heat is blown on the part to dry it before packaging.

Action

Finally, the part drops off the end of the conveyor, activating a photo-electric switch, which counts the part, and a pick-and-place robot. The pick-and-place robot takes the part and places it in a plastic carton.

When all 9 puzzle parts are placed in the carton, a plastic wrap is placed over the finished product and it is sealed by an electronic heat seal device.

Conclusion

Conclusions to brief descriptions of operation are optional. At times, writers follow the description with a discussion of the advantages and disadvantages of the process or with a brief summary. If you have written a relatively brief, well-constructed description, you do not need a summary.

Planning the Description of a Human System

A human system is a sequence of actions in which one or more people act in specified ways. If you describe in general what happens in such a system, you have a process description. But if you tell people how to act, you have a procedure or a set of instructions. The description of the product tester selection policy in the next section describes actions that the members of the group usually take on. A set of instructions, by contrast, would tell a person exactly what to do, in detail, for each step.

You plan for such a document as for a regular process description (see pp. 235–237). Like the process description shown earlier (pp. 237–240), this document contains an introduction, which defines the process and its major sequences, and the body, which describes the process in detail.

Writing the Description of a Human System

The following section shows the usual form for writing a description of a human system. Include an introduction and a body. The conclusion is optional.

Introduction

In the introduction, orient the reader to the process. The following introduction states the purpose of the document, explains why the reader needs the information, and lists the major steps in the process (Product Tester).

PRODUCT TESTER SELECTION POLICY

This document explains Illuminated Ink's product tester selection policy, which determines which individuals will be asked to test our newest products. Applicants who are interested in improving their chances of being selected as a product tester should read this document carefully. The three steps involved in this process are:

1. Applying for Product Tester Position
2. Responding to Surveys
3. Selecting Product Testers

If you have questions or concerns regarding this policy, please contact us at writeus@IlluminatedInk.com.

Body: Sequence of a Person's Activities

In the body, describe each step in sequence. Present as much detail as necessary about the quality and quantity of the actions (Product Tester).

APPLYING FOR PRODUCT TESTER POSITION

All applicants must submit a valid Product Tester Application form. This form may be completed online at www.IlluminatedInk/ProductTester Application.com, or it may be filled out in person at any Catholic conference at which Illuminated Ink is vending. A current list of the conferences that we plan to attend may be found at www.IlluminatedInk/Conference Schedule.com.

It is extremely important that applicants fill out their application as completely as possible. *Applicants will automatically be rejected if they do not provide:*

- a valid e-mail address
- a valid snail mail address
- the name, birth date, and interests of at least one child under the age of 18

RESPONDING TO SURVEYS

Once a month, all the product tester applicants entered into our database are sent an e-mail with a request to visit a specific website to view and respond to our latest Product Proposal(s) via an on-line survey. *Applicants who choose NOT to respond to this request will not be eligible to test the new product being proposed.*

SELECTING PRODUCT TESTERS

Actual product testers will be selected from the pool of Product Proposal surveys that we receive in a timely manner. Applicants are first checked to see if their profile meets our needs (e.g., female child between the ages of 6 and 12), and then the field is narrowed based on the quality of each response we receive. An individual who has taken the time to really think about how our idea could be improved upon will be given greater weight than someone who does not offer any constructive comments.

Please note: *Any Product Tester who does not submit the results of the product test within 30 days of receiving the product will automatically have their application voided for all future product tests.*

Conclusion (Optional)

A conclusion is optional. If you choose to include one, you might discuss a number of topics, depending on the audience's needs, including the advantages and disadvantages of the process.

Worksheet for Planning a Description

□ Name the audience for this description.

□ Estimate the level of their knowledge about the concepts on which this description is based and about the topic itself.

□ Name your goal for your readers.
Should readers know the parts or steps in detail or in broad outline?
Should they focus on the components of each step or part, or on the effect or significance of each step or part?
Should they focus just on the machine or process, or grasp the broader context of the topic (such as who uses it, where and how it is regulated, who makes it, and its applications and advantages)?

□ Select an approach.
What will you do first in each paragraph?
In what sequence will you present the explanatory detail?

☐ **Plan a visual aid.**
   What is your goal with the visual aid? to provide a realistic introduction? to give an overview? to be the focus of the text? to supplement the text?
   Will you have one visual aid for each step or part, or will you use just one visual and refer to it often?

☐ **Choose the type of visual aid. Use a visual that will help your reader grasp the topic.**

☐ **Decide on the visual aid's size (not too large or too small) and placement (for example, after the introduction).**

☐ **Construct a rough visual now. Finish it later.**

☐ **Devise a style sheet. Decide how you will handle heads, margins, paragraphing, and visual aid captions.**

☐ **To write a description of mechanisms**
   Name each part.
   Name each subpart.
   Define each part and subpart.
   List details of size, weight, method of attachment, and so forth.
   Tell its function.

☐ **To write a description of processes**
   Name each step.
   Name each substep.
   Tell its end goal.
   List details of quality and quantity of the action.
   Tell significance of action.

Worksheet for Evaluating a Description

1. For each part of the report—introduction, body, visual aid—answer the following questions:
   • Is it appropriate to the goal for the audience?
   • Is it consistent with all the other parts and its own subparts?
   • Is it clear to the audience and faithful to the reality?
2. Do the visual aid and the text helpfully, actively interact with each other?
3. Does the report build a mental model that an audience can use for future action?
4. Does the writer convince you that he or she is believable? (Consider two dimensions: statement of background and method of presentation.)

(continued)

(continued)

5. Check these items for a **mechanism description.**
   Introduction
   - Purpose statement is present and clear.
   - Cause of your writing is present and clear.
   - Preview of the paper is present.
   - Topic is named and defined.
   - Each item in list is a thing (not an action).
   Body
   - Each new section has effective use of keyword.
   - Either first or second sentence defines each part.
   - Any necessary background is given.
   - Subparts are listed early in the paragraph.
   - Each part has enough details (size, shape, material, location, method of attainment).
   Format
   - Heads have consistent format.
   - Visual aid is clearly drawn.
   - Visual aid has clear, correct caption at the bottom.
   - Visual aid has callouts.
   - Callouts are keywords in text.
   Style
   - Sentences use active voice if possible.
   - All technical and jargon terms are defined.
   - No spelling or grammar mistakes remain.

6. Check these items for a **process description.**
   Introduction
   - Purpose statement is present and clear.
   - Cause of your writing is present and clear.
   - Preview of the paper is present.
   - Topic is named and defined.
   - Each item in list is an action (not a thing).
   Body
   - Each new section has effective use of keyword.
   - Either first or second sentence defines each step (check "no" if any of the steps are not defined).
   - Any necessary background is given.
   - End result or overall goal of each step is explained.
   - Each step has enough details (specific actions, substeps, quantity and quality of actions)
   Format
   - Heads have consistent format.
   - Visual aid is clearly drawn.

- Visual aid has clear, correct caption at the bottom.
- Visual aid has callouts.
- Callouts are keywords in text.

Style
- Sentences use active voice if possible.
- All technical and jargon terms are defined.
- No spelling or grammar mistakes remain.

Examples

Examples 10.1 to 10.5 describe a mechanism and four processes. Examples 10.1, 10.2, 10.4, and 10.5 appear as if they were parts of longer documents. Example 10.3 is a memo. The form (memo or part of a longer document) depends solely on the situation.

Example 10.1

Description of a Mechanism

SKINFOLD CALIPER

The following information explains the skinfold caliper and its individual parts. The skinfold caliper (see Figure 1) is an instrument used to measure a double layer of skin and subcutaneous fat (fat below the skin) at a specific body site. The measurement that results is an indirect estimate of body fatness or calorie stores. The instrument is approximately 10 inches long, is made of stainless steel, and is easily held in one hand. The skinfold caliper consists of the following parts: caliper jaws, press and handle, and gauge.

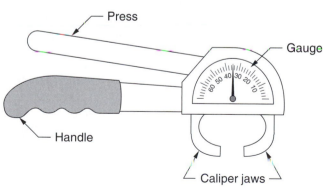

Figure 1
Skinfold Caliper

(continued)

Example 10.1

(continued)

CALIPER JAWS

The caliper jaws consist of two curved prongs. Each prong is approximately ¼ inch long. The prongs project out from the half-moon-shaped gauge housing. They are placed over the skinfold when the measurement is taken. They clasp the portion of the skinfold to be measured.

PRESS AND HANDLE

The press is the lever that controls the caliper jaws. Engaging the press opens the caliper jaws so they can slip over the skinfold. Releasing the press closes the jaws on the skinfold, allowing the actual measurement. The press is 4.5 inches long and .5 inch thick. It is manipulated by the thumb while the fingers grip the caliper handle. The caliper handle is 6 inches long and .5 inch thick. The outside edge of the handle has three indentations, which make the caliper easier to grip.

GAUGE

The gauge records the skinfold measurement. It is white, half-moon shaped, with 65 evenly spaced black markings and a pointer. Each marking represents 1 centimeter. The pointer projects from the middle of the straight edge of the half-moon-shaped gauge to the black markings. When the jaws tighten, the pointer swings to the marking that is the skinfold thickness.

Example 10.2

Description of a Process

DIAGNOSING NECK PAIN

Determining the source of the pain is essential to recommend the right method of treatment and rehabilitation. Therefore a comprehensive examination is required to determine the cause of neck pain.

Your orthopaedist will take a complete history of the difficulties you are having with your neck. He or she may ask you about other illnesses, any injury that occurred to your neck and any complaints you have associated with neck pain. Previous treatment for your neck condition will also be noted.

Next, your orthopaedist will perform a physical examination. This examination may include evaluation of neck motion, neck tenderness, and the function of the nerves and muscles in your arms and legs.

X-ray studies often will be done to allow your orthopaedist to look closely at the bones in your neck. These simple diagnostic techniques often help orthopaedists to determine the cause of neck pain and to prescribe effective treatment.

Source: "Diagnosing Neck Pain." *Neck Pain.* Park Ridge, IL: American Academy of Orthopaedic Surgeons, 1989.

Example 10.3

Description of a
Process

PROCESS DESCRIPTION: IDENTIFICATION OF UNKNOWN CHEMICALS

Date: March 28, 2006
To: Auguste Dupin
From: Kris-Jilk, Lab Research Department
Subject: Identification of Unknown Chemicals

This memo is to familiarize you with one of the most common procedures done in the research laboratory at ACME Pharmaceuticals, Inc. Mr. Dupin, have you ever come across a container in your household cleaning cupboard and, because the label had fallen off, had absolutely no idea of what it was? Well, even though we in the lab do not solicit door to door for work, this whole concept of taking an unknown compound and identifying it is one of the most important aspects of our research lab. We deal almost exclusively with medicinal agents, but are often called on by the local police department to identify their seized unknown chemicals and by the hospital to work on unidentifiable agents found in blood and tissue samples. Figure 1 lists the six steps we use to identify an unknown chemical.

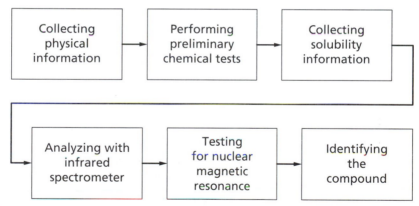

Figure 1
Process of Identification

COLLECTING PHYSICAL INFORMATION

Collecting physical information about an unknown compound gives general information pertaining to the overall chemical. Taking note of the physical state, color, smell, melting point, and boiling point of the compound gives you important basic information that deals with the most fundamental chemical properties of whatever it is you are working with.

PERFORMING PRELIMINARY CHEMICAL TESTS

Running universal preliminary chemical tests on the unknown identifies what major compound classification it falls under. All organic compounds

(continued)

Example 10.3

(continued)

can be classified into approximately 15 different categories, and running some very basic tests helps us to place the unknown into its general category. For example, if the ignition test produces black, sooty smoke, then it is evident that the general classification of the compound is an aromatic compound.

COLLECTING SOLUBILITY INFORMATION

Collecting information on the solubility of unknown compounds identifies the properties they exhibit when they are introduced to other compounds. By observing how an unknown compound reacts with a variety of other compounds, it is possible to gain insight into some of the compound's specific chemical structure. If, for example, the unknown is soluble in water, then it is clear that it is polar.

ANALYZING WITH INFRARED SPECTROMETER

Analysis of the infrared spectrometer test shows what functional groups are attached to the carbon "backbone" of the compound. By knowing the kinds of chemicals attached to the parent chemical, it is possible to begin sketching a picture of what the unknown is. These attached chemicals are mainly responsible for how the unknown reacted in the solubility tests.

TESTING FOR NUCLEAR MAGNETIC RESONANCE

Running the nuclear magnetic resonance test on an unknown compound gives the essential information about the parent carbon structure (this is the compound that is at the base of the molecule and is often called the *parent compound*). This test produces evidence relating to the number of carbons present, how they are bonded to one another, and how they interact with the attached hydrogens. This information is especially important because it shows what is at the core of what may be a very big structure.

IDENTIFYING THE COMPOUND

Identification of the unknown is now possible. All the information from the tests run by the researcher can be compiled, and in most cases, the unknown compound can be identified. One of the exceptions to this process is that, even with all this data, more extensive tests need to be completed before the compound is identifiable. The other exception is that the compound is common and could be identified after doing the preliminary chemical tests.

The process by which unknown compounds are identified begins by collecting very general information and continues until the information that is collected is very specific. These tests are run daily and are a vital function of the work done in the lab.

Example 10.4

Description
of a Process

Source: From *Ultimate
Chocolate* by Patricia
Lousada. Copyright
© 1997 Dorling
Kindersley. Reprinted
by permission of
Dorling Kindersley.

MAKING CHOCOLATE

Chocolate, like coffee, originates in a bean, but one that grows on a tree, not a bush. The exotic cocoa tree produces a blizzard of pink and white flowers, green unripe fruit, and bright golden cocoa pods, all at the same time. Encased in the pod is the dark little cocoa bean. Put through an intricate production process, the bean is transformed into cocoa mass, which ultimately becomes that magically pleasurable ingredient, chocolate (Lousada 28).

THE COCOA TREE

The region between the twentieth parallels, with the exception of parts of Africa, is the home of the cocoa tree. The tree begins to bear fruit once it is four years old and has an active lifespan of at least sixty years. Its fruit grows directly out of the older wood of the trunk and main branches, reaching the size of a small football and ripening to a rich golden color. Inside the ripe pods, purplish-brown cocoa beans are surrounded by pale pink pulp. After the pods are cut from the tree, beans and pulp are left to ferment together. The beans turn a dull red and develop their characteristic flavor. After fermentation, the beans are dried in the sun, acquiring their final "chocolate" color. They are now ready for shipping to the manufacturing countries. Various bean types have been bred and experts take pride in their ability to distinguish chocolate made from Criollo or Trinitario beans from that made from Forastero.

PROCESSING THE BEANS

When the dried beans reach the processing plant, they are cleaned and checked for quality, and then roasted. Roasting, an important stage in the manufacturing process, develops the flavor of the beans and loosens the kernels from the hard outer shell. Each chocolate manufacturer has its own roasting secrets, which contribute significantly to the chocolate's flavor. After roasting, the beans have a distinctive chocolaty smell. The next step is to crack the beans open, discarding their shells and husks, to obtain the kernels, called nibs. It is the processing of these small, brown nibs that gives us chocolate. The roasted nibs, which contain on average 54 percent cocoa butter, are ground into a dark, thick paste called cocoa mass or solids. When more pressure is applied to the cocoa mass, the resulting products are cocoa butter and a solid cocoa cake. From the cocoa cake, when it is crushed into cocoa crumbs and then finely ground, comes cocoa powder.

CONCHING

Chocolate is generally cocoa solids and sugar, with added cocoa butter (in the case of semisweet chocolate), or milk (in the case of milk chocolate), plus vanilla and other flavorings. Conching, in which the chocolate mixture is

(continued)

Example 10.4

(continued)

heated in huge vats and rotated with large paddles to blend it, is the final manufacturing process. Small additions of cocoa butter and lecithin, an emulsifier, are made to create the smooth, voluptuous qualities essential to the final product.

SWEETENING CHOCOLATE

Baking or bitter chocolate is simply cocoa solids and cocoa butter. To produce the great range of chocolates, from bittersweet to semisweet to sweet, more cocoa butter plus varying amounts of sugar, vanilla, and lecithin are added. The flavor and sweetness of a chocolate will be unique to its maker, with one brand's bittersweet tasting like another brand's semisweet. Changing your usual brand of chocolate can make a difference in the flavor of a favorite recipe. To make milk chocolate, milk solids replace some of the cocoa solids. White chocolate is not, in fact, a real chocolate since it is made without cocoa solids; brands containing all cocoa butter, rather than vegetable oil, are best (Lousada 28–29).

Example 10.5

Writing the Process Description

CONCRETE, SLUMP, AND COMPRESSIVE STRENGTH

Concrete is one of the most universal construction materials in the world because its component raw materials are inorganic, noncombustible, highly versatile, and relatively low in cost in comparison to other materials. It is rated by its compressive strength after a 28-day curing period. Specified compressive strengths of concrete are produced by varying the proportions of cement, sand (a.k.a. fine aggregate), course aggregate, and water combined to make the concrete paste. One of the most important factors affecting the strength of concrete is the water-to-cement ratio, expressed in pounds or gallons of water per sack of cement. The sections that follow will explain the nature of slump tests, discuss the testing process, and end with slump test indications.

NATURE OF SLUMP TESTING

Concrete must always be made with a workability, plasticity, and consistency suitable for job conditions. Workability is a measure of how easy or difficult it is to place, consolidate, and finish concrete. Plasticity determines concrete's ease of molding. Consistency is the ability of freshly mixed concrete to *flow*. In general, the higher the slump, the wetter the mixture; however, too much water in the mix may cause segregation of the mid-components (aggregates, etc.), producing nonuniform concrete.

Concrete for buildings is usually mixed in a transit mix truck and tested in the field (i.e., on the job site) for the water-to-cement ratio by means of a slump test. The slump test is a measure of concrete consistency. This test is performed by rodding concrete from the transit mix chute into a conical steel

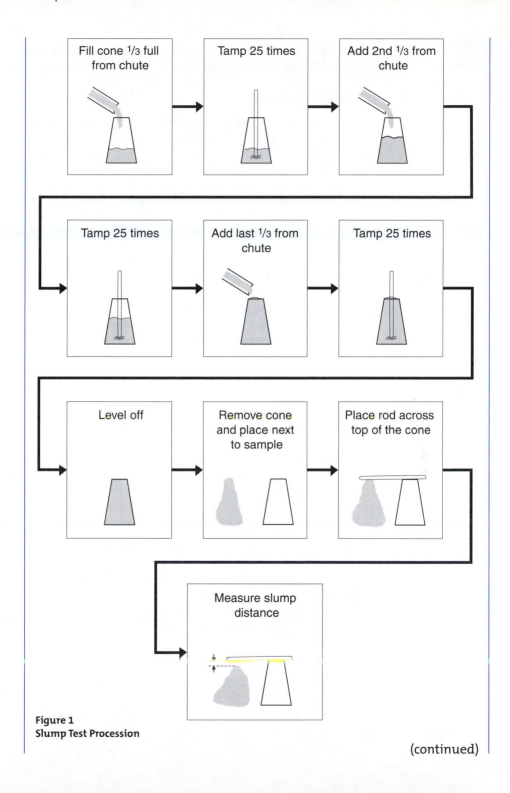

Figure 1
Slump Test Procession

(continued)

Example 10.5

(continued)

form. When the form is removed, the amount of "drop" or "slump" is measured. Figure 1 illustrates this method of determining concrete consistency.

TESTING PROCESS

A slump test is made to ensure the concrete conforms to specifications and has the flowability required for placing. The test is carried out as soon as a batch of concrete is mixed, and a standard slump cone and tamping rod are required to carry it out. The cone is made of sheet metal, 4 inches in diameter at the top, 8 inches in diameter at the bottom, and 12 inches high. The rod is a $\frac{5}{8}$-inch bullet-nosed rod about 24 inches long.

Obtaining a Sample

Concrete is poured from the transit mix truck chute directly into the slump test cone as soon as the batch is mixed in order to allow for the most accurate, efficient test results. If testing is done at a later point, the concrete will give incorrect slump measurements due to partial setup (hardening) or "bleeding" caused by separation of components. The cone is filled in three equal layers, each tamped 25 times with the rod. This is done to compact the sample so excess air and voids caused by larger course aggregates are eliminated, and a more accurate slump measurement is given.

Form Removal

After the third layer is in place and has been tamped, the concrete is struck off level, the cone is lifted carefully and set down beside the slumped concrete, and the rod is laid across the top of the cone. If the cone is not carefully removed from the sample, the test results will be inaccurate because of a shift of the concrete.

Slump Measurement

The distance from the underside of the rod to the average height of the top of the concrete is measured and registered as the amount of slump in inches. Different slumps are needed for various types of concrete construction. Slump is usually indicated in the job specifications as a range, such as 2 to 4 inches, or as a maximum value not to be exceeded. Table 1 shows slump values for unspecified jobs according to general industry recommendations. When making batch adjustments, the slump can be increased by about 1 inch by adding 10 pounds of water per cubic yard of cement.

Table 1 Recommended Slumps for Various Types of Construction

Concrete Construction	Slump (inches)	
	Maximum	Minimum
Reinforced foundation walls and footings	3	1
Plain footings, caissons, and substructure walls	3	1
Beams and reinforced walls	4	1
Building columns	4	1
Pavements and slabs	3	1
Mass concrete	2	1

TEST INDICATIONS

Slump is indicative of workability when assessing similar mixtures. However, it should not be used to compare mixtures of totally different proportions. When used with different batches of the same mixture, a change in slump indicates a change in consistency and in the characteristic of materials, mixture proportions, or watery content.

Exercises

▶ You Create

1. In class, develop a brief mechanism description by brainstorming. Hand in all your work to your instructor. Your instructor may ask you to perform this activity in groups of three or four.

   - Brainstorm the names of parts and subparts.
   - Choose the most significant parts.
   - Arrange the parts into a logical pattern.
   - Name and define each part in the first sentence.
   - Describe each part in a paragraph.
   - Create a visual aid of your mechanism, complete with appropriate callouts (use keywords from the text as callouts).

2. In class, develop a mechanism description through a visual aid. Hand in all your work to your instructor. Your instructor may ask you to perform this activity in groups of three or four.

   - Draw a visual aid of your mechanism. Use callouts to point out key parts.
   - Name each part that the audience needs to understand.

- Select a logical pattern for discussing the parts.
- Name and define each part in a sentence.
- Describe each part in a paragraph; be sure to discuss each part named in the callouts.

3. In class, develop a process description through brainstorming. Assume that you need either to demonstrate that a problem exists or to provide cause-effect theoretical background. Hand in all your work to your instructor. Your instructor may ask you to perform this activity in groups of three or four.

- Brainstorm the names of as many steps and substeps as you can.
- Arrange the steps into chronological or cause-effect order.
- Define the end goal of each step in one sentence.

4. In class, develop a brief process description through visual aids. Assume that your audience needs a basic understanding of the process in order to discuss it at a meeting. Hand in all your work to your instructor. Your instructor may ask you to perform this activity in groups of three or four.

- Draw a flow chart of the process, or else a diagram of the parts interacting, as in the following diagram:

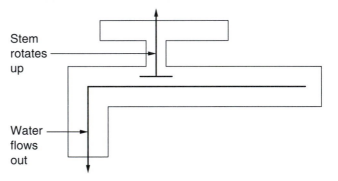

Stem rotates up

Water flows out

- Write a brief paragraph for each step.

▶ You Revise

5. Either singly or in groups of two to four, analyze these paragraphs for consistency in presentation. Rewrite the paragraphs to eliminate passive voice. What identity does the author of this text appear to express? Do you like that? Rewrite the paragraphs to achieve a different identity.

     Date: March 28, 2007
     To: Dan Riordan
     From: Tadd Hohlfelder
     Subject: To give an overview of the Norclad case-out process

As the new supervisor it is important for you to be familiar with the Nor-clad case-out process. I compiled this information while reviewing the department last week.

The purpose of the process is to construct complete window units from separate clad awnings, casement, and picture assemblies. After construction, they are tagged and moved to shipping.

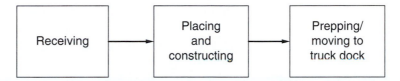

When subassemblies are received in case-out they must be checked in to be sure all are accounted for. Each of the four tables has a stack of orders which need to be built. There is not one correct method for checking in assemblies. For example, some workers check their subassemblies separately as they build them. Others choose to check them all in at the beginning of the shift.

Once they are checked in, the subassemblies for a particular order are placed on a table. All components of the unit are also placed on the table. The components are listed on the order sheet. One order might call for a $6\frac{3}{4}$-inch extension with screens and storm panes. Another may just have the standard $4\frac{9}{16}$-inch extension without other options. After correct placement, the unit can be built. We use several types and sizes of air-powered nail guns to construct our units.

After construction, the unit must be prepped and sent to shipping. We are currently using plastic strips to protect the units while shipping. The strips are placed on the facing edges of the units. Cardboard pieces are then stapled on the corners of the unit. On the work order for each unit there are three tags attached: green, white, and red. The green tag must be attached to the finished unit to identify it for shipping. The remaining tags are left on the order, which is placed in a bin on your desk upon completion. The units are loaded on racks and moved to the truck dock using either pallet jacks or fork trucks.

6. Analyze the strategy of the "help" example in these two paragraphs. Do you like the way it is introduced in each section? Do you like using the same example throughout the description? Decide what kind of identity this author achieves by presenting the example as she does. Does she make you feel confident? Rewrite the paragraph so that you create a different identity for the author.

OBTAINING BACKGROUND INFORMATION

Obtaining background information means to gather information from the customers in order to get a feeling of their overall knowledge of computer software. This information is usually gathered through interviews or questionnaires. The information we get is used to design a user interface that will be easy for the customer to learn.

We can use the simple example of getting help when trying to understand what a certain function does. For example, if the customer is familiar with making the help text appear on the screen by typing "help," we design ours to work in a similar fashion.

DEFINING THE SEMANTICS

Defining the semantics of the user interface means to come up with a clear view of all the tasks the customer must do to make the system perform. When performing these tasks, two customers will, more than likely, take two different paths through the interface to obtain the same outcome. We must discuss each of these paths for each task, then create and describe them within our system.

If you refer back to the help text example, you may understand this better. If two users are looking for help on a particular topic, one may go to a help "index" to find it, and the other straight to the topic by typing in "help <subject>." Both customers will get the same result, but each did it in his or her own way.

7. Redo this paragraph so that your supervisor can take it to a committee that needs to know what happens in this step.

HANDLE

The handle is made from kiln dried walnut rough cut lumber $1'' \times 8'' \times 11'$. Lumber is cropped to $30''$ lengths. Cut to $30''$ lengths. Jointed to get a true edge. Ripped to $1\frac{1}{4}''$ widths. $1\frac{3}{4}''$ sections are cropped off starting $\frac{1}{4}''$ from one end leaving $\frac{1}{16}''$ between marks. 15 sections are turned down to $\frac{3}{4}''$ diameter one end and $\frac{5}{8}''$ diameter other end. Diameter is 100% inspected and lengths are marked off. Sections are cut to $1\frac{3}{4}''$ lengths. $\frac{5}{16}''$ diameter hole is drilled $1\frac{1}{4}''$ deep. Drill depth is inspected randomly. All surfaces are sanded and 100% inspected.

▶ You Analyze

8. Compare these two versions of the same paragraph. Which version gives you more confidence in the writer? In groups of three or four, discuss the stylistic features that cause confidence.

A. To affix the bacteria means to "glue" them to the slide so that they are permanently mounted. Basically there are three steps. First, the lab assistant lights the Bunsen burner and grasps the slide on each end. Second, the assistant dips the loop of the sterile poker into the culture and smears the liquid onto the center of the slide. Third, the assistant passes the slide (wet side up) several times through the flame.

B. Affix the bacteria to the slide. First, the lab assistant places a sterile poker in the culture of bacteria. Then he or she places the loop on the center of the clean glass slide and smears it around in a tiny circle. The lab assistant then heat fixes the slide with the bacteria on it by passing it through the flames of the Bunsen burner two or three times.

9. In class (or in small groups if your instructor prefers), compare a paragraph from the document on the product tester selection policy (pp. 241–242) with a paragraph from the ColorScript Laser process (pp. 234–235). How are the paragraphs organized? What is the function of the first sentence of each? Which paragraph seems to more effectively convey its message to the audience? Be prepared to briefly present your findings to the class.

Writing Assignments

1. Assume that you must describe a problem with a process at your workplace. Describe the process in detail, and then explain the problem and offer a solution. Use a memo format with heads and a visual aid. As you work, you can use the worksheet in this chapter. Use Exercise 4 or 5 (above) to start your work.

2. Write a brief description of the steps you took to solve a problem. Assume that your audience is someone who must be assured that your solution is based on credible actions, but who does not know the terms and concepts you must use. For instance, you could explain the process you used to test an object or the process you used to select a vendor for a product your company must purchase.

3. Write an article for a company newsletter, describing a common process on the job. Use a visual aid. Sample topics might include the route a check follows through a bank, the billing procedure for accounts receivable, the company grievance procedure, the route a job takes through a printing plant, or the method for laminating sheets of materials together to form a package. Fill out the worksheet in this chapter. Use Exercise 4 or 5 (above) to start your work. Your article should answer the question "Have you ever wondered how we . . . ?"

4. Write several paragraphs to convince an audience to purchase a mechanism or to implement a process. The mechanism might be a machine, and the procedure might be a system, such as hiring new personnel. Describe the advantages that this mechanism or process offers over the mechanism or process currently in use. Fill out the worksheet in this chapter. Use a visual aid. Use Exercise 2 or 3 (above) to start your work. Choose a mechanism or process you know well, or else choose from this list (for X, substitute an actual name).

the lens system of brand X camera

the action of brand X bike gear shift

the X theory of product design

the X theory of handling employee grievances

the X retort process

how brand X air conditioner cools air

how brand X solar furnace heats a room

5. Expand into a several-page paper one of the brief descriptions you wrote in Exercises 1–4. Your audience is a manager who needs general background. (Alternate: your audience is a sixth-grade class.) Bring a draft of the paper to class. In groups of two to three, evaluate the draft in terms of these concerns:

   a. Does the introduction present the purpose of the mechanism process and provide a basis for your credibility?
   b. Does the introduction present a preview of the paper?
   c. Does each new section start with an effective keyword?
   d. Are the details sufficient to explain the part or step to the audience?
   e. Is the visual aid correctly sized, clear, and clearly referred to?
   f. Do callouts in the visual aid duplicate key terms in the text?
   g. Is the style at a high enough quality level?

6. Write a learning report for the writing assignment you just completed. See Chapter 5, Writing Assignment 7, pages 134–135, for details of the assignment.

Web Exercises

1. Describe a Web browser homepage (Netscape, Internet Explorer, one of the many search engines) as if it were a mechanism. Name and explain the function of each part of the page.

2. Search the Web using the keywords "process description" (or, to change your results slightly, add a company name—"process description" Ford). Analyze the descriptions you find both for the way they are organized and the role they play in the website or document. Then, following your instructor's directions, use the example you found to create a process description. (Alternate: write an analytical report explaining the organizations and roles you found.)

3. Describe the process of finding some type of information (for instance, air fares or technical data relevant to your major or job focus). Name and explain the sequence of steps that a person must follow in order to find results efficiently. (Note continuation of this exercise in Chapter 11, Web Exercise 2.)

Works Cited

How Things Work in Your Home. New York: Holt, 1987.

Jordan, Michael P. *Fundamentals of Technical Description*. Malabar, FL: Robert E. Krieger, 1984.

Lousada, Patricia. *Chocolate*. New York: DK Publishing, 1999.

"Product Tester Selection Policy." *Illuminated Ink* (10 Nov. 2003): 10 Nov. 2003. <www.IlluminatedInk.com>.

"QMS ColorScript Laser 1000." *Macworld* (July 1994): 75.

Section

2

Technical Communication Applications

Chapter

11 Sets of Instructions

Chapter 11
In a Nutshell

The goal of a set of instructions is to enable readers to take charge of the situation and accomplish whatever it is that they need to do.

Introduction

▶ Tell the end goal of the instructions (or do that in the title).

▶ Define any terms the readers might not know; if necessary, explain the level of knowledge you expect.

▶ List tools they must have or conditions to be aware of.

Body Steps

▶ Explain one action at a time.

▶ Tell the readers what they need to know to do the step, including warnings, special conditions, and any "good enough" criteria that allow them to judge whether they have done the step correctly.

Format

▶ Use clear heads.

▶ Number each step.

▶ Provide visuals that are big enough, clear enough, and near enough (usually directly under or next to) the appropriate text.

▶ Use lots of white space that clearly indicates the main and the subordinate sections.

▶ Write the goal at the "top" of the section—so the readers can skip the rest if they already know how to do that.

Tone

▶ Be definite. Make each order explicit. If the monitor "must be placed on top of the CPU," don't say "should."

▶ Discover what readers feel is arbitrary by asking them in a field test.

Sets of instructions appear everywhere. Magazines and books explain how to canoe, how to prepare income taxes, and how to take effective photographs; consumer manuals explain how to assemble stereo systems, how to program VCRs, and how to make purchased items work. On the job you will write instructions for performing many processes and running machines. This chapter explains how to plan and write a useful set of instructions.

Planning the Set of Instructions

To plan your instructions, determine your goal, consider your audience, analyze the sequence, choose visual aids, and follow the usual form. In the following discussion, the subject is exposing Dylux paper in order to make a "proof" or "review copy" of a color print that many people will review for accuracy and effectiveness.

Determine Your Goal

Instructions enable readers to complete a project or to learn a process. *To complete a project* means to arrive at a definite end result: The reader can complete a form or assemble a toy or make a garage door open and close on command. *To learn a process* means to become proficient enough to perform the process without the set of instructions. The reader can paddle a canoe, log on to the computer, or adjust the camera. In effect, every set of instructions should become obsolete as the reader either finishes the project or learns to perform the process without the set of instructions.

Consider the Audience

When you analyze your audience, estimate their knowledge level and any physical or emotional constraints they might have.

Estimate the Audience's Knowledge Level

The audience will be either absolute beginners who know nothing about the process or intermediates who understand the process but need a memory jog before they can function effectively.

The reader's knowledge level determines how much information you need to include. Think, for instance, about telling beginners to turn on a computer. They will not be able to do this because they will not know where to look for the power switch. For an intermediate, however, "turn it on" is sufficient.

Globalization and Instructions

Knowing your international audience means knowing their cultural norms. You need to know how to present ideas to them. Martin Schell warns writers that in some countries the phrase "Turn Power On" is an essential first step for any set of instructions, whereas in other cultures this step is assumed and unnecessary, and, in some cases, it is insulting to include such an obvious step.

The same rules of clear and concise writing that you apply to other documents are especially important in writing instructions. Keep your prose simple and succinct. As much as possible, use the present tense; stay away from contractions; pare down your writing to one idea per sentence; avoid using jargon and clichés (Dehaas).

When writing the actual instructions, understand that someone will read them very literally. Use the active voice. Instead of "The lever should be positioned at the 'on' position," write "Put the lever in the 'on' position." The passive voice makes the sentence vague and unclear; the reader may not know what is expected of him or her, especially as many languages do not use passive voice. By changing the instructions to the active voice, you will make the instructions more precise (Dehaas).

Consider the word order in your sentences; each idea should build on the one before it. If you are describing the steps in a task, write them in the order in which they are performed. It will be easier for someone with limited English skills to follow short, numbered steps than to try to understand long, complex sentences (Dehaas).

Be aware of the kinds of prepositions that you use. These parts of speech are often very hard to learn. If you write a complex sentence that uses a lot of prepositions ("Open the program that is on the desktop or in the start menu under program"), the reader may miss one or more relationships and will not be able to perform the task.

For further reference:
Nancy Hoft Consulting at www.world-ready.com is a training and consulting firm that specializes in effective techniques for communicating with multilingual and multicultural audiences. Look here for advice on writing and designing for a global community and for links to other helpful sites.

Works Cited

Dehaas, David. "Say What You Mean." *Occupational Health and Safety Magazine.*
    31 Jan. 2004 <www.ohscanada.com/Training/saywhatyoumean.asp>.
Schell, Martin A. "Frequently Asked Questions About Globalization and Local-
    ization." *American Services in Asia.* 31 Jan. 2004 <www.globalenglish.info/faq
    .htm#two>.

Identify Constraints

Emotional and physical constraints may interfere with the audience's attempts to follow instructions. Many people have a good deal of anxiety about doing something for the first time. They worry that they will make mistakes and that those mistakes will cost them their labor. If they tighten the wrench too hard, will the bolt snap off? If they hit the wrong key, will they lose the entire contents of their disk? To offset this anxiety, include tips about what should take place at each step and about what to do if something else happens. Step 5 in the example "Making a Dylux Proof," page 273, explains what it means when the overhead light turns off—nothing is wrong; the process is finished.

The physical constraints are usually the materials needed to perform the process, but they might also be special environmental considerations. A Phillips screw cannot be tightened with a regular screwdriver; a 3-pound hammer cannot be swung in a restricted space; in a darkroom, only a red light can shine. Physical constraints also include safety concerns. If touching a certain electrical connection can injure the reader, make that very clear. Step 1 in the example on page 272 tells readers that it doesn't make any difference how they lay the Dylux into the frame—either way achieves the same result.

Examples for Different Audiences

To see how the audience affects the set of instructions, compare the brief version below with "Exposing a Single Color" (pp. 272–273). The section on pages 272–273 explains the steps in detail, assuming that the beginner audience needs detailed "hand-holding" assistance. The brief example below, designed for an intermediate audience, simply lists the sequence of steps to jog the reader's memory.

Instructions for an Intermediate

1. Lay the Dylux and black film in the frame.
2. Select the channel and key it in. Enter.
3. Close and latch. Open after the light goes out.

Analyze the Sequence

The sequence is the chronological order of the steps involved. To analyze the sequence, determine the end goal, analyze the tasks, name and explain the tasks, and analyze any special conditions. (See the sample flow chart of an analysis in Figure 11.1, p. 266.)

Determine the End Goal

The end goal is whatever you want the reader to achieve, the "place" at which the user will arrive. This goal affects the number of steps in your sequence because different end goals will require you to provide different sets of instructions, with different sections. In the preceding example, the end goal is "The user will finish exposing a single color," and the document ends at that point. Other end goals, however, are possible. For instance, if the goal were "The user will finish exposing four colors," the sequence would obviously include more steps and sections.

Figure 11.1

Flow Chart of an
Analysis

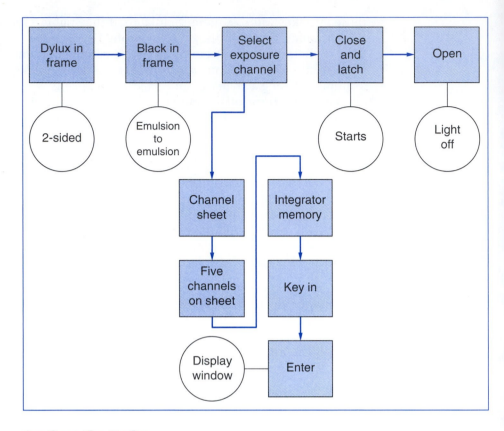

Analyze the Tasks

You have two goals here: determine the sequence and name the steps. To determine the sequence, you either go backward from the end goal or perform the sequence yourself. If the end goal is to remove the exposed film, the question to ask is "What step must the user perform before removing the exposed film?" The answer is "Close and latch the frame." If you continue to go backward, the next question is "What does the user do before latching the frame?" As you answer that question, another will be suggested, and then another—until you are back at the beginning, walking into the room with the film. You can also do the process yourself; as you do it, record every act you take. Then perform the task a second time, following your written notes exactly. You will quickly find whether or not you have included all the steps.

Name and Explain the Tasks

Having decided on the sequence, you name each task and explain any subtask or special information that accompanies it. The Dylux example has five subtasks labeled a to d under the task "Select the exposure channel," and many of the steps include explanations for the audience. For instance, step 3d (p. 273) tells the user that channel 1 is for the black exposure.

Analyze Conditions

You must also analyze any special conditions that the user must know about. For instance, step 2 explains that the films must be emulsion to emulsion. Safety considerations are very important, and safety warnings are an essential part of many instructions. *If it will hurt them or the machine, tell the audience.* Warn the user not to touch a hot bulb and to turn off the machine before working on it.

Example of Process Analysis.

An easy way to conduct your analysis is to make a flow chart of the process. Put the steps in boxes and any notes in circles (see Figure 11.1).

Choose Visual Aids

Visual aids either clarify or replace the prose explanation. Figure 2 in the Dylux example (p. 273) *replaces* text; to describe the position of the Memory button in words would take far more than the seven words used to convey the instruction.

The figure below *clarifies* the text—and reassures the readers that their actions are correct—by showing what the screen will look like as the actions occur on a computer.

> At the TO: prompt type their NAME, not their real name but their e-mail user name, and then press enter. The SUBJ: prompt will appear. (See Figure 1—note that you do not have to type in all capital letters, though you

```
mail>SEND
TO: SMITHJ
SUBJ: Learning e-mail
Enter your message below
```

Figure 1
On-Campus E-Mail Address

may.) On our system, the user name is generally the last name and the first one or two letters of the first name.

Here are a few guidelines for choosing visual aids:

▶ Use a visual aid to orient the reader. For instance, present a drawing of a keyboard with the return key highlighted.
▶ Use a visual aid to show the effect of an action. For instance, show what a screen looks like *after* the user enters a command.
▶ Decide whether you need one or two visual aids for the entire process or one visual aid per step. Use one per step if each step is complicated. Choose a clear drawing or photograph. (To determine which one to use, see Chapter 7.)
▶ Place the visual aid as close as possible to the relevant discussion, usually either below the text or to the left.
▶ Make each visual aid large enough. Do not skimp on size.
▶ Clearly identify each visual aid. Beneath each one, put a caption (e.g., *Figure 1. E-Mail Address* or *Fig. 1. E-Mail Address*).
▶ Refer to each visual aid at the appropriate place in the text.
▶ Use *callouts*—letters or words to indicate key parts. Draw a line or an arrow from each callout to the part.

Follow the Usual Form for Instructions

The usual form for a set of instructions is an introduction followed by a step-by-step body. The introduction states the purpose of the set of instructions, and the steps present all the actions in chronological order. The models at the end of this chapter illustrate these guidelines. Make a style sheet of all your decisions.

For steps and visual aids, use these guidelines:

▶ Place a highlighted (underlined or boldfaced) head at the beginning of each section.
▶ Number each step.
▶ Start the second and following lines of each step under the first letter of the first word in the first line.
▶ Use margins to indicate "relative weight"; show substeps by indenting to the right in outline style.
▶ Decide where you will place the visual aids. Usually place them to the left or below the text.
▶ Use white space above and below each step. Do not cramp the text.

For columns, the decisions are more complex. Basically, you can choose one or two columns, but their arrangement can vary, and each will have different effects on the reader. Figure 11.2 presents several basic layouts you can choose. You can place visual aids below or to the right or left of the text. To the left and below are very common places. Generally, you place to the left (text or visuals) whatever you want to emphasize.

Figure 11.2

Different Column Arrangements for Instructions

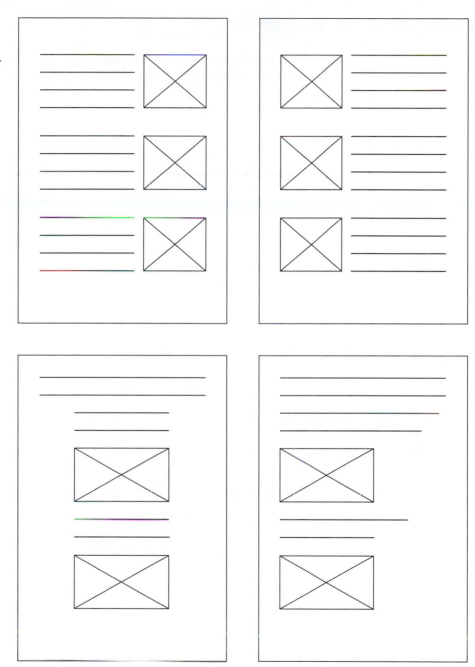

Writing the Set of Instructions

A clear set of instructions has an introduction and a body. After you have drafted them, you will be more confident that your instructions are clear if you field-test them.

Write an Effective Introduction

Although short introductions are the norm, you may want to include many different bits of information, depending on your analysis of the audience's knowledge level and of the demands of the process. You should always

▶ State the objective of the instructions for the reader.

Depending on the audience, you may also

▶ Define the process.
▶ Define important terms.
▶ List any necessary tools, materials, or conditions.
▶ Explain who needs to use the process.
▶ Explain where and/or when to perform the process.
▶ List assumptions you make about the audience's knowledge.

A Sample Introduction to a Set of Instructions

In the following introduction, note that the writer states the objective ("These instructions enable you to make a single- or multicolored Dylux"), defines the topic, lists knowledge assumptions, and lists materials.

MAKING A DYLUX PROOF

INTRODUCTION

End goal
Background

These instructions enable you to make a single- or multicolored Dylux. A Dylux proof is a single-color proof that uses different shades of blue to represent each of the process colors. A Dylux is used to check for copy content, layout, and position.

Knowledge
assumption
Materials list

These instructions assume you know how to operate the integrator. Before you start, you need regular Scotch tape and films to proof.

Write an Effective Body

The body consists of numbered steps arranged in chronological order. Construct the steps carefully, place the information in the correct order, use imperative verbs, and do not omit articles (*a, an,* and *the*) or prepositions.

TIP
Two Style Tips for Instructions

1. Use imperative verbs.

   An imperative verb gives an order. Imperative verbs make clear that the step must be done. Notice below that "should" introduces a note of uncertainty about whether the act must be performed.

   Say

   Turn on the exposure frame.

   Rather than

   You should turn on the exposure frame.

2. Retain the short words.

   Use *a, an, the* in all the usual places. Eliminating these "short words" often makes the instructions harder to grasp because it blurs the distinction between verbs, nouns, and adjectives.

   No short words:

   Using register marks on Dylux film, register film with image on Dylux.

   Short words added:

   Using the register marks on the Dylux film, register the film with the image on the Dylux.

Construct Steps Carefully

To make each step clear, follow these guidelines:

- Number each step.
- State only one action per number (although the effect of the action is often included in the step).
- Explain unusual effects.
- Give important rationales.
- Refer to visual aids.
- Make suggestions for avoiding or correcting mistakes.
- Place safety cautions before the instructions.

Review the Examples on pages 277–283 to see how the writers incorporated these guidelines. An example of how to write the body follows.

Sample Body

Here is the body of the set of instructions that follows the introduction on page 270.

Head for sequence

PREPARATION

1. Turn on the exposure frame. (See Figure 1.)

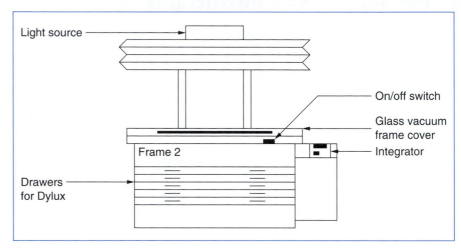

Light source

On/off switch

Glass vacuum
frame cover

Integrator

Frame 2

Drawers
for Dylux

Figure 1
Exposure Frame

Instruction

Explanatory
comment

Special condition

Action

Action

2. Lay your films on the work counter (emulsion down; emulsion side is
the dull side) in the order in which you will proof them (first film to be
exposed on top, and the last film to be exposed on the bottom—usually black, yellow, magenta, cyan).
3. Clean both sides of the glass on the vacuum frame using the glass
cleaner and cheesecloth (located on a shelf to the left of frame 2).
4. Raise the glass cover on the vacuum frame.
5. Get one sheet of Dylux, large enough to hold the entire image. The
Dylux paper can be found in the drawers under frame 2.

EXPOSING A SINGLE COLOR (OR THE FIRST COLOR IN A SERIES)

1. Lay the Dylux in the vacuum frame. The Dylux paper is two-sided, so it
does not matter which side is up.
2. Place the film to be exposed on top of the Dylux. Black is usually the
first color to be exposed. Film must be emulsion down, so Dylux and
film are emulsion to emulsion.
3. Select the exposure channel for the first exposure.

Note that use of
substeps keeps the
number of main
steps small

   a. Refer to the sheet listing the channels and what material each is set
   to expose. This sheet is posted above the integrator.
   b. Locate the proper channel on the sheet. There should be five channels on the Dylux: one channel for each of the four process colors
   (black, yellow, magenta, cyan) and one for clearing.
   c. Push the **Memory** button on the integrator. (See Figure 2.)

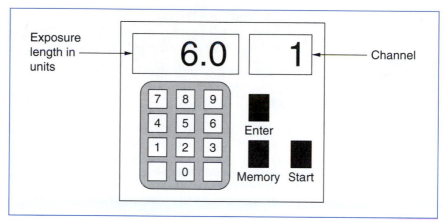

Figure 2
Integrator

    d.   Key in the appropriate channel number on the number pad. (For example, channel 1 is for the black Dylux exposure.)

    e.   Press the **Enter** button on the integrator. After you press **Enter,** the exposure time in units will appear in the top left display window.

4. Close and latch the vacuum frame.
    Note: The exposure starts when the frame is latched.

5. When the exposure is complete (when the overhead light goes out), open the vacuum frame and remove the film.

    Note: After this exposure there will be an image on the Dylux paper.

Special note

Head for special condition

If You Have a Single-Color Proof

Remove the Dylux from the vacuum frame and skip to the "Clearing a Dylux" section.

Head for special condition

If You Have a Multicolor Proof

Leave the Dylux in the frame and continue with the next section, "Exposing Additional Colors."

EXPOSING ADDITIONAL COLORS

Introduction to subsection

The first exposure will image register marks onto the Dylux. From this point on, each film must be lined up (registered) to the register marks on the Dylux.

1. Place the next negative film (probably yellow) over the Dylux paper. Using the register marks on the Dylux and the film, register the film with the image on the Dylux.

2. Tape the Dylux and film together.

3. Select the proper exposure channel for the color you are exposing.

4. Close and latch the vacuum frame.
5. When the exposure is complete, open the vacuum frame and remove the film, leaving the Dylux in the frame.

Repeat steps 1–5 until all colors have been exposed.

CLEARING A DYLUX

Clearing is the final exposure. The clearing exposure is made without films. The Dylux is exposed to the light through a clearing filter, which moves into place when the clearing channel is selected.

1. After all colors have been exposed, remove the Dylux from the vacuum frame and close the vacuum frame.
2. Place the Dylux on top of the glass.
3. Select the channel for clearing a Dylux.
4. Push **Start** to clear the Dylux. (See Figure 2.)
5. Once the Dylux has been cleared, trim it according to the specifications on the job ticket.

One sentence instruction avoids lengthy repetition

All main heads emphasize action

Field-Testing Instructions

A field test is a method of direct observation by which you can check the accuracy of your instructions. To perform a field test, ask someone who is unfamiliar with the process to follow your instructions while you watch. If you have written the instructions correctly, the reader should be able to perform the entire activity without asking any questions. When you field-test instructions, keep a record of all the places where the reader hesitates or asks you a question.

Worksheet for Preparing Instructions

☐ **Assess the audience for these instructions.**
  • Estimate the amount of knowledge the audience has about the process. Are they beginners or intermediates?
  • What will you tell the readers in the introduction? What will you assume about them? What do they need to know? What can they get from your instructions? How do they decide if they want to read your instructions? What will make them feel you are helpful and not just filling in lines for an assignment? How will you orient them to the situation?

☐ **What is the end goal for your readers?**

☐ Analyze the process.
  • Construct a flow chart that moves backward from the end goal.
  • Use as many boxes as you need.

☐ List all the conditions that must be true for the end goal to occur. (For instance, what must be true for a document to open in a word processing program? The machine is turned on, the disk is inserted, the main menu appears, and the directory appears.)

☐ List all the words and terms that the audience might not know.

☐ List all the materials that a person must have in order to carry out the process.

☐ Where do the readers need a visual aid to "give them permission," or to orient them to the situation, or to show them something quickly that is easy to see but hard to describe in words?

☐ Draw the visual aids that will help readers grasp this process. Use visuals that illustrate the action or show the effect of the action.

☐ How will you arrange this material on the page so that it is easy for readers to read quickly, but also to keep their place or find it again as they read?

☐ Construct a style sheet. Choose your head system, margins, columns, method of treating individual steps, and style for writing captions.

☐ Convert the topic of each box in the flow chart into an imperative instruction. Add cautions, suggestions, and substeps. Decide whether a sequence of steps should be one step with several substeps or should be treated as individual steps.

☐ How will you tell them each step? How—and where—will you tell them results of a step? How—and where—will you tell them background or variations in a step?

☐ Why should you write them a set of instructions in the first place? Why not write them a short report or an article? A report tells the results of a project, an article informally explains the concepts related to a project, and a set of instructions tells how to do the project.

TIP
Information Order in a Step

If your step contains more than just the action, arrange the items as action-effect. In the following example, the first sentence is the action, the second sentence is the effect.

Press **Enter** on the integrator.
After you press **Enter**, the exposure time in units will appear in the top left display window.

If your step contains a caution or warning, place it first, before you tell the audience the action to perform.

1. CAUTION: DO NOT LIGHT THE MATCH DIRECTLY OVER THE BUNSEN BURNER!

Light the match and slowly bring it toward the top of the Bunsen burner.

Worksheet for Evaluating Instructions

☐ **Evaluate your work. Answer these questions:**
- Does the introduction tell what the instructions will enable the reader to do?
- Does the introduction contain all the necessary information on special conditions, materials, and tools?
- Is each step a single, clear action?
- Does any step need more information—result of the action, safety warning, definitions, action hints?
- Do the steps follow in a clear sequence?
- Are appropriate visual aids present? Does any step either need or not need a visual aid?
- Are the visual aids presented effectively (size, caption, position on page)?
- Does the page layout help the reader?
- Are all terms used consistently?

Examples

The three examples that follow exemplify sets of instructions.

Example 11.1

Instructions for a
Beginner

INSTRUCTIONS: HOW TO USE THE MODEL 6050 pH METER

Introduction

This set of instructions provides a step-by-step process to accurately test the pH of any given solution using the pH Meter Model 6050. The pH meter is designed primarily to measure pH or mV (millivolts) in grounded or un-grounded solutions. This set of instructions assumes that the pH meter is plugged in and that the electrode is immersed in a two-molar solution of potassium chloride.

Materials Needed

- Beaker containing 100 ml of 7.00 pH buffer solution
- Beaker containing 100 ml of 4.00 pH buffer solution
- Thermometer
- Squeeze bottle containing distilled water
- Four squares of lint-free tissue paper

How to Program the pH Meter

1. Press the button marked pH (A in Figure 1) to set the meter to pH mode.
2. Set pH sensitivity by pushing the pH sensitivity button down to .01 (B in Figure 1).
3. Gently remove the pH electrode (C in Figure 1) from the plastic bottle in which it is stored, and rinse it gently with distilled water from your squeeze bottle.

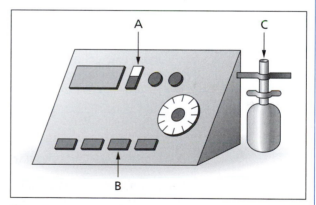

Figure 1
Sargent-Welch pH Meter Model 6050

4. Carefully lower the electrode into the beaker containing the pH 7.00 buffer solution.
5. Set temperature control.
   a. Using the thermometer, take the temperature of pH 7.0 buffer solution.
   b. Turn the temperature dial (D in Figure 2) to the temperature reading on the thermometer in degrees Celsius.

(continued)

Example 11.1

(continued)

6. Set electrode asymmetry (intercept) by rotating the dial marked "intercept" (E in Figure 2) until the digital display (F in Figure 2) reads 7.00.

7. Raise the electrode from the 7.00 pH buffer solution, rinse gently with distilled water from your squeeze bottle, and dry tip of the electrode using lint-free tissue paper.

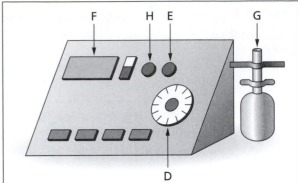

Figure 2
Sargent-Welch pH Meter Model 6050

8. Lower the electrode (G in Figure 2) into the buffer solution of pH 4.00 to set the lower pH limit.

9. Set the response adjustment (slope) by rotating the dial marked "slope" (H in Figure 2) until the digital display reads 4.00.

10. Raise the electrode from the 4.00 pH buffer solution.

11. Rinse the electrode gently with distilled water from your squeeze bottle.

12. Dry the tip of the electrode using lint-free tissue paper.

Example 11.2

Instructions for a Beginner

HOW TO ADD BACKGROUND SOUND TO A WEBPAGE

The process of adding background sound to a website will allow you to hear sound clips when your page is opened in a browser. By adding background sound to your webpage, you will add excitement to your site! Consider using background sound that enhances the information on your site and is interesting and pleasing for the listeners. There are many different sound file types that you can use. Some of these file types include MIDI, RMF, RMI, WAV, and MOD. Prior to adding sound to an HTML document, you must have a sound file and an existing HTML document saved in the same directory.

To Add Background Sound to a Webpage

1. Open Notepad by selecting **Start/Programs/Accessories/Notepad**.

2. Open an existing HTML file by selecting **File/Open** and then finding the HTML file to open.

3. Add the following code to the **BODY** section of the HTML file to add the music file **cheers.mid** to the background:

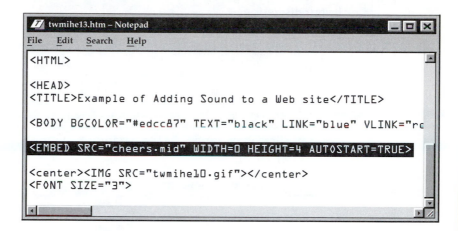

```
twmihe13.htm – Notepad
File   Edit   Search   Help
<HTML>

<HEAD>
<TITLE>Example of Adding Sound to a Web site</TITLE>

<BODY BGCOLOR="#edcc87" TEXT="black" LINK="blue" VLINK="re

<EMBED SRC="cheers.mid" WIDTH=0 HEIGHT=4 AUTOSTART=TRUE>

<center><IMG SRC="twmihe10.gif"></center>
<FONT SIZE="3">
```

The **WIDTH** and **HEIGHT** parameters are set to small numbers so there will be no visual changes to the webpage. The **"AUTOSTART=TRUE"** statement is added so the sound will begin playing as soon as the webpage is opened in a browser.

4. Select **File/Save** to save the modified HTML document to your disk.

To View Your Modified File in Internet Explorer

1. Double-click on the **My Computer** icon on the desktop.
2. Double-click on the **3 ½ Floppy (A)** icon in the My Computer window.
3. Double-click on the **HTML file name** you modified to view your changes in Internet Explorer.

Example 11.3

Multiple-Section Instructions

Indentation Instructions for FrontPage

INTRODUCTION

This document is a set of instructions for using indentation in Microsoft FrontPage 2000. In order to use these instructions effectively, you must know how to start up FrontPage, create a new webpage, enter text into that webpage, and open menus such as "File," "Format," and "Help." The first instruction set, "Accessing a paragraph menu for a specific paragraph," is a required

(continued)

Example 11-3

(continued)

precursor to using the final three instruction sets. Read it first if you are not familiar with opening the paragraph menu for a specific paragraph. All italicized text is supplementary to the main instruction set and can be read at your discretion.

ACCESSING THE PARAGRAPH MENU

Method 1

1. Place the cursor within the paragraph you wish to perform indentation on.
2. Open the "Format" menu.
3. Left click "Paragraph . . .". (The paragraph menu shown in Figure 1 will open.)

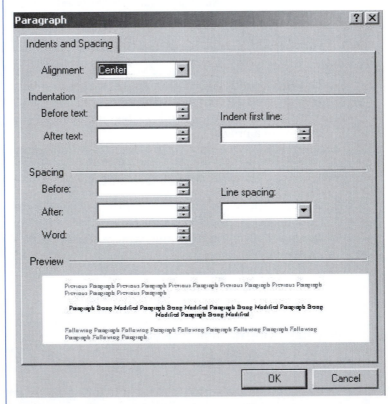

Figure 1
The FrontPage Paragraph Menu

Method 2

1. Right click within the paragraph on which you wish to perform indentation.

2. Left click "Paragraph . . ." from the pop-up menu. (The paragraph menu shown in Figure 1 will open.)

It is possible that the values seen in the entry boxes of your paragraph menu are different from those shown in Figure 1. This is not a problem, and quite normal.

LEFT-INDENTING THE FIRST LINE

Method

1. Access the paragraph menu for the paragraph you wish to perform indentation on. (The paragraph menu will open.)

2. In the "Indentation" subsection of the paragraph menu, locate the "Indent first line" box. (See Figure 2.)

3. Enter the size (in *points*) of the indentation into this box. (A number of points will now be shown in this box.)

4. Left-click the "OK" button on the paragraph menu. (The first line of the paragraph will be left-indented.)

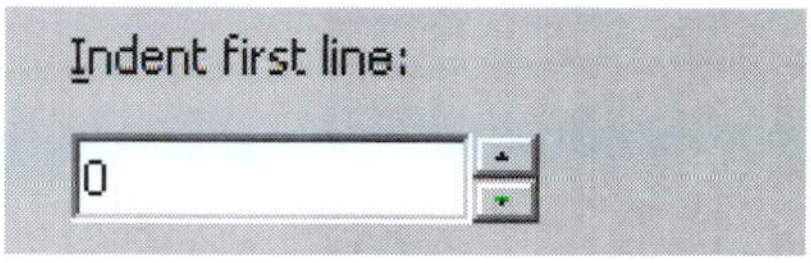

Figure 2
Indent First Line Box

When entering the value into the box, you can see an example of the effect in the "Preview" section of the paragraph menu shown in Figure 1.

LEFT-INDENTING A PARAGRAPH

Method

1. Access the paragraph menu for the paragraph you wish to perform indentation on.

2. In the "Indentation" subsection of the paragraph menu, locate the "Before text" box. (See Figure 3.)

3. Enter the size (in *points*) of the indentation into this box. (A number of points will now be shown in this box.)

4. Left-click the "OK" button on the paragraph menu. (The paragraph will be left-indented.)

Figure 3
Before Text Box

When entering the value into the box, you can see an example of the effect in the "Preview" section of the paragraph menu shown in Figure 1.

(continued)

Example 11.3

(continued)

RIGHT-INDENTING A PARAGRAPH

Method

1. Access the paragraph menu for the paragraph you wish to perform indentation on.
2. In the "Indentation" subsection of the paragraph menu, locate the "After text" box. (See Figure 4.)
3. Enter the size (in points) of the indentation into this box. (A number of points should now be in this box.)
4. Left-click the "OK" button on the paragraph menu. (The paragraph will be right-indented.)

Figure 4
After Text Box

The results of a right indent may not be apparent if your text does not extend to the right margin.

When entering the value into the box, you can see an example of the effect in the "Preview" section of the paragraph menu shown in Figure 1.

Example 11.4

Professional Instructions

Source: Elizabeth Castro, *HTML4 for the World Wide Web: Visual Quickstart Guide,* p. 38, © 1998. Republished by permission of Pearson Education, Inc. Publishing as Peachpit Press.

Understanding the HEAD and BODY

Most webpages are divided into two sections: the HEAD and the BODY. The HEAD section provides information about the URL of your webpage as well as its relationship with the other pages at your site. The only element in the HEAD section that is visible to the user is the title of the webpage (*see page 39*).

```
                    code.html
<!DOCTYPE HTML PUBLIC "-//W3C//DTD HTML 4.0
Transitional//EN"><HTML>
<HEAD>

</HEAD>
<BODY>



</BODY>
</HTML>
```

Figure 2.18
Every HTML document should be divided into a HEAD and a BODY.

To Create the HEAD Sections

1. Directly after the initial !DOCTYPE and HTML tags (*see page 37*), type **<HEAD>**.
2. Create the HEAD section, including the TITLE (*see page 39*). Add META information (*see pages 290–293*) and the BASE (*see page 113*), if desired.
3. Type **</HEAD>**.

The BODY of your HTML document contains the bulk of your web-page, including all the text, graphics, and formatting.

To Create the Body

1. After the final</HEAD> tag and before anything else, type **<BODY>**.
2. Create the contents of your webpage.
3. Type **</BODY>**.

Tip

For pages with frames, the BODY section is replaced by the FRAMESET.

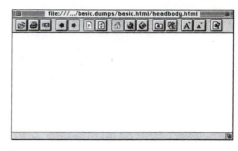

Figure 2.19
With no title and no contents, a browser has to scrape together a little substance (in the form of a title) from the file name of the HTML document.

Exercises

▶ You Create

1. Construct a visual aid that illustrates an action. For instance, show a jack properly positioned for changing a tire. Then write the instructions that would accompany that visual.

2. Write a set of instructions for a common activity, such as wrapping a package, tying a shoe, or programming a telephone. Choose one of the columnar formats shown in Figure 11.2 on page 269. Have a classmate try to perform the process by following your instructions. Discuss with the class the decisions you had to make to write the instructions. Consider word choice, layout, visual aids, sequence of steps, etc.

3. Make a flow chart or decision chart of a process. Choose an easy topic, such as a hobby, a campus activity, or some everyday task. In class, write the instructions that a person would need to perform the process. Depending on your instructor's preferences, you may either use your own chart or exchange charts with another student and write instructions for that student's chart.

▶ You Revise

4. Rewrite the following steps from the instructions for changing a car's oil:

    1. Get drainage pan and place it under the oil pan of the car.
    2. Grab a crescent wrench and locate the oil plug, on one side of the oil pan.
    3. Use the crescent wrench to turn the plug counterclockwise (ccw) until it comes out and oil drains out.
    4. While this is draining, grab a filter wrench and locate oil filter.
    5. Turn the oil filter counterclockwise with the filter wrench until it comes off and the oil drains into the drainage pan.

5. Rewrite the following steps from instructions for controlling text on a webpage. Draw visuals that would clarify the instructions. If you write these instructions on your computer, create screen captures that illustrate the steps.

    1. Pick the webpage that you want to do this to.
    2. Click on Table symbol, scroll down to darken in one box, and then click on that box.
    3. Double-right-click on the table anywhere and another window should pop up like the one below. On this table you want to click on Table Properties . . .

       Once you have completed step three you will see the box below.
    4. In this box you need to make the border Size 0 like I have done here and click "Apply." This will make your table look like this.
    5. Highlight your **whole** page by putting your curser at the very top and scrolling down selecting everything. This should change some of the background color and shows how much you have highlighted.
    6. Click on "Edit" and scroll down to "Copy". After you click on this put your curser into the table that you made after step four.
    7. Go back to "Edit" and this time select "Paste."

       This will place your whole webpage inside of the table and you will not be able to see the boarder of the table on the web like you can see it now.
    8. Erase your old webpage and you are done!

6. Convert the following paragraphs into a set of instructions:

    First I went to the website www.uwstout.edu/place/studentwebregistration .html to read the general instructions given by the Co-op and Placement Office to reach the new website and log in to my profile. I was directed to www.uwstout.edu/place/ and was instructed to click on the *"Students create or update your eRecruiting profile"* link to enter the log-in screen. The username is the last 5 digits of my ID number with the word *"stout"* added on the end, and the password is the last 4 digits.

I next chose to enter my personal information by clicking on the *"Edit your profile"* link. There are then options to update *"Personal Info, Academics, Future Plans,"* and *"Administration"* links to choose. In each of these, I either typed in information about myself, my education, my qualifications, and plans or chose information from drop-down lists containing possible options. As each section was completed, more information about me was saved in my profile.

After finishing entering personal information, I chose to post my résumé online. I clicked on the *"Documents"* link and read the instructions there to upload files. After clicking on *"Upload Documents"* and choosing *"Résumé,"* I was instructed to type in the name of the file containing my résumé or search for it among the folders on my computer and push *"Upload."*

7. Rewrite all of the items in Exercise 4 from the point of view of a "chatty help" columnist in a newspaper. Use paragraphs, not numbered steps.

8. Rewrite Example 11.3 as a report structured in paragraphs.

▶ Group

9. Compare Example 11.2 with Example 12.2 (pp. 309–310). In groups of three or four, discuss the differences in tone and in the presentation of the action. Report to the class which document you prefer to read and why.

10. For Writing Assignment 1 or 2, construct a flow chart of the process. Explain it to a small peer group who question you closely, causing you to explain the steps in detail. Revise the chart based on this discussion.

11. For Writing Assignment 1 or 2, create a template for your instructions, including methods for handling heads, introduction, steps, visual aids, captions, and columns. Review this template with your peer group, explaining why you have made the choices you have. Your peer group will edit the template for consistency and effectiveness.

12. For Writing Assignment 1 or 2, bring the final draft of your instructions to your peer group. Choose a person to field test. With your instructor's permission, field-test each other's instructions. Note every place where your classmate hesitates or asks a question, and revise your instructions accordingly.

13. Bring an article instruction to class. Computer and household magazines offer the best sources for these articles. In groups of three or four, analyze the models and decide why the authors decided to use the article method. Depending on your instructor, either report your analysis to the class, or as a group rewrite the instructions, either with a different tone (say, as a coach) or as a numbered set.

Writing Assignments

1. Write a set of instructions for a process you know well. Fill out the worksheet and then write the instructions. Use visual aids and design your pages effectively, using one of the columnar formats shown on page 269. Pick a process that a beginning student in your major will have to perform or choose something that you do as a hobby or at a job, such as waxing skis, developing film, ringing up a sale, or taking inventory.

2. Divide into groups of three or four. Pick a topic that everyone knows, such as checking books out of the library, applying for financial aid, reserving a meeting room, operating an LCD projector, replacing a lost ID or driver's license, or appealing a grade. Then each team should write a set of instructions for that process. Complete the worksheet on pages 274–275. When you are finished, decide which team's set is best in terms of design, clarity of steps, and introduction.

3. Write a learning report for the writing assignment you just completed. See Chapter 5, Writing Assignment 7, pages 134–135, for details of the assignment.

Web Exercises

1. Give instructions to a beginner on how to create a webpage using a wizard, template, or Web-authoring tool with which you are familiar.

2. Convert the process paper that you wrote for Chapter 10, Web Exercise 3, into a set of instructions for a beginner.

Work Cited

Castro, Elizabeth. *HTML for the World Wide Web*. Berkeley, CA: Peachpit, 1998.

Chapter 12

Memorandums and Informal Reports

Chapter 12
In a Nutshell

Memos. A memo is any document (regardless of length) that has memo heads (Date, To, From, Subject) at the top. The subject line should relate the contents to the reader's needs.

Informal reports. Informal reports are usually short (1 to 10 pages). Their goal is to convey the message in an understandable context, from a credible person, in clear, easy-to-read text.

Informal report structure. The informal report structure is the IMRD (Introduction, Method, Results, and Discussion).

The *Introduction* explains your goal and why this situation has developed.

The *Method* outlines what you did to find out about the situation. It establishes your credibility.

The *Results* establish what you found out, the information the reader can use.

The *Discussion* describes the implications of the information. It gives the reader a new context.

Informal report strategies. Key strategies include

▶ Explain your purpose—what your reader will get from the report.

▶ Use a top-down strategy.

▶ Develop a clear visual logic.

▶ Provide the contents in an easy-to-grasp sequence and help the reader out by defining, using analogies, and explaining the significance to the person or organization.

T he day-to-day operation of a company depends on memos and informal reports that circulate within and among its departments. These documents report on various problems and present information about products, methods, and equipment. The basic informal format, easy to use in nearly any situation, has been adapted to many purposes throughout industry.

This chapter explains the elements of memos, the elements of informal reports, and the types of informal reports, including analytical reports, IMRD reports, progress reports, and outline reports.

The Elements of Memos

Memos are used to report everything from results of tests to announcements of meetings. In industry you must write memos clearly and quickly. Your ability to do so tells a reader a great deal about your abilities as a problem solver and decision maker. This section explains memo headings and provides a sample memo report.

Memo Headings

The memo format consists of specific lines placed at the top of a page: *To, From, Subject,* and *Date* lines. That's all there is to it. What follows below those lines is a memo report. Usually such a report is brief—from one or two sentences to one or two pages. Theoretically there is no limit to a memo's length, but in practice such reports are seldom longer than four or five pages.

Follow these guidelines to set up a memo or memo report:

1. If using a preprinted form, fill in the blanks; if not, follow guidelines 2–5.

2. Place the To, From, and Subject lines at the left margin.

3. Place the date either to the right, without a head, or at the top of the list with a head (Date:).

4. Follow each item with a colon and the appropriate information.

5. Choose a method of capitalization and placement of colons (see examples).

6. Name the contents or main point in the subject line.

7. Place the names of those people who are to receive copies below the name of the main recipient (usually with the head cc:).

8. Sign to the right of your typed name.

Memo Format: Example 1

Date on far right   February 14, 2007

To:	E. J. Mentzer
cc:	Jane Thompson
From:	Judy Davis

Copy line
Signature

Subject: Remodeling of Office Complex

Subject line—only
first letters
capitalized

Memo Format: Example 2

DATE:	February 14, 2007
TO:	E. J. Mentzer
FROM:	Judy Davis

Date line
Memo heads in
all caps
Signature

SUBJECT: REMODELING OF OFFICE COMPLEX

Subject line
capitalized for
emphasis

Memo Format: Example 3

March 29, 2007

| To: | E. J. Mentzer |
| From: | Judy Davis |

Memo heads
aligned on colons

Subject: Remodeling of Office Complex

A Sample Memo Report

A memo can contain any kind of information that your audience needs. The following memo is a recommendation based on criteria.

April 1, 2007

To:	Bill Foresight
From:	Carol Frank, Food Service Director
Subject:	Purchase of an open-top range

Purpose of memo
Credibility of writer

Basic conclusion first

Data to support
conclusion
Four criteria: cost,
energy efficiency,
rating, design
features

Here is a preliminary recommendation on which brand of open-top range to purchase for the Food Service Department. After comparing the specification sheets of several brands, I found that two brands satisfy our needs: Montague and Franklin, but Montague is the better choice.

The Montague is cheaper ($499 vs. $512). It is more energy efficient: it has an overall rating of 103,000 BTU/hour, whereas the Franklin has a rating of 138,000 BTU/hour. The Montague has several design features not found on the Franklin, including a 3-position rack, a removable oven bottom, a continuous-cleaning oven, and a solid hot top. I will provide a detailed report next week.

The Elements of Informal Reports

Informal reports are those that will not have wide distribution, will not be published, and are shorter than 10 pages (General Motors). This kind of report follows a fairly standard format that can be adapted to many situations, from presenting background to recommending and proposing. The format basically has two parts: an introduction and a discussion.

Introduction

Introductions orient readers to the contents of the document. You can choose from several options, basing your decision on the audience's knowledge level and community attitudes. To create an introduction, you can do one of three things: provide the objective, provide context, or provide an expanded context.

Provide the Objective

The basic informal introduction is a one-sentence statement of the purpose or main point of the project or report, sometimes of both. This type of introduction is appropriate for almost all situations and readers.

Objective of the project

To evaluate whether the customer service counter should install an Iconglow personal computer system

Objective of the report

To report on investigation of the feasibility of installing an Iconglow personal computer system at the customer service counter

If this statement is enough for your readers, go right into the discussion. If not, add context sections as explained below.

Provide Context

To provide *context* for a report means to explain the situation that caused you to write the report. This type of introduction is an excellent way to begin informal reports. It is especially helpful for readers who are unfamiliar with the project. Include four pieces of information: cause, credibility, purpose, and preview. Follow these guidelines:

▶ Tell what caused you to write. Perhaps you are reporting on an assignment, or you may have discovered something the recipient needs to know.
▶ Explain why you are credible in the situation. You are credible because of either your actions or your position.
▶ State the report's purpose. Use one clear sentence: "This report recommends that customer service should install an Iconglow computer system."
▶ Preview the contents. List the main heads that will follow.

Here is a sample basic introduction.

Cause for writing

Source of
credibility

Purpose

Preview

> I am responding to your recent request that I determine whether cus-
> tomer service should install an Iconglow computer system. In gathering
> this information, I interviewed John Broderick, the Iconglow Regional
> Sales Representative. He reviewed records of basic personnel activities.
> This report recommends that customer service install an Iconglow sys-
> tem. I base the recommendation on cost, space, training, and customer
> relations.

Special Case: Alert the Reader to a Problem. Sometimes the easiest way to provide context is to set up a problem statement. Use one of the following methods:

- Contrast a general truth (positive) with the problem (negative).
- Contrast the problem (negative) with a proposed solution (positive).

In either case, point out the significance of the problem or the solution. If you cast the problem as a negative, show how it violates some expected norm. If you are proposing a solution, point out its positive effect. Here is a sample problem-solution introduction.

Negative problem
and its significance

Proposed solution

Positive significance

Purpose of report

> Processing customers at the service desk is a time-consuming process.
> The service representative fills out three different forms while the cus-
> tomer and those in line wait, annoyed. Some customers go elsewhere to
> shop. An Iconglow computer system would eliminate the waiting, cutting
> average service time from 10 minutes to 1 minute. This report recom-
> mends that we purchase the Iconglow system.

Provide an Expanded Context

To provide an expanded context, create a several-paragraph introduction. You must include a purpose or an objective, and then add other sections that you might need: a summary, a background, a conclusions/recommendation section, or a combination of them.

The *summary*—also called an "abstract" or sometimes "executive summary"— is a one-to-one miniaturization of the discussion section. If the discussion section has three parts, the summary has three statements, each giving the major point of one of the sections. After reading this section, the reader should have the gist of your report.

The *background statement* gives the reader a context by explaining the project's methodology or history. If the report has only an objective statement, this section orients the reader to the material in the report.

Writers present the *conclusions and/or recommendations* early in the report because this section contains the basic information that readers need. It can provide information that differs from the summary, but it replaces the summary.

Examples of Introductory Options

VERSION 1

Objective ·|· Objective
To evaluate whether to install an Iconglow system at the customer service counter.

Conclusions and Recommendation

Context ·|· 1. The system will pay for itself in one year.
2. The office area contains ample space for the system.
3. The system will not interfere with attending to customers.

I recommend that we install the Iconglow System.

Background

Context ·|· Customer service proposed installation of an Iconglow system to handle all updating functions. The system would reduce the number of employee hours required to complete these functions. The system includes two computers, a printer, programs, and cables. I reviewed personnel figures and discussed the proposal with Iconglow's sales representative.

VERSION 2

Introduction

Cause for writing ·|· I am responding to your recent request that I determine whether customer service should install an Iconglow computer system. In gathering
Source of credibility ·|· this information, I interviewed John Broderick, the Iconglow Regional Sales Representative. He reviewed records of basic personnel activities.
Purpose ·|· This report recommends that customer service install an Iconglow sys-
Preview ·|· tem. I base the recommendation on cost, space, training, and customer relations.

Conclusions

1. The system will pay for itself in one year.
2. The office area contains ample space for the system.
3. The system will not interfere with attending to customers.

Discussion

The discussion section contains the more detailed, full information of the report. Writers subdivide this section with heads, use visual aids, and sometimes give the discussion its own introduction and conclusion. If you write the introduction well, your reader will find no surprises, just more depth, in the introduction.

Two format concerns that arise in planning the discussion section are pagination and heads.

Pagination

Paginate informal reports either with just a page number or with header information as well. Follow these guidelines:

- If you use just page numbers, place them in the upper right corner or in the bottom center.
- If you use header information, arrange the various elements across the top of the page. Generally, the page number goes to the far right and other information (report title, report number, recipient, and/or date) appears to the left.

Iconglow Recommendation   12/24/07   2

Heads

Informal reports almost always contain heads. Usually you need only one level; the most commonly used format is the "side left." Follow these guidelines:

- Place heads at the left margin, triple-spaced above and double-spaced below. Use underline or boldface.
- Capitalize only the first letter of each main word (do not capitalize *a, an, the,* or prepositions).
- Do not punctuate after heads (unless you ask a question).
- Use a word or phrase that indicates the contents immediately following.
- At times, use a question for an effective head.

Side left, boldface   **Will the New System Save Money?**
Double-space

The new Iconglow system will pay for itself within 6 months. Currently, employees spend 87 hours a month updating files. The new system will reduce that figure to 27, a savings of 60 hours. These 60 hours represent a payroll savings of $435.00 a month. Because the new system costs $2450, the savings alone will pay for the system in 6 months ($435 \times 6 = 2610$). This amount of time is under the 1-year period allowed for recovery.

Triple space

Double-space   **Is There Enough Space for the System?**

The computers will easily fit in their allocated spaces. One computer and the printer will occupy space at the refund desk, and the other computer will sit on the customer service counter. Both areas were reorganized to accommodate the machine and allow for efficient work flow.

Will the Computers Affect Customer Relations?

The Iconglow system will allow employees to process customer complaints more quickly, reducing a 15-minute wait to seconds. This speeded-up handling of problems will eliminate customer complaints about standing in line.

Types of Informal Reports

Writers use informal reports in many situations. This section introduces you to several variations.

IMRD Reports

An IMRD (*I*ntroduction, *M*ethodology, *R*esults, *D*iscussion) report is a standard way to present information that is the result of some kind of research. This approach can present laboratory research, questionnaire results, or the results of any action whose goal is to find out about a topic and discuss the significance of what was discovered. The IMRD report causes you to tell a story about your project in a way that most readers will find satisfying. This kind of report allows you to provide new knowledge for a reader and to fit that knowledge into a bigger context. Your research project started out with some kind of question that you investigated in a certain way. You found information, and you explain that the information is important in various ways.

▶ For the *introduction,* present the question you investigated (the goal of the project) and the point of the paper. It is helpful to give a general answer to the question. Consider these questions:
  • What is the goal of this project?
  • What is the goal of this report?
▶ For the *methodology section,* write a process description of your actions and why you performed those actions. This section establishes your credibility. Explain such things as whom you talked to, and describe any actions you took and why. This description should allow a reader to replicate your actions. Consider these questions:
  • What steps or actions did you take to achieve the goal or answer the questions? (Explain all your actions. Arrange them in sequence, if necessary.)
  • Why did you perform those actions?
▶ For the *results section,* tell what you discovered, usually by presenting a table or graph of the data. If a visual aid is all you need in this section, combine it with the discussion section. If you add text, tell the readers what to focus on in the results. Honesty requires that you point out material that might contradict what you expected to discover. Consider these questions:
  • What are the results of each action or sequence?

- Can I present the results in one visual aid?

▶ In the *discussion section,* explain the significance of what you found out. Either interpret it by relating it to some other important concept or suggest its causes or effects. Relate the results to the problem or concerns you mentioned in the introduction. If the method affects the results, tell how and suggest changes. Often you can suggest or recommend further actions at the end of this section. Consider these questions:

- Did you achieve your goal? (If you didn't, say so, and explain why.)
- What are the implications of your results? for you and your goals? for other people and their goals?
- What new questions do your results raise?

THE HISTORY OF HTML

IMRD Report

INTRODUCTION

The goal of this IMRD Report is to describe my research for my history of HTML (*HyperText Markup Language*) research report. The goal of this project is to produce an HTML document that describes the history of HTML. This report will inform the reader of the methods used to obtain the information for the report, the results of the application of the methods, and questions that arise because of the results.

Importance of HTML's History

The history of HTML is important to anyone who wants to know more about how and why the Internet works. HTML is the language per se of the Internet. Documents on the Net are written using HTML or using an editor that places the HTML into the document. Knowing why HTML was created, how it has developed, and why it has changed are all important points in understanding why the Internet has developed as it has. The functionality and limits of HTML dictate how we can illustrate our ideas on the Internet and how we view the ideas of others.

METHODS

Internet Searches

The methods used in my research consisted exclusively of Internet searches, and my goal was to collect enough information to write a document on the history of HTML. I chose to use the Internet because HTML has been related to the World Wide Web since its inception, and, hence, much of its history can be found on-line. The primary search engines I used were Google and Metacrawler. Search terms/phrases used to locate information in conjunction with both of these search engines were

"HTML History," "History of HTML," "W3C," and "CERN." Using these search terms in both search engines produced a multitude of results, of which I had to choose the most pertinent.

Drilling for Information

After using these search engines to locate pertinent sites, I had to drill down through these sites to locate more specific information. This consisted of looking for specific dates, places, and names that had to do with the development of HTML. Each time I found a page that gave me specific information that I thought would be useful, I used Internet Explorer's Favorite menu to bookmark the page for later use in writing my report.

RESULTS

The application of my methods yielded a plethora of information on the history of HTML. Specific information on how HTML originated at CERN (Conseil Européen pour la Recherche Nucléaire) and how HTML was originally proposed by Tim Berners-Lee was helpful in formulating the early history of HTML. I was also able to find two different time lines of the development of HTML on two independent sites. All of the pages that I found and used for information are listed in Figure 1. The webpage title,

Webpage	Description
The Early History of HTML	This page covers HTML history from 1990 to 1992.
HTML History	This page covers all of HTML's specifications as well as cascading style sheets.
The History of HTML	This page covers basic terminology associated with HTML and the origin of HTML at CERN.
HTML Overview	This page contains an excellent glossary of terms as well as an HTML time line and narrative on the early development of HTML. It also contains links to the different HTML specifications.
Some Early Ideas for HTML	This page contains information on hypertext systems that came before HTML and gives a description of what hypertext is. It is published by the current HTML regulatory agency, the W3C (World Wide Web Consortium).
The History of HTML	This page contains a description of each version for HTML, what changed from version to version, and why.
What Is HyperText	This page contains a description of hypertext.
Quick HTML History	This page contains a small time line of HTML's history.

Figure 1
Useful Webpages

which is also a link to that page, as well as a short description of each page, can be found there.

DISCUSSION

The results above show I have met my goal for information collection, and I am now prepared to write my HTML history. My information, while answering many of my questions, raised more. These questions need further study: Why did HTML became so popular, while other hypertext systems did not? Why did certain browsers such as Mosaic outclass other browsers to such a great degree?

Brief Analytical Reports

Brief analytical reports are very common in industry. Writers review an issue with the goal of revealing important factors in the issue and of presenting relevant conclusions. The two reports below illustrate varied uses of this form.

CREDIBLE RESOURCES AVAILABLE FOR USE BY DIETITIANS

INTRODUCTION

The use of computers and the Internet has become a part of the daily life of most Americans. As dietetics professionals, being aware of the resources available on the World Wide Web is necessary. Dietitians can access nutrition education materials, current legislation information, job opportunities, government programs, disease/disorder information, and more with the click of a button.

FINDING CREDIBLE RESOURCES

When doing a simple Internet search using the search engine Google and using the keywords "dietetics" or "dietitian," several Web matches appear. When utilizing Web resources, it is important to use the CAR (Credibility, Accuracy, and Reliability) method to determine whether the information provided is appropriate for use in one's practice. The CAR method was utilized reviewing websites related to dietetics. All websites described in this report passed the CAR examination. A summary of the findings are given in Table 1.

WEB RESOURCES AVAILABLE

Government Sites

The most detailed and reliable information found on the Internet came from government resources. The website www.nutrition.gov is the official nutrition site for the United States government. This site links to all nutrition-related information, ranging from food safety and security to diabetes and disease management. The best part of this site is that it links to all federal nutrition programs. These sites give important information as well as provide the ability to download and print forms. Nutrition education, tools and resources are also available on this site. The second government site, agriculture.senate.gov, provides archived federal bills as well as current legislation regarding nutrition, forestry, and agriculture.

Professional Sites

Professional websites that provided many links and extremely reliable information include www.webdietitian.com and www.eatright.org. Both websites contain a wide variety of resources beneficial to dietitians, ranging from current nutrition issues to patient education materials.

The website dietetics.co.uk is a message board forum for dietitians across the world. It is based in the United Kingdom. Dietetics professionals can post and reply to message boards dealing with all aspects of the dietetics profession, including enteral/parenteral feedings, professional issues, nutrition assessment and screening, freelance and private practice dietetics, and more.

Dietetics.com houses links to state and national dietetic associations, antiquackery information, and other basic information to help the dietetic professional.

Dietetic Career Searches

The American Dietetic Association has a career link page that is a national database of current openings in the field of dietetics. A search can be narrowed by choosing an area of discipline or choosing by location. However, the career link is not extensive at this point and does not offer many positions.

The site Jobs in Dietetics, jobsindietetics.com, offers a nationwide career search. There is a membership charge for utilizing this Web resource. Its member-only approach makes it impossible to summarize its usability or quality.

RESULTS OF QUERY

Table 1 summarizes the above paragraphs. The websites were placed into three categories: professional, government, and career search. It was noted whether the website provided outside links. The availability of links on pages was taken into consideration in the overall rating of the quality and usefulness of the websites. Based on the CAR analysis, each site was scored with a 1 to 5 rating with 5 being the highest quality and most beneficial to dietetic practitioners.

Table 1

Summary of Dietetics Resources on the Web

Name of Site	Web Address	Category	Links Available?	Rating (5 = best)
American Dietetic Association	www.eatright.org	Professional	Yes! Many	5
Web Dietitian	webdietitian.com	Professional	Yes! Many	5
U.S. Senate Committee on Agriculture, Forestry, and Nutrition	agriculture.senate.gov	Government	Yes! Some not nutrition related.	4
Nutrition. Gov	www.nutrition.gov	Government	Yes! Many	5
American Dietetic Association Career Link	www.adacareerlink.org	Dietetic Career Search	Yes! Other medical career searches	3
Jobs in Dietetics	jobsindietetics.com	Dietetic Career Search	No	2
Dietetics.Com	dietetics.com	Professional	Yes	3
Nutrition and Dietetics Forum	dietetics.co.uk	Professional	No	3

CONCLUSION

There are many beneficial resources available to dietetic professionals on the World Wide Web. When viewing websites; it is important to keep the CAR (Credibility, Accuracy, and Reliability) method in mind. Professional, government, and dietetic career search capabilities can be found and utilized easily using the Internet. Becoming familiar with this process will enhance the dietetic professional.

Resources

American Dietetic Association. 5 Nov. 2002 <www.eatright.org> (November 5, 2002).
American Dietetic Association Career Link. 7 Nov. 2002 <www.adacareerlink.org> (November 7, 2002).
Dietetics.Com. 5 Nov. 2002 <http://dietetics.com> (November 5, 2002).
Jobs in Dietetics. 5 Nov. 2002 <http://jobsindietetics.com> (November 5, 2002).
Nutrition and Dietetics Forum. 6 Nov. 2002 <http://dietetics.co.uk> (November 6, 2002).
Nutrition.Gov. 6 Nov. 2002 <http://nutrition.gov> (November 6, 2002).
U.S. Senate Committee on Agriculture, Forestry, and Nutrition. 7 Nov. 2002 <http://agriculture.senate.gov> (November 7, 2002).
Web Dietitian. 5 Nov. 2002 <www.webdietitian.com> (November 5, 2002).

CHIPPEWAVALLEYHELPWANTED.COM VERSUS MONSTER.COM

Content Analysis

INTRODUCTION

This section deals with the content of the job search websites Chippewa ValleyHelpWanted.com and Monster.com. I researched the number of ways to apply for a job on each website, the number of available job results for several cities, and the general length of the job descriptions from these results.

WAYS TO APPLY

I found that the two websites had similar methods for applying for jobs. Monster.com has one more option, as shown in the following table, but otherwise the sites were comparable in this aspect. There is more information about some companies listed in Monster.com that isn't included in the other website that may be helpful in deciding whether a company would be a desirable place to apply and work for the future.

	Post Résumé for Free	Automatically Send Résumé to Chosen Employers	Company E-Mail Addresses/ Web Links	View All Company Job Opportunities	Learn About Company
ChippewaValley HelpWanted.com	X	X	X	X	
Monster.com	X	X	X	X	X

NUMBER OF JOBS LISTED

The following table illustrates the number of jobs that are listed for certain cities in the central Wisconsin area as well as the total number of jobs listed for the area. Monster.com had many more jobs by far, but often didn't have jobs listed in smaller cities that ChippewaValleyHelpWanted.com did include. The larger database of jobs would be beneficial for those looking for jobs in larger cities or just in the area in general, while listings for smaller cities would be helpful to those searching for a job in a specific area that may not be as populated.

	Menomonie	Eau Claire	Chippewa Falls	Madison	Total
ChippewaValley HelpWanted.com	1	19	2	1	32
Monster.com	0	74	1	>400	654

LENGTH OF JOB DESCRIPTIONS

I looked at postings for the same job on both websites and compared the information that the posting included and found that the basic information was the same, but Monster.com had more specific details, such as salary ranges, position types, and a reference code to easily distinguish the particular job from others listed by the same company. Depending on the focus of a person's job search, this information may be a key factor in deciding whether to apply for a job. Once again, if the major deciding factor in applying for a job is based on location or job type, ChippewaValley HelpWanted.com provided this information and would be an adequate resource for job searching.

CONCLUSION

I have found ChippewaValleyHelpWanted.com and Monster.com to provide much of the same information in their job search functions. Monster.com contained more information about companies and a larger listing of available jobs, but the majority of jobs were located in larger cities. ChippewaValley HelpWanted.com seemed to be targeted more for people looking for jobs specifically in this area and included job listings for smaller towns that were not included in the other website.

Sources

ChippewaValleyHelpWanted.com
Monster.com

Progress Reports

Progress reports inform management about the status of a project. Submitted regularly throughout the life of the project, they let the readers know whether work is progressing satisfactorily—that is, within the project's budget and time limitations. To write an effective progress report, follow the usual process. Evaluate your audience's knowledge and needs. Determine how much they know, what they expect to find in your report, and how they will use the information. Select the topics you will cover. The standard sections are the following:

- Introduction
- Work Completed
- Work Scheduled
- Problems

In the Introduction, name the project, define the time period covered by the report, and state the purpose: to inform readers about the current status of the project. In the Work Completed section, specify the time period, divide the project into major tasks, and report the appropriate details. In the Work Scheduled section, explain the work that will occur on each major task in the next time period. In the Problems section, discuss any special topics that require the reader's attention.

PROGRESS REPORT

Date: March 29, 2006
To: Dan Riordan, Practicum Manager
From: Julia Seeger
Subject: Progress Report on Construction Manual

SUMMARY

I am working on developing a user manual for the Universal Test Machine, TestWorks QT, for the UW-Stout Construction Lab. Tests have been modified for the students' use, which will require a new set of instructions. The instructions will be designed to help guide the student through machine setup, starting up the TestWorks QT software program, running the test, and proper shutdown.

WORK COMPLETED

The client and I have decided that the manual will be hard copy, 5.5″ × 8″, bound with a plastic spiral. Each step of the process will include an illustration. The client decided that the manual would include instructions for three different types of tests: Bending, Compression, and Tensile. I have written instructions for the Compression test. Additional information will be added after the client reviews the instructions. Thursday, March 28, I met with the client in the construction lab where approximately 15 photos were taken for the Tensile, Compression, and Bending tests. All photos will use JPG format.

WORK SCHEDULED

Digital photos of the Bending, Compression, and Tensile tests will be viewed and enhanced using Photoshop. I plan to have the photos prepared and sent to the client by Wednesday to view and approve.

The next client meeting occurs on April 6. After that meeting I will develop the written instructions. The client will receive the rough draft of the instructions by April 15.

PROBLEMS

The Universal Test machine is scheduled to be used for classes during the only hours I have free to conduct usability testing in the instructions. At present no students have agreed to serve as usability testers. I will work with the lab instructor to resolve these issues.

Outline Reports

An expanded outline is a common type of report, set up like a résumé, with distinct headings. This form often accompanies an oral presentation. The speaker follows the outline, explaining details at the appropriate places. Procedural specifications and retail management reports often use this form. The brevity of the form allows the writer to condense material, but of course the reader must be able to comprehend the condensed information. To write this kind of report, follow these guidelines:

▶ Use heads to indicate sections *and* to function as introductions.
▶ Present information in phrases or sentences, not paragraphs.
▶ Indent information (as in an outline) underneath the appropriate head.

Sample Outline Report

REPLACE GYMNASIUM FLOORING?

December 10, 2007

Researcher: Aaron Santana

Purpose: Evaluate whether Athletic Department should install new flooring in the Memorial Gymnasium.

Side heads serve as introductions to each section

Method: Interview Athletic Director, Athletic Trainer, vendors. Used these criteria:

Essential material presented in phrases and lists

• Cost—not to exceed 250K
• Time—less than 3 months
• Benefits—personal health and overall usage must be impacted

Conclusions: Gym floors meet all criteria

• Cost is less than 250K allocated
• Time to install is less than 3 months
• Benefits—fewer injuries; likelihood of increased usage with greater durability.

Recommendation: Install the new floor.

Worksheet for Planning a Project

☐ Write the question you want answered.

☐ Create a research plan.
   a. List topics and keywords that might help you find information on your question

(continued)

(continued)

    b. List a method for finding out about those topics. Tell which specific acts you will undertake. E.g., "Explore Compendex and Ebscohost using X and Y as keywords." or "Talk to all employees affected by the change, using questions X and Y."

☐ **Carry out your plan.**

Worksheet for IMRD Reports

☐ **Write an introduction in which you briefly describe the goal of your project and your goal in this report. Give enough information to orient a reader to your situation.**

☐ **Write the methods statement.**
- Name the actions that you took in enough detail so that a reader could replicate the acts if necessary.
- Use terms and details at a level appropriate to the reader, but necessary for the subject.
- Explain *why* you chose this strategy or actions.

☐ **Name the actual results of the actions. This section might be very short.**

☐ **Tell the significance of your actions.**
- What do the results indicate?
- How do the results relate to the audience concerns?
- What will you do next, as a result of this project?
- Did you accomplish your goals?
- How is this important to your classmates, in this class?

☐ **Develop a style sheet for your report.**
- How will you handle heads?
- How will you handle chunks?
- How will you handle page numbers?
- How will you handle visual aids?
- How will you handle the title/memo heads?

☐ **Develop an idea of how you will present yourself.**
- Will you write in the first person?
- Will you call the reader "you"?
- Will you write short or long sentences?
- Are you an expert? How do experts sound? What will you do to make yourself sound like one?
- *Key question:* Why should I believe you? Why are you credible?

Worksheet for Informal Reports

☐ Identify the audience.
  • Who will receive this report?
  • How familiar are they with the topic?
  • How will the audience use this report?
  • What type of report does your audience expect in this situation (lengthier prose? outline? lots of design? just gray?)

☐ Determine your schedule for completing the report.

☐ Determine how you will prove your credibility.

☐ Outline the discussion section. Will this section contain background? Divide the section into appropriate subsections.

☐ For IMRDs, outline each of the three body sections.
  • Clearly distinguish methods and results.
  • In the discussion, relate results to the audience's concerns.

☐ Prepare the visual aids you need.
  • What function will the visuals serve for the reader?
  • What type of visual aid will best convey your message?

☐ Prepare a style sheet for heads (one or two levels), margins, page numbers, and visual aid captions.

☐ Select and write the type of introduction you need
  • To give the objective of the report.
  • To provide brief context.
  • To provide expanded context.

☐ Select the combination of introductory elements you will use to give the gist of the report to the reader. Write each section.

Worksheet for Evaluating IMRDs

Read your report or a peer's. Then answer these questions:

☐ Introduction
  • Is the goal of this project clear?
  • What is the basic question that the writer answers?

☐ Methods
  • Does the writer tell all steps or actions that he or she followed to achieve the goal or answer the question? Often there are several sequences of them.
  • Is it clear why the writer took those steps or actions?

☐ Results
  • Does this section present all the things that the writer found out?

(continued)

(continued)

     • If there is a visual aid, does it help you grasp the results quickly?

☐ **Discussion**
     • Does the discussion answer the question or explain success or failure in achieving the project's goal?
     • What are the implications of the results? Implications mean (1) effects on various groups of people or their goals, or (2) perceptions about the system (e.g., Web search engines, Web authoring programs) discussed in the report.
     • What new questions do the results cause?

Examples

Examples 12.1–12.4 show four informal reports. These reports illustrate the wide range of topics that the informal report can present. Note the varied handling of the introduction and of the format of the pages. The goal in all the reports is to make the readers confident that they have the information necessary to make a decision.

Example 12.1

IMRD Report

EVALUATION OF COLLABORATIVE SOFTWARE APPLICABILITY AT UW–STOUT BY JAMES J. JANISSE

INTRODUCTION

Technical professionals are faced with an increasing number of tools intended to help their job productivity. Technology has also changed the environment and context of their work. Traditional teamwork can now be performed in an electronic or virtual manner and with collaborative software tools purported to significantly enhance job performance. A key part of UW–Stout's continuing mission is to monitor and evaluate these industry tools.

    Groove (version 1.3) is considered the leading-edge collaboration software package in the industry and was developed by Groove Networks. This privately held company was created in 1997 and its headquarters are in Beverly, Massachusetts. The founder, Ray Ozzie, is best known as the creator of Lotus Notes, the world's leading groupware product. (There are more than 75 million users worldwide.)

    The purpose of this report is to evaluate the applicability and feasibility of using Groove collaboration software in selected technical areas at the University of Wisconsin–Stout (UW–Stout).

METHODOLOGY

This cursory evaluation targeted two areas of technical study at UW-Stout: the Technical Writing Practicum and development of the new, on-line

Industrial Management Case Studies Course within the College of Technology, Engineering, and Management (CTEM). The selection was based on the strong interest and support from groups and the fact they represented different parts of the technical spectrum. There was a special consideration for the CTEM evaluation of Groove. Any new software tool must provide some instructional delivery capability like the incumbent tool, BlackBoard.Com. Therefore, from a CTEM perspective, Groove was essentially evaluated against the test criteria and an existing software tool.

The evaluation of the Groove software was performed by

1. identifying a subject matter expert in each area. Professor Dan Riordan and Chairperson Donna Stewart represented Technical Writing and CTEM, respectively.
2. downloading the latest demonstration version of the Groove software from www.groove.net/downloads/ to test laptop and desktop computers.
3. developing a functional profile of the Groove software that would serve as an evaluation checklist. This technique is commonly used in software development in an activity called the Application Walk-Through. Evaluators list all major software categories and functions in a table format to establish the beginning of a simple but powerful scorecard.
4. defining typical test group project activities, scenarios, and testing media to be addressed by each item in the evaluation checklist.
5. conducting a series of tests, with each expert assessing the applicability of Groove functions against the Technical Writing Practicum and Industrial Case Studies Course evaluation checklist.
6. summarizing the evaluation findings by jointly grading the fit or feasibility of each functional area for the two target groups.

RESULTS

Evaluators assessed each criterion by determining the level of

- functional fit to test scenarios
- performance
- ease-of-use
- intangibles such as departmental considerations, software personality, etc.

The range of subjective ratings was noted as

- low (L)—limited function or no match
- medium (M)—moderate function or adequate match
- high (H)—high function or significant match
- not applicable (N/A)

Additionally, each test group designated three criteria as critical or must-have, and noted these ratings with asterisks.

(continued)

Example 12.1

(continued)

The findings of the Groove collaborative evaluation are summarized in Table 1. Nine of the fourteen criteria for Technical Writing were low, including several functions needed for dynamic and joint editing. The Industrial Case Studies ratings included six lows, two not applicable ratings, and a note about prohibitive costs.

Table 1

Ratings of Groove Functional Applicability to Test Groups

Groove Functional Area	Technical Writing Practicum	Industrial Management Case Studies Course
Shared workspaces	H	H
Collaboration—Common File Directory with version control	M	L**
Collaboration—Discussion Space	L–M**	M** No statistics of readers or number of visits to the discussion thread
Collaboration—Notepad	L**—Didn't permit the desired dynamic editing; slow to refresh	L
Collaboration—Outliner	L**—Performance didn't permit reasonable joint editing	N/A
Collaboration—Pictures	L—FYI only	L
Collaboration—Sketchpad	L—No practical use found	L
Collaboration—common Web browser	L	N/A
Productivity—calendar	L	L
Productivity—contacts	L	M
Productivity—course delivery and administration	N/A	L**
Messaging—instant text	M	L
Messaging—asynchronous audio recordings	L	M
Security and privacy	H	H
Costs	M—Might be for a small workgroup	L—Cost prohibitive for desired class size of 20+ students ($49 each)

DISCUSSION

The results of the collaborative evaluation show that Groove would have limited applicability and value to either the Technical Writing Practicum or Industrial Case Studies Course. Existing or alternative software solutions should be persued.

From a Technical Writing Practicum perspective, Groove had some interesting functions. Initial impressions anticipated dynamic editing, but testing the software in actual project scenarios proved otherwise. Groove did not provide the right functionality to permit a team to rapidly perform quality creation or editing of text. For example, evaluators had a difficult time converting a table of text into a paragraph using Groove Outliner, Notepad, and Discussion Space functions. While some other tests yielded moderate ratings, Groove did not address the most important criteria for the Technical Writing Practicum.

Similarly, Groove did not score high for the key criteria established by the Industrial Case Studies team. Neither the file directory nor collaborative functions provided the interactive discussion and analysis needed for students to perform case study work. Although there might be some potential for interaction with Groove, there were deficiencies in delivering basic course activities and addressing course administration as a standalone product.

It is worth noting that Groove should still be monitored as a tool for educational delivery. The April issue of *InfoWorld* previewed an impending release of Groove 2.0 with extensive enhancements and a total integration partnership with Microsoft.

Example 12.2

IMRD Report

IMRD: ADDING BACKGROUND SOUND TO WEBPAGES

INTRODUCTION

I set out to add background sound to a webpage. My resources included an IBM-compatible PC with Windows 98 and software that included Notepad and Internet Explorer 4.0. Computers with these resources can be found in the Main Lab of the Micheels Hall Computer Lab. Before I began, I downloaded a sound file from the Internet and had an existing HTML document to which I wanted to add the sound file. These two files were in the same directory on my disk. Some of the different sound file types that you can use include .midi, .rmf, .rmi, .wav, and .mod.

METHOD

I opened Notepad by selecting Start/Programs/Accessories/Notepad. I then opened up an existing HTML file that I wanted to add sound to by selecting File/Open and then finding the HTML file I wanted to open.

(continued)

Example 12.2

(continued)

Once the file was open, I added the following code to the body section of my HTML file to add the music file, *cheers.mid,* to the background:

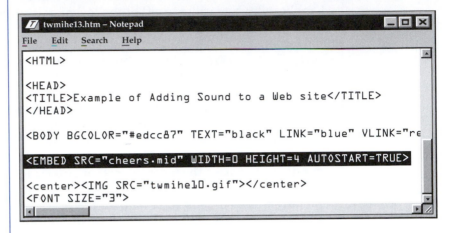

```
<HTML>

<HEAD>
<TITLE>Example of Adding Sound to a Web site</TITLE>
</HEAD>

<BODY BGCOLOR="#edcc87" TEXT="black" LINK="blue" VLINK="re

<EMBED SRC="cheers.mid" WIDTH=0 HEIGHT=4 AUTOSTART=TRUE>

<center><IMG SRC="twmihe10.gif"></center>
<FONT SIZE="3">
```

I then selected File/Save to save the modified HTML document.

Once the file was saved, I wanted to view the modified page in Internet Explorer. I did this by first double-clicking on the My Computer icon. Then I double-clicked on the 3½ Floppy (A) icon. I finally opened the file by double-clicking on the filename.

RESULT

In the HTML code that I inserted in my file, the WIDTH and HEIGHT parameters are set to small numbers so there will be no visual changes to the webpage. AUTOSTART = TRUE is used so that the sound file will start playing as soon as the page is opened.

Once I modified and saved my HTML document in Notepad, I had a current copy of the document on my disk in the same directory as the sound file.

When I double-clicked on the My Computer icon, a list of the available drives appeared in the window.

When I double-clicked on the 3½ Floppy (A) icon, a list of the files on my disk appeared in the window.

When I double-clicked on the HTML filename that I modified, Internet Explorer opened with my HTML file and the *cheers.mid* sound file playing in the background. There were no visual changes to the webpage.

DISCUSSION

My webpage was made more interesting by following this easy process of adding music to the background.

If you are working at a computer that has a sound card, but no speakers, you can still hear the audio by plugging headphones into the computer.

Example 12.3

Analytical Report

Date: April 29, 2006
To: Joseph King
From: Chris Lindblad
Subject: Purchase of a function generator for the control module tester

The module test area will be conducting the testing and troubleshooting of the SSD control module upon receiving the control module tester. The testing of this module will require the use of a function generator, which the test area does not currently have.

I have talked with numerous sales representatives and have discussed the purchase with other technicians in the test area. Basing my criteria on budget allowance, operating features, ease of operation, and future applications, I recommend the purchase of the Tektronix model AFG 5101 function generator.

Table 1
Cost of Function Generator, Options, and Accessories

Item	Tektronix AFG 5101	Philip PM 5192
Function generator	$3695	$4050
Options	350	425
Accessories	55	45
Total	4100	4520

TEKTRONIX IS WITHIN THE BUDGET

Currently the budget allows up to $6000 for the purchase of a function generator for the control module tester. As you can see from Table 1, both function generators are priced below the allowed budget. Not reflected on the total purchase price is a 10% discount the company currently receives on the purchase of electronic test equipment from Northern States Electronics Inc., the regional distributor of Tektronix Inc. With this discount included, the purchase price of the Tektronix model would drop to a total of $3690.

TEKTRONIX HAS BETTER OPERATING FEATURES

The Tektronix model AFG 5101 function generator has the ability to produce the required 10-MHz clock sine wave along with four other signal wave forms in a frequency range from .012 to 20 MHz. The Philips model PM 5192 function generator has the ability to produce the required clock signal and four other signal wave forms but only in the range of .1 to 20 MHz. The Tektronix model also has the feature of changeable pods, which can be quickly replaced if the unit should fail, whereas the Philips model would have to be returned to a service center for repair, which would result in downtime on the tester.

(continued)

Example 12.3

(continued)

TEKTRONIX HAS BETTER OPERATIONAL SUPPORT

Both function generators are relatively easy to operate and are fully programmable. Both allow for the programming of selected wave forms, eliminating the need to lead information into the function generator before each test is run. A Tektronix representative will present a one-day training session to the module test technicians and will be available by phone for any further questions. The Philips model is accompanied by a manual and a 20-minute training video.

TEKTRONIX WILL UPGRADE EASIER

With the future module designs that will be coming into the test area and the faster speeds in which they will operate, the Tektronix model AFG 5101 has a larger operating frequency range, and the changeable pod feature will allow it to be upgraded for possible future uses. The Philips model has a lower frequency range and cannot be upgraded.

    If you require any further information or documentation on my recommendation, please contact me at the module test department.

Example 12.4

Memo
Recommendation

Date: July 17, 2006
To: Marcus Hammerle
From: John Furlano
Subject: Recommendation of tooling schedule

This letter is a follow-up of our discussion pertaining to the Storage Cover tooling schedule, which we discussed briefly during your visit at MPD on Thursday, June 27.

    There are three main options available to expedite the pilot run date from October 2 up to the week of October 14.

1. Postpone the tool chroming until after the pilot run.
2. Postpone any major tool modifications until after the pilot run.
3. Expedite the tool building.

I believe that option 1 is the best choice. There will not be any additional costs for this option, and quality parts will still be produced for the pilot run. Option 2 may not be a reliable option because we cannot judge until after sampling what modifications may be necessary. Option 3 will carry additional cost due to overtime labor.

    I am also optimistic, yet concerned, about the September 9 sample date (two samples at your facility, hand drilled and bonded, no paint). My concern

> is with delays through customs for shipping parts from the tool shop in Canada to the United States for assembly and back to Canada.
>
> We are taking every step possible to stay on schedule with the Draft #3 tooling schedule, which you have a copy of. The enclosed tooling schedule (Draft #4) shows the projected pilot run date if option 1 above is employed. I welcome your comments or suggestions on any of the above issues. Thank you.
>
> enc: tooling schedule draft #4

Exercises

▶ You Create

1. Create an objective/summary introduction for the Iconglow report on pages 293–294.

2. Create a different introduction for the analytical report on purchasing a function generator on pages 311–312.

3. Write a methodology statement that explains how you recently went about solving some problem or discovered some information. When you have finished, construct a visual aid that shows the results of your actions. Compare these statements and visuals in groups of two to three.

4. Write the introduction for the material you wrote in Exercise 3.

5. Write the discussion section for the material you wrote in Exercise 3.

▶ You Analyze

6. Because introductions imply a lot about the relationship of the writer to the reader, analyze the introductions of the reports in Examples 12.1–12.3 to determine what you can about the audience-writer relationship. How is that relationship affected when you change the introduction as you did in Exercise 2?

7. Read the introduction and body of another student's paper from one of the Writing Assignments below. Does the discussion really present all the material needed to support the introduction? Are the visual aids effective? Is the format effective?

▶ Group

8. In groups of three or four, analyze the sections of the IMRD report on collaborative software on pages 306–309. How is the introduction related to the discussion? Do you feel that you know everything you need to know after reading the first few paragraphs? What does the discussion section add to the report?

9. In groups of three, read the introduction of each person's paper from one of the Writing Assignments below. Decide whether to maintain the current arrangement; if not, propose another.

10. a. In groups of two or three, decide on a question that you will find the answer to. A good example is how to use some aspect of e-mail, the library, or the Web. Before the next class, find the answer. In class, write an IMRD that presents your answer.

    I = question you wanted to answer and goal of this paper
    M = relevant actions you took to find the answer
    R = the actual answer
    D = the implications of the answer for yourself or other people with your level of knowledge and interest

    b. In groups of two or three, read each other's IMRD reports. Answer these questions:

    Do you know the question that had to be answered?
    Could you perform the actions or steps given to arrive at the answer?
    Is the answer clear?
    Is the discussion helpful or irrelevant?

11. In groups of two or three, compare the differences in Example 11.2 (pp. 278–279) and Example 12.2 (pp. 309–310). Focus on differences in tone and in presentation of the actions. Report to the class which document you prefer to read and why. What principles would affect your choice to write an IMRD or a set of instructions?

Writing Assignments

1. Write an informal report in which you use a table or graph to explain a problem and its solution to your manager. Select a problem from your area of professional interest—for example, a problem you solved (or saw someone else solve) on a job. Consider topics such as pilferage of towels in a hotel, difficulties in manufacturing a machine part, a sales decline in a store at a mall, difficulties with a measuring device in a lab, or problems in the shipping department of a furniture company. Use at least one visual aid.

2. Write an IMRD report in which you explain a topic you have investigated. The report could be a lab report or a report of any investigation. For instance, you could compare the fastest way to reproduce a paper, by scanning or retyping, or give the results of a session in which you learned something about navigating on the Internet, or present the results of an interview you conducted about any worthwhile concern at your school or business. Your instructor may combine this assignment with Writing Assignment 3.

3. Bring a draft of the IMRD you are writing to class. In groups of two or three, evaluate these concerns:

a. Is the basic research question clear?

b. Does the method make you feel like a professional is reporting?

c. Could you replicate the actions? Could other people?

d. Does only method—and not results—appear in the method section?

e. Is the method statement written like instructions or a process description? Which is best for this situation?

f. Are the results clear? Are they a clear answer to the original question?

g. Does every topic mentioned in the discussion section have a clear basis of fact in the methods or results section?

h. Is the significance the writer points out useful?

i. Does the visual aid help you with the methods or result section? Would it help other people?

j. Is the tone all right? Or is it too dry? too chatty? too technical?

k. Does the formatting of the report make it easy to read?

4. Rewrite the IMRD from Writing Assignment 2 from a completely different framework—for instance, a coach explaining the subject to a high school team. After you complete the new IMRD, in groups of three or four, discuss the difference "author identity" makes and create questions to tell writers how to choose an identity.

5. Convert your IMRD report from Writing Assignment 2 into an article for a newsletter.

6. Convert your IMRD report from Writing Assignment 2 into a set of instructions. After you complete the instructions, in groups of three or four, construct a list of the differences between the two, especially the method statement. Alternate: In groups of three or four, construct a set of guidelines for when to use instructions and when to use IMRD. Hand this list in to your instructor.

7. Write an outline report in which you summarize a long report. Depending on your instructor's requirements, use a report you have already written or one you are writing in this class.

8. Write a learning report for the writing assignment you just completed. See Chapter 5, Writing Assignment 7, pages 134–135, for details of the assignment.

Web Exercise

Write an IMRD that explains a research project on the effectiveness of a search strategy on the Web. Choose any set of three words (e.g., plastic + biodegradable + packaging). Choose any major search engine (Yahoo!, AltaVista). Using the "advanced" or "custom" search mode, type in your keywords in three sequences—plastic + biodegradable + packaging, packaging + plastic + biodegradable, biodegradable + packaging + plastic. Investigate the first three sites for each search. In the IMRD, explain your method and results and discuss the effectiveness of the strategy and of the search engine for this kind of topic.

Large group alternative: Divide the class into groups of four. All members of the class agree to use the same keywords, but each group will use a different search engine. After the individual searches are completed, have each group compile a report in which they present their results to the class orally, via e-mail, or on the Web.

Works Cited

General Motors. *Writing Style Guide*. Rev. ed. GM1620. By Lawrence H. Freeman and Terry R. Bacon. Warren, MI: Author, 1991.

Keppler, Herbert. "SLR: Going Where No Macro Has Dared to Go Before." *Popular Photography* (August 1997).

Focus on
E-Mail

E-mail has become a major way by which people communicate, sending everything from birthday greetings to intergovernmental communications. E-mail has revolutionized interpersonal communications, so much so that a problem has arisen: How do we deal with the glut of e-mail? Steve Gilbert, who has run a LISTSERV for many years, says, "I try to limit the number of messages to a few each week to avoid making even worse the e-mail glut many of my friends and colleagues are experiencing." How do you fashion your e-mail so it doesn't sit unread in your recipient's in-box or, worse, is dumped altogether?

Here are a few tips that will help you write effective, clear e-mail messages that will be read and acted upon in a timely manner.

Write a Clear Subject Line Experts who have studied e-mail find that the subject line is the most important item when trying to connect with the intended reader. Messages are often displayed in a directory that lists the sender's name, the date, and the subject. Many readers choose to read or delete messages solely on the basis of the subject line, because they can't possibly take the time to respond to so much mail. Your message will more likely be opened if the subject line connects with the reader's needs. If the reader is not engaged by the subject line, he or she will often simply delete the message unread. Here are some tips:

- Start with an information-bearing word. Say

  "Budget Request—meeting scheduled"
  rather than "budget meeting" or "meeting."
  Or
  "Hi—meet me after your class?"
  rather than "Hi."
- Keep the subject line relatively short. This tip could conflict with the previous one, so be judicious in your phrasing of the information-bearing word or phrase.
- People often open messages with RE in the subject line (so don't change the subject when

you reply). In a subject line, state content— "Response to your 7-25 budget request."
- Make the subject line a short summary of your message. (Nielsen; Rhodes; "ITS")

Use the To and CC Lines Effectively The To line should contain only the names of persons who you are asking to do something. In the CC line, list people who should know about the message, or who are getting the e-mail simply for information purposes ("ITS").

Check Addresses Many e-mail addresses are remarkably similar. It is quite easy to make a typing mistake, so that the e-mail intended for jonessu goes instead to joness or jonesd. Although this is often a minor annoyance, it can be a major embarrassment if the content is sensitive or classified ("ITS").

Consider Whether to Send an Attachment Attachments take more time to download and often easily become separated from the original e-mail. In addition, many attachments can't be opened at all by the receiver, especially if they were created in another platform or by an application not owned by the receiver.

If you do send an attachment, be sure that the document contains such information as a title and the name of the person who sent it. Sometimes this information appears only in the e-mail; if the e-mail is deleted, the attachment becomes difficult to make meaningful. If the attachment is long, consider posting it on a website (if that option is easily available to you) and sending your recipients an e-mail with the RRL to that space ("ITS"). In order to avoid "losing" an attachment, or to ensure that there are no problems opening the document, paste the contents directly into the e-mail. Note, however, that this strategy makes the e-mail long, so in the introduction establish the context for the content. Be sure to give the attachment a meaningful filename. If the attachment is opened

(continued)

(continued)

directly from the e-mail, the context for it is clear. But if the e-mail is gone and the attachment resides in a directory with many other files, the filename must be meaningful. Say "jonesresume" rather than "resume," or "ABCapplicationform" rather than "ABCaf."

Keep Messages Short and to the Point Research has established that readers categorize e-mails. "To-do" messages require some action from the recipient. Often, these messages stay in in-boxes as a reminder to the recipient of work to do. "To read" messages usually are long documents that take time and effort to read. Although the content could be important, the length causes recipients to delay reading them. "Indeterminate" messages are those whose significance is not clear to the reader. Like long messages, these messages are usually not read, but left in in-boxes so that when there is time enough, the reader will make the effort to read the message and determine the significance (Rhodes).

Establish the Context In the body of the e-mail, repeat questions or key phrases. Briefly explain why you are writing, then go on with your message. If a person has sent out 20 messages the day before, he or she might not easily remember exactly what was sent to you. Offer help. Remember that you are not in a dialogue in which the other person can respond instantaneously to your statements, so avoid the temptation to use one-line speeches. For instance, don't just write one word—"No"—but explain the topic you are saying "No" to. One respondent to an e-mail survey said, "X is unbelievable in that he never puts in the context of what he is replying to. He always comes up with these one-line responses, and I have no idea what it is that he's talking about" (Rhodes).

Remember to Use Paragraphs E-mail's format has a kind of hypnotic quality that encourages people to write as if they were speaking. And, of course, in speech there are no obvious paragraphs. However, remember that e-mail is text that a person reads, so chunk into manageable paragraphs. Use keywords at the beginning of units in order to establish the context of the sentence or paragraph that follows.

Signal the End Because e-mail exists in scrolling screen form, there is no obvious cue to its end, unlike a hard copy where you always know when you are on the last page. Therefore, signal the end by typing your name, with or without a closing. You may also use the words "the end" or a line of asterisks.

Avoid Mind Dumps The point of e-mail is to satisfy the reader's needs as concisely as possible. Do not ramble. Plan for a moment before you start to write. If you have "on-line fear," the same strange emotional response that often makes people give awkward, rambling messages on an answering machine, type your message first on a familiar word processing program, when you have time to gather your thoughts and get them down coherently. Edit in the word processor, then upload and send (know the capabilities of your system).

Don't Type in All Caps The lack of variation in letter size makes the message much harder to grasp and gives the impression that you're shouting.

Get Permission to Publish E-mail is the intellectual property of its creator. Do not publish an e-mail message unless the creator gives you permission.

Be Prudent Technically (and legally), the institution that provides you the e-mail service (such as your university or employer or governmental agency) owns the e-mail you are sending and receiving. As a result, any number of people can access individual e-mails if they have some reason to. Be careful about sending sensitive or personal information. In addition, remember that any e-mail is easily forwarded—and so is any attachment—

without your knowledge. Although you might think that the sensitive meeting notes that you send to the committee chair will remain only on her computer, or that the personal comments you make about another person will stay buried in an in-box, it is all too easy for these messages to be forwarded, deliberately or accidentally, to others.

Works Cited

Gilbert, Steve. "Personalizing Pedagogy with E-Mail." *Syllabus Magazine* March 2003. 13 Oct. 2003 <www.syllabus.com/mag.asp?month=3&year=2003>.

"ITS Email Tips." Email and Network Services, Yale University 17 Aug. 2002. 13 Oct. 2003 <www.yale.edu/email/emailtips.html>.

Nielsen, Jakob. "Microcontent: How to Write Headlines, Page Titles, and Subject Lines." *Useit.com* September 1998. 13 Oct. 2003 <www.useit.com/alertbox/980906.html>.

Rhodes, John S. "The Usability of Email Subject Lines." 8 Feb. 2001. *Webword.com* 13 Oct. 2003 <http://webword.com/moving/subjectlines.html>.

Ethics and E-Mail

E-mail is a form of written electronic communication only a little over ten years old. Being aware of the ethical guidelines for conduct when sending communications over the Internet is important in remaining professional and courteous.

Frequently, the recipient of an e-mail passes on all or part of its content to other recipients. The professional is conscientious of attributing true sources of such material, even in a medium as informal as e-mail. The following excerpt from an e-mail policy contains guidelines for being professional and sensitive to the concerns and contributions of others when forwarding material from a previous sender:

Good practice [in the use of quotations in e-mail] includes:
- quoting only that part of another message that is relevant
- including enough context when quoting so that the recipient will not be misled as to the meaning or intentions of the person quoted
- attributing quotations to the person quoted
- not using quotations in messages to someone who was not a recipient of the original message, unless you have the permission of the original sender, or unless you can be reasonably satisfied that the original sender would not object to being quoted. (Lynmar Solutions)

Perhaps the most controversial topic concerning ethics and e-mail is the practice of "flaming"—sending angry e-mails inciting "flame wars" (Shapiro and Anderson).

The absence, in e-mail, of the facial expression, tone of voice, and feedback in face-to-face conversations, together with the speed of response that is possible with e-mail, can lead to "flaming" and "flame wars."

To avoid these, it is good practice:
- to avoid *ad hominem* expressions such as "you must be stupid if you don't understand that . . ." or "only an idiot would think that . . ."
- to allow yourself a "cooling off" period, before responding to e-mail that annoys you, and to be temperate in your response; you should be particularly careful if your response will go to more than just the original sender, e.g., to an entire mailing list
- to make sure when appropriate, for instance by the addition of conventional symbols such as ":-)", that humorous remarks cannot be taken seriously. (Lynmar Solutions)

Works Cited

Lynmar Solutions. "E-mail Ethics and Good Practice: A Sample Policy." 2004 March 1 <www.lynmarsolutions.co.uk/files/emailethics.doc>.

Shapiro, Norman Z., and Robert H. Anderson. "Toward an Ethics and Etiquette for E-mail." July 1985. *Rand.* 1 March 2004 <www.rand.org/publications/MR/R3283/#rece>.

Chapter 13 Developing Websites

Chapter 13
In a Nutshell

Websites and Web documents are important methods of conveying information. Creating effective websites requires careful planning, drafting, and testing.

To plan effectively, you need to consider your audience. Determine who they are. Of course, on the Web they could be anyone in the world, but that's too broad. A helpful way to create a sense of your audience is to define a role for them, as if they were actors in your "Web play." Are they customers? students? curiosity surfers?

In addition to considering your audience, you need to plan a flow chart and a template. The flow chart is a device that indicates how you will link your material together. For instance, if you have four files and if you want your reader to link from any one to any other one, your flow chart would look like Figure 1. Each line is a link and each box is a webpage.

Your template is a design of your site's look. It shows how you will place various kinds of information (title, text, links, visuals) so that your reader can easily grasp the sense of your site.

Websites create special concerns for writing. Good Web text is scannable (easy to find key ideas), cor-rect (no spelling, grammar mistakes), and consistent (all items treated in a similar fashion).

Visuals must be legible, but not so large that they take up most of the screen or take a long time to load.

Websites must be tested to make sure links work, visuals appear, and the site displays consistently in various browsers.

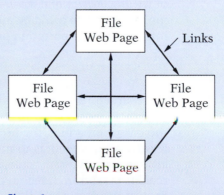

Figure 1
Sample Flow Chart

The Web is one of the primary means of communication today. Millions of people use it every day to find information, to purchase items, and to entertain themselves. Because it is so easy to use, the Web has changed the method of disseminating information. In the past, vital information (of whatever kind—from research data to sale items) was printed on paper (as a report or a catalog, for instance) and sent to intended audiences. Now the vital information is "posted," and the intended audience must search for it. Universities, corporations, organizations of all types, and private individuals all maintain large websites to make information available to viewers.

Because of this shift, technical communicators must know how to create documents that are both clear and easy to read on screen. Creating such documents requires the same general process of planning, drafting, and finishing as in creating any document, but with special considerations for the on-line situation.

This chapter covers basic Web concepts, planning, drafting, and testing Web documents. Examples show Web reports and Web instructions.

Basic Web Concepts

Three basic concepts that will help you create effective websites and documents for readers are hierarchy, Web structure, and reader freedom.

Hierarchy

Hierarchy is the structure of the contents of a document. All websites and Web documents have a hierarchy, that is, levels of information. The highest level is the *homepage,* a term that can apply either to an entire site or to a document. Lower levels are called *nodes*; the paths among the nodes are the *links.*

Figure 13.1 shows three levels in the hierarchy, each giving more detail. *Writing* is the most general category. *Technical* and *fiction* are the two subcategories of writing. *Reports* and *novels* are subdivisions, respectively, of their types of writing. More levels could be added. For instance, reports could be broken down into feasibility reports and proposals.

Web Structure

"Web structure" means that the document contains hyperlinks (or "links") that allow readers to structure their own reading sequence. When the reader clicks the cursor on a link, the browser opens the screen indicated by the link. This feature allows readers to move to new topics quickly and in any order. This arrangement is a radical departure in organizing strategy. The author gives the readers maximum freedom to choose the order in which they will view the site or read the document.

Figure 13.1

Hierarchy

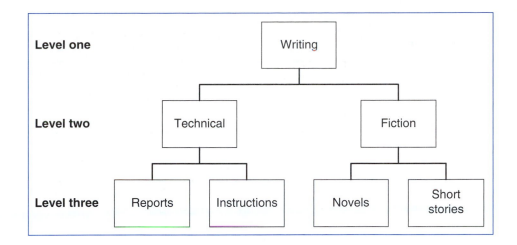

To see the difference between traditional and Web structures, consider these two examples. If a document has seven sections and a traditional (or "linear") structure, then a reader will progress through the sections as shown in Figure 13.2. But if the same seven sections have a Web structure, then they would look like Figure 13.3 (p. 324), with each line a link. Once readers arrive at the start, or home, screen, they may read the document in any order they please.

The two Web homepages in Figures 13.4 and 13.5 show two ways that authors used Web structure. Both have a lengthy report, which they want the readers to be able to read without having to scroll through various screens. Maertens (Figure 13.4, p. 324) has divided the report into five linkable sections. The homepage provides an index and brief abstract of each section. The reader can choose any section and link to it. Currier (Figure 13.5, pp. 325–326) has provided "anchor links" that move the screen to that part of the document without scrolling.

To "think Web" is a radical departure in organizing strategy. You can give readers maximum freedom to choose in what order they read your document.

Figure 13.2

Linear Structure

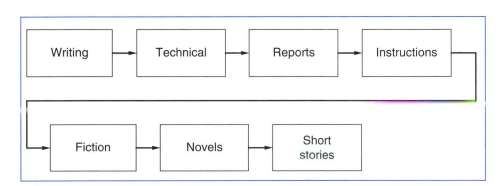

Figure 13.3

Web Structure

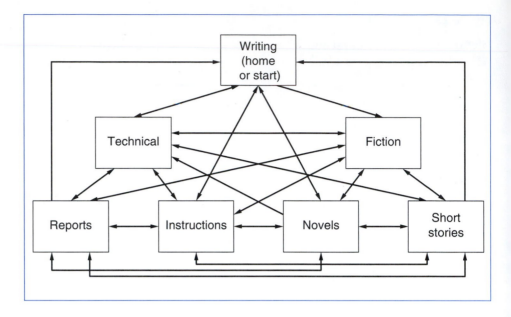

Figure 13.4

Linked Section Report

Clicking on a link causes that page to open on the screen.

Tips for Making a Good Website

The following report contains some ideas that I feel are important to consider if one is going to produce a good website. These are just some of the ideas I have learned through my Technical Writing class. Remember, you are free to do whatever you want, but the following ideas may be helpful in achieving your goal.

Go to

- **Planning Is Key**-Planning is essential in order for you to produce a good..........
- **Establishing a Purpose for Your Website**-Before you embark on this voyage..........
- **Implementing Links in Your Website**-In order to make your site easy to use..........
- **Handling Visuals and Text**-These seem to spice up your website, but..........
- **Interactive Websites**-People do not just want to see a webpage, they want to..........

Return to

- **Matt Maertens Homepage**
- **Technical Writing Index Page**
- **English Department Student Projects Page**

Let me know what you think! e-mail me **maertensm@uwstout.edu**

Figure 13.5

Anchor Link
Strategy

How to Create a Great Webpage

HTML stands for Hypertext Markup Language. This language allows the computer to read your document and makes it appear in Web format. Once you get the hang of this computer language, the possibilities are endless. However, good quality writing and construction of your website could determine how many people will stop and stay long enough to check out your site. The following areas are important to consider in order to create a great website:

Before You Begin|Here Goes|Introduction|Chunks and Heads|Graphics

Links to anchor.
Clicking on the link
causes that text to
appear at the top of
the screen.

BEFORE YOU BEGIN
Before you start typing anything into the computer, sit down and plan out what you are going to do. Designing a website is 90% planning and 10% actual work. Figure out the purpose of your website. Things can get out of control fast, so knowing your boundaries is essential.

Click on "Top" to
jump back to the first
paragraph.

It is incredibly helpful to draw a map of your website. When creating links, things can get confusing. Good links enable you to get in, through, and out of the document easily. Nobody likes to get trapped in a webpage with no way out. Having a map of links in front of you can make linking your pages easier. You will save yourself a lot of time and hassle, if you work out what you are going to do before you do it. **Top**

Anchor

HERE GOES!
Now that you know what you are going to do, it is time to acquire the things necessary to get it done. You will need a computer, a simple word processing program, a list of HTML commands and what they do, and a browser. Type your information in a simple word processor, such as Simple Text, or save it in your present word processing program as a text file. You will need knowledge of HTML commands to convert your page into a language the computer can read. A browser allows you to open up your document on the Web to see what it looks like. **Top**

Anchor

Introduction
An introduction is an important part of the website. The introduction contains a lot of necessary information, such as the reason for the site and why it would be of interest to the viewer. In the introduction, the purpose of the website should be clearly stated. If viewers don't know what the site is about or how it will benefit them, they are not going to stick around. **Top**

Anchor

Chunks and Heads
People like information presented in small chunks. It is easier for them to digest. It also gives you a better chance of holding their attention. People don't have the desire or patience to read through a lot of unbroken text. It is also a good idea to use heads. They are also useful in breaking up the page. Heads inform the reader about each section's content, giving them the option of whether they want to read it. **Top**

Anchor

GRAPHICS
When incorporating graphics into your webpage, there are several things to consider. Determine how long it takes the graphic to load. If it takes too long, try to make the graphic smaller. If that doesn't help, you should cut it out. People don't like to wait very long for a graphic to

(continued)

Figure 13.5

(continued)

load. Another thing to consider about graphics is if they will appear when the document is load-ing or if the viewer will have to click on a graphics icon to bring up the picture. Either one is ac-ceptable. It is up to the designer which way to go. The Web can provide additional options for incorporating a graphic into you site.   <u>**Top**</u>

<u>My Set of Instructions\How to Reply in Eudora</u>

<u>Nikki's Index\Technical Writing Home Page</u>

Reader Freedom

Reader freedom is the degree to which the reader of the website can easily se-lect the order in which he or she will read sections of a document. Whereas hierarchy imposes control on the reader's freedom, Web structure provides freedom. The Web author must find a way to combine the two. Figures 13.6 to 13.8 demonstrate how authors can use the two concepts to affect the way readers view the document or site.

In Figure 13.6, the reader starts at the home page and can progress only to one of the two level 2 nodes, and then to one of the level 3 nodes. To arrive at *reports,* the reader must click to *technical* and then to *reports.* To get to *instruc-tions* from *reports,* the reader must first go back to *technical.* To get to *novels* from *reports* the reader must follow the path back to *writing* and then click forward to *novels.*

Figures 13.7 and 13.8 show progressively less control by the author and more control for the reader. In the hierarchy shown in Figure 13.7, the reader can move directly from *reports* to *instructions* without clicking back to *technical.* In the hierarchy shown in Figure 13.8, the reader can move to any document from any other document. A reader could click from *reports* to *fiction* and then to *instructions.*

Figure 13.6

Little Reader Freedom

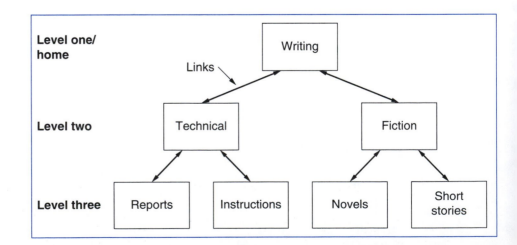

Figure 13.7

Moderate Freedom

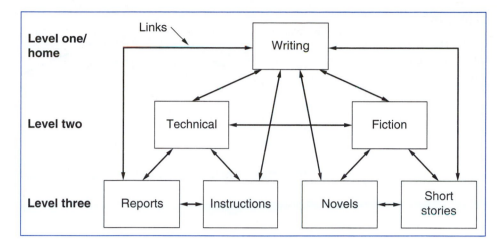

Figure 13.8

Absolute Freedom

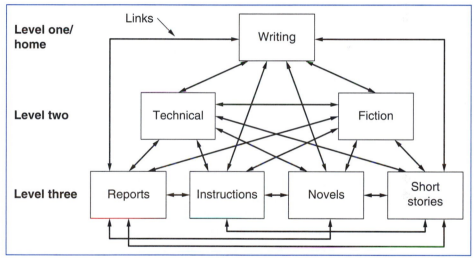

Figure 13.9

Controlled Hierarchy

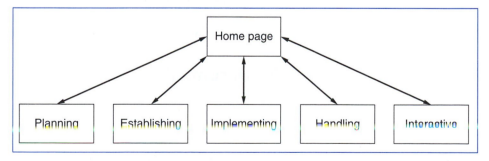

The Maertens model in Figure 13.4 appears to have a tightly controlled hierarchy, as shown in Figure 13.9. But if he would supply links between each section, then the document would have a hierarchy of little control and great reader freedom, as in Figure 13.10 (p. 328).

Figure 13.10

Web Structure
Allowing Reader
Freedom

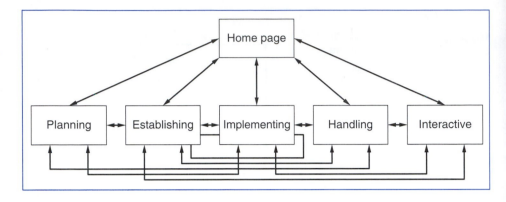

Guidelines for Working with Web Structure

The two possibilities of rigid hierarchical organization and loose Web organization mean that the writer must choose to insert enough links to be flexible,
but not so many that the reader is overwhelmed with choices. William Horton
suggests that an effective strategy is to "layer" documents, "designing them so
that they can serve different users for different purposes, each user getting the
information needed for the task at hand" (178).

Horton suggests that each level of the hierarchy is a layer and that each layer
provides more detailed information. Use this principle in these ways:

▶ In higher layers, put information that everyone needs. To learn about writing, everyone needs the definitions on the homepage, but only a few readers need the concepts explained on the instructions page.
▶ Control reader's paths. Figure 13.6, for instance, indicates that readers have
complete control in either of the two major categories, technical writing and
fiction, but that accessing the material in the other category will require the
reader to "start over." The author has assumed that readers in fiction will
want to know more about that category, so it should be easy to get around
in it. However, it is less likely that they would want to compare items in fiction with the items in technical writing, so the path to it is restricted.

Planning a Website or Web Document

In the planning stage, consider these four aspects (based on December; Horton; Hunt; Wilkinson): Decide your goal, analyze your audience, evaluate the
questions the audience will ask, create a flow chart, and create a template.

Decide Your Goal

Your site or document should have a "mission statement"—for instance, "To explain the purposes and services of the campus antique auto club." Make this

statement as narrow and specific as possible. It will help you with the many other decisions that you will have to make.

Analyze Your Audience

Ask the standard questions: Who is the audience? How much do they know? What is their level of expertise? (See Chapter 2 for more information on audience analysis.)

Do not answer: Anyone who comes onto the Web. Such a broad answer will not allow you to make decisions. Narrowly focus these answers—they are people who are interested in antique cars, the university in general, or clubs in particular. They know a little or a lot; they will have experience or not. A website aimed at an audience of people who have restored antique cars is quite different from a website aimed at students who have some interest in old cars and wish to join a university group.

One helpful way to think about audience is to think about the site or document as a stage (Coney and Steehouder). Your audience member is a member of a cast and thus has a "role" in the site. For instance, the audience can assume the role of "antique car experts" or "students who want to join a club." Most audience members find it easy to adopt such a role if it is clear to them. If you understand the role you want your audience members to play, you will be able to make better decisions about how to present your site or document to them.

Evaluate the Questions the Audience Will Ask

Speculate on general questions: What is the purpose of the club? When does the club meet? What are the club activities? What antique cars does the club have? Can I learn about where to purchase an antique car? Can I learn how to restore an antique car? What are the bylaws of the club?

You can decide which of these questions you want to, or should, answer. Your audience decisions will help you. For instance, if you feel that only car buffs will look at the site, then there is no need to provide rudimentary information about cars.

Create a Flow Chart of Your Site

A flow chart indicates the site's or document's nodes, their hierarchy, and the degree of freedom that the reader will have. The flow chart gives a visual map of your site, allowing you to control the creation of the links. It shows you how much control you will allow your audience, and because it is a blueprint of the site, it gives you a method to check whether you have inserted all your links. In the flow chart in Figure 13.11, the arrows indicate a link from one file to another, and the curved lines indicate escape links back to the homepage from each of the files. Notice that this structure is much like the one shown in Figure 13.7 (p. 327). The reader has total control inside each section of the hierarchy, but must return at least to the second level in order to transfer to another category.

Follow these guidelines for degrees of control (Figure 13.11 illustrates all of these points):

▶ Provide a link from the homepage to all nodes.
▶ Provide a link from the node to all subparts of the node.
▶ Provide a link between nodes.
▶ Provide shortcut links from higher levels to key data at lower levels. Note the two lines that run from the homepage to the items marked "K."
▶ Provide shortcut or "escape" links from all levels back to the home page. Escape links are shown by the lighter curved lines in Figure 13.11.
▶ Provide enough links for multiple paths. Unless you have a compelling reason to prevent random navigation, readers should be able to read the site and the documents in it in any order. Use links to create paths that will allow this type of reader freedom.

Create a Template

Your site or document must have a consistent visual logic. Select a background color, a font (but be aware that individual users can change the font that appears

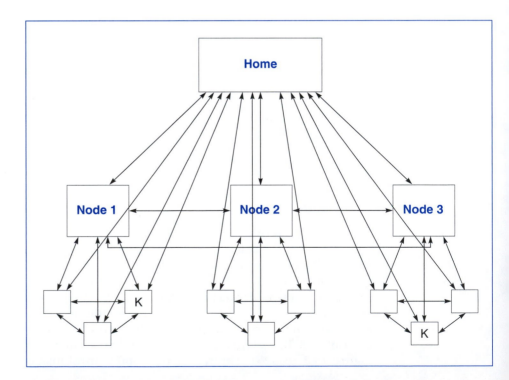

Figure 13.11

Full Flow Chart
of a Website

on their screen), and a consistent spot to place titles, introductions, lists, return links, and e-mail links. The information on each screen will change, but the way it is presented will remain the same.

The easiest way to create a template is to make a sample page and keep a record of each of your decisions. As you make the template, include all of these items:

▶ Title
▶ Introductory text
▶ List of nodes (actually a table of contents for the document or site)
▶ Shortcut links
▶ Escape links
▶ Color/font/size of heading
▶ Color/design of background
▶ Color/font/size of text
▶ Placement of blocks of similar types of text

Consider, for instance, Figures 13.12 and 13.13 (p. 332), pages from the Geology node of the Arches National Park website. Each page looks the same. Similar elements appear in the same position, color, and font. Notice all of the following:

1. Title of the section—flush left, sans serif font, rule beneath

2. Links to other nodes—left-hand column, serif font

3. Title of the page—flush left in right-hand column, sans serif font, black

4. Heads—flush left in right-hand column, serif font, color

5. Text—block paragraphs, double-space between paragraphs, serif font

6. Shortcut links—2 lines below text, serif font

7. Privacy/author information—3 lines at bottom left, serif font

8. Two-column format

9. Visually separated elements. Note that white space and color bands clearly separate the various elements of the page. Each element is said to be "grouped," a visually effective method of helping the reader grasp the information.

Drafting

Creating a website or document takes several drafts. Creating clear content, effective structures for reader freedom, and accessible pages seldom happens in one draft. As you create your site or document, orient your reader, write in a scannable style, establish credibility, and use visuals effectively (Nielson; Spyridakis; Williams).

Figure 13.12

Arches National
Park Geology
Introduction
Home Page

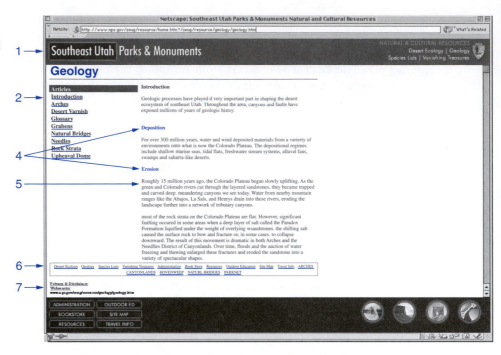

See webpage at <www.nps.gov/seug/resource/home.html>.

Figure 13.13

Arches National
Park Desert
Varnish Web
Page

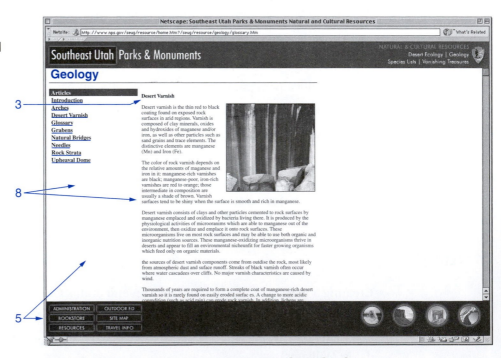

See webpage at <www.nps.gov/seug/resource/home.html>.

Orient Your Reader

As readers surf through websites and documents, they lose track of where they are. This disorientation causes confusion and diminishes the ability to draw meaning from a page. Three key methods to orient the reader are shown in Figures 13.12 and 13.13:

▶ Provide an informative title at the top of the page.
▶ Provide an introductory sentence that either announces or defines the topic under discussion.
▶ Repeat key information at consistent spots on the page.

Write in a Scannable Style

A scannable style is one that presents information by highlighting key terms and concepts and placing them first in any sequence. Because reading from a screen is a physically difficult task, a scannable style often makes text easier to read (Nielson). Several strategies will help you make the page scannable:

▶ Use chunks. Create smaller chunks; use more short paragraphs. Notice the short paragraphs in Figures 13.12 and 13.13.
▶ Use headings. Place heads throughout the text to help readers grasp the overall structure. Heads function as an "in-text" outline, which helps readers orient themselves (see Figure 13.12).
▶ Use bulleted lists. Notice the "Go To" list in Figure 13.4. It is much easier to read than this linear version: Go to: Planning Is Key; Establishing a Purpose for Your Website; Implementing Links in Your Website; Handling Visuals and Text; Interactive Websites.
▶ Add abstracts after the link. If the word in the link is not self-evident, add a brief description. For instance,

Instructions: The steps you need to build a better telescope.

▶ Make a link-title connection. Use a word in the link that repeats a word in the title of the section where the reader will arrive. If the link is *Planning,* the title of the page that appears should include *Planning.*
▶ Use one idea per paragraph. Notice the two paragraphs under "Erosion"in Figure 13.12. Each deals with only one idea: the creation of canyons and the creation of spectacular shapes.
▶ Use the inverted pyramid style. The term *inverted pyramid* is a synonym for top-down. Put the key idea of the paragraph first, and then give supporting detail. Notice in the "concise" example below that the first sentence gives the main idea ("prepares students") and the rest of the paragraph supplies details on how the program actually prepares students.
▶ Use introductions that tell the purpose of the screen, especially if it is a node screen. See Figure 13.4 or Figure 13.12.

▶ Write concisely. Because it is difficult to read text on a screen, excess wording simply compounds the problem. Jakob Nielson suggests that writing can be *promotional* (overblown, to be avoided), *concise,* or *scannable.* Writers should try for the latter two.

PROMOTIONAL

This fun program leads young scholars into the exciting world of technical communication. Scholars have the unprecedented opportunity to study multimedia in all the state-of-the-art software and hardware. They can also, in a particularly innovative aspect, develop an Applied Field that will allow them to interact with specialists in other areas. And they can, as students in many majors cannot, take a course that readies them to enter the challenging world of work.

CONCISE

This program prepares students for the exciting world of Technical Communication. The program features state-of-the-art multimedia software and hardware; Applied Fields that develop expertise in a specialty; and a professional development course which facilitates beginning a career.

SCANNABLE

Three key features of the new technical communication programs are

- State-of-the-art multimedia hardware and software.
- Applied Fields to develop specialized expertise.
- Professional development to facilitate finding a job.

Establish Credibility

Because anyone can post anything on a website, readers look for, and are reassured by, some proof of credibility (Coney; Nielson; Spyridakis). Some features that enhance a site's credibility are

▶ Information about the author, including name, e-mail address, and organizational affiliation. A website on effective dieting is more credible if the author is a person with an e-mail address who is an associate professor in a dietetics department at a university.

▶ The date that the site was posted or updated.

▶ A statement about privacy. Many websites include a "Privacy" link to a file that explains their policy.

▶ No typos or spelling or grammar mistakes. These mistakes seem even more glaring on the Web. If all the world can see your material, shouldn't you care enough to get it correct?

▸ Links to other sites. These links show that the authors know the field. Be careful of such links because they lead viewers off your site. Careful Web authors often add a statement near an "off-site" link, telling the readers that they are about to leave the site and reminding them to use their browser's Back button or Go menu to return. Check such links regularly to make sure that they work.

▸ High-quality graphics. The ability to present quality graphics shows that the writer knows how to use software and hardware, as well as that the writer wants to make a clear contact with the reader. The clear visual in Figure 13.13 makes the reader confident that the author has a high level of expertise.

Use Visuals Effectively

Visual aids in a Web document perform all the same functions as in a paper document (see Chapter 7). Visuals summarize data, allow readers to explore data, provide a different conceptual entry point into a report, and engage expectations. Used well, visuals enhance webpages; but used poorly, the visuals are annoying. Visual aids must be sized correctly and interact with the text (Horton).

Correct size can be electronic size or physical size or both. Electronic size is the number of kilobytes (K) that the visual uses. The larger the K, the slower the image loads. A site with several large-K (above 150K) color images will take a long time to load on a 56K modem.

Follow these guidelines to control the electronic size of your visuals:

▸ Use fewer visuals.
▸ Use a software program to compress visuals to reduce their electronic size.
▸ Use a "thumbnail" linked to a larger version (see Figure 13.14).

Figure 13.14

Use of
Thumbnails

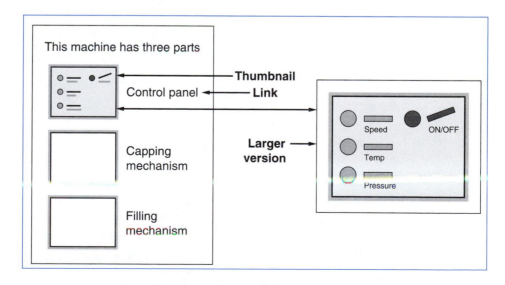

Figure 13.15

Effective Sizes of
Visuals on Screen

Physical size is related to the number of pixels occupying the screen. The average screen is 15 inches with 480 × 640 pixels. If the visual aid occupies most of that pixel space, there is nothing left for text (see Figure 13.15).

Learn to manipulate image size. All programs that help you produce images (such as Photoshop) allow you to alter dimensions. Often, if you make an image smaller, you make it more illegible. If you capture the image of a screen, for instance, and then reduce it to one-fourth its original size, the text on the screen will probably be illegible. To fix this problem, Web authors *resample* the image, a process that restores legibility. Consult the help menu or manuals of programs like Photoshop or Front Page to learn how to resample.

Notice Figures 1 and 2 in "Focus on HTML" (pp. 353–354). The image of the cabin, as scanned in, originally was approximately four times as large as it is on screen. Notice in Figure 1 that the writer has sized the image by inserting *width* and *height* commands into the code.

Follow these guidelines for the best presentation of physical size:

▶ If the text and visual complement each other (as in a step in a set of instructions), use no more than half the screen width for the visual.
▶ If you reduce a visual in size, resample it.
▶ Use visuals on screen to clarify your message for your readers.

Testing

· · · · · ·

Your Web document must be usable. Readers must be able to navigate the site easily and access the information they need. In order to ensure easy navigation and access, you must test your document. Either perform the test yourself or have another person do it. Testing consists of checking for basic editing, audience effectiveness, consistency, navigation, the electronic environment, and clarity.

Basic Editing

Web documents include large amounts of text that must be presented with the same exactness as text in a paper document. If the editing details are handled effectively, the credibility of the site is increased.

Check your site for stylistic elements.

▶ Spelling
▶ Fragments, run-ons, comma splices
▶ Overuse of *there are, this is*
▶ Weak pronoun reference
▶ Scannable presentation, including use of the inverted pyramid style

Audience Effectiveness

Your site must put the audience into a role. Checking for audience role is highly subjective.

▶ Is it apparent what role the audience should assume, either from direct comments or from clear implications in the way the audience is addressed?
▶ Do all parts of the site help the audience assume that role?

Consistency

Readers find sites easier to navigate and access if all items are handled consistently (Pearrow). The document has many text features that should be repeated consistently in order to establish visual logic. These features include font, font size, color, placement on the screen, and treatment (bold, italics, all caps). In addition, a document has many visual features that must remain consistent in order to establish the visual logic. These features include size and placement on the screen.

Check these textual items for font, font size, color, placement on the screen, and treatment:

▶ Titles
▶ Headings
▶ Captions for figures
▶ Body text
▶ Lists
▶ Links

Check these visual items for size and placement on the screen:

▶ Clip art
▶ Photographs
▶ Tables and graphs
▶ Screen captures

Ethics and Websites

For websites, as for all sources of information, technical communicators must act ethically. As with all documents, it is unethical to manipulate data, use deliberately misleading or ambiguous language, exaggerate claims, or conceal information that your users need in order to make good decisions (Summers and Summers 127).

However, communicators must also act ethically in regard to all the elements of websites: their design, code, graphics, and text. A major issue is plagiarism, using someone else's work without permission or acknowledgment. To state it succinctly, all aspects of websites are protected by copyright law. You are not permitted to copy anything from a website and use it on your website unless you either have permission or clearly acknowledge the source. As one expert says, "A work is copyrighted by the author at the moment of creation." A copyright notice or symbol, or "official" registration of a copyright, is not necessary—the rights automatically exist (Pfaff-Harris).

Another expert says this: "Simply put, by the time you see a webpage posted on the Web or a Usenet message posted in a newsgroup, it . . . is protected by copyright. You are not permitted to copy it, even if there is no copyright notice on the page or message" (Bunday). If you do post without permission or acknowledgment, you are not only violating the copyright law, "you are acting unethically because you are benefiting from someone else's work, without permission or due compensation. You are taking something that does not belong to you" (Murtaugh).

The best way for a communicator to ensure that she or he is using resources ethically is simply to ask the administrator of a website or the author of an e-mail message for permission to use the material and to give credit where credit is due. Often an e-mail request is all that is needed.

From another point of view, creators of websites can encourage ethical action by placing copyright notices on their sites, even though such notice is not required in order to be protected by the copyright law. Such a notice includes a copyright symbol, the years of creation and last modification, the name of the copyright holder, and the phrase "All rights reserved" (Murtaugh).

Another, somewhat different ethical strategy is to help viewers situate themselves. Summers and Summers suggest that you should include information on the creator and purpose of the site, and on the date it was last updated. This type of information will "help users evaluate the quality of the content you have provided" (Summers and Summers 130).

For a helpful overview of websites and ethical issues, see the article by Pfaff-Harris.

Works Cited

Bunday, Karl M. "Building Better Web Sites." 2000. Learn in Freedom. 24 April 2004. <http://learninfreedom.org/technical_notes.html>.

Murtaugh, Tim. "Pointing and Laughing: FAQs." 4 April 2002. New Press: Pirated-Sites.com. 24 April 2004. <www.pirated-sites.com/faqs/>.

Pfaff-Harris, Kristina. "Copyright Issues on the Web." The Internet TESL
    Journal II.10 (October 1996). 24 April 2004. <http://iteslj.org/Articles/
    Harris-Copyright.html>.
Summers, Kathryn, and Michael Summers. *Creating Websites That Work*. Boston:
    Houghton Mifflin, 2005.

Navigation

To check for navigation is to investigate whether all the links work and whether
the path through the material makes the material accessible. To check whether
all the links work is easy. Simply try each link. Make note of any that are "broken," that is, do not lead to any document. To determine if the path through the
material makes the material accessible is more subjective. An effective way to
investigate accessibility is to ask questions of yourself or your tester (Pearrow):

▶ Starting at the homepage, can you find information X quickly?
▶ From any point in the site or document, can you easily return to the home-
   page or top?
▶ Does the homepage give the reader an overview of the purpose and con-
   tents of the site or document?
▶ Does the title of any page repeat the wording of the link that led to it?
▶ At any time is the user annoyed? For instance, does a visual take a long
   time to open?

The Electronic Environment

The electronic environment of your site or document is the way in which it interacts with the reader's viewing equipment—modem, computer, and browser
software. The basic guideline is that all your material should appear on screen
quickly and be designed as you intended.

Check these electronic aspects:

▶ How long does the site or document take to load? Ten seconds appears to be
   the limit readers will wait before they get annoyed and click away to another site. Answer this question by using different access methods. Load the
   site over a 56K modem and over a T1 line. The differences are often large.
   Some programs (such as Front Page) provide a menu item that gives this
   information.
▶ Does the browser used affect whether the features appear? The two major
   browsers are Internet Explorer and Netscape Communicator. In most cases,
   but not all, a site will appear exactly the same regardless of which browser
   the viewer uses. For various program coding reasons, more sophisticated
   elements such as tables, frames, and videos sometimes work in one browser
   but not in the other. Checking a site with both browsers will ensure that

readers see the items that you intended in the manner in which you intended.

If you must include features that only one browser supports, alert viewers to that effect ("Use Internet Explorer in order to view all the features of this site."). If you are using an application that viewers can download a version of for free (e.g., Shockwave), include a link to that site.

Clarity

The site or document must appear clearly on the screen. A viewer must be able to read all the elements.

To check for clarity, answer the following questions:

▶ Do all the visual aids appear? If not, edit the Web document to make sure that they do.

▶ Can all the text be read? Sometimes, inexperienced Web authors use color combinations that make text hard to read (black text on a blue background, yellow text on a white background). Revise the color (see "Focus on Color," pp. 166–173).

▶ Are the visual aids clear? If visual aids are fuzzy or illegible, edit them in a software program (such as Photoshop) that allows you to resample the image.

Worksheet for Planning a Website or Document

☐ Identify the audience and the role they will play.

☐ Identify questions the audience will have about the content.

☐ Identify probable nodes for the site or document.

☐ Create a flow chart that indicates hierarchy and paths.

☐ Plan paths that give readers the freedom they need.

☐ Plan features of site (both screen and text items) that will facilitate the way readers find the information that they need.

☐ Create a screen template that groups similar information into distinct locations, including placement of visual aids.

☐ Choose font and color for heads and text.

☐ Determine which visual aids you need to convey your information.

☐ Choose a neutral background (light blues are good).

☐ Write text in manageable chunks.

Worksheet for Evaluating a Website

Homepage
- Does the title make the content clear?
- Does the introduction tell you the purpose of the site?
- Can the homepage fit on one screen only?
- Did the site load quickly?

Navigation / Links
- Does every link work?
- Does every link have the same wording as the title of the page it links to?
- Are links coded or designed so that similar links look the same?
- Are all links of the same type always in the same location on each page?
- Do readers have sufficient freedom to access sections?
- Is the level of freedom too restrictive or annoying?
- Did you get lost navigating the site? If so, where?

Style
- Does the word choice indicate an exact awareness of the audience's knowledge and expectations?
- Are all the words spelled correctly?
- Is the grammar correct?

Text/Screen
- Is the background the same color in each document?
- Is the title of the page in the largest type on that page?
- Are all similar objects placed in the same place and do they have the same size?
- Are items on the page "clumped" so that the most important are together, set in bigger type and placed higher up?
- Can an audience easily read each document or is the font too small or too big or too busy?
- Does every document have the same font, including size?
- Is the text always aligned left?
- Is the format of each page consistent with every other page in the site?

Information
- Can readers easily figure out your plan of organization?
- Is the plan easy to find in the visual design you present?
- Does each section use appropriately convincing examples to inform the reader?
- Does each section start with a clear introduction that lists both the purpose and the parts of that section?
- Does each section contain a list of references with title of page, URL, and date visited?

(continued)

(continued)

> **Visual Aid Design**
> - Does each image load?
> - Does the picture or table add a dimension of detail or interest not available in words?
> - Is each visual aid in the same relative place on the page?
> - Is the visual aid one that really helps a reader (and is not a waste of space and reader time)?
> - Is there a clear cross-reference to each visual aid from the text?
> - Does each visual aid have a caption?
> - Is each visual aid roughly the same size?

Examples

Here are the report sections from the homepage presented in Figure 13.4, another informational Web report, and two sets of instructions. Note that Example 13.3 contains links to many other sections that are lower in the report's hierarchy. Those sections are not reproduced here.

Example 13.1

Report Sections That Can Be Linked to Homepage

> PLANNING IS THE KEY TO SUCCESS
>
> Much like any other project that you may tackle, creating a good website involves a little bit of planning. Many things have to be taken into account when you begin writing your own webpage.
>   The following should be considered:
>
> 1. Appearance
> 2. Actual size
> 3. Links
> 4. Visuals
> 5. Audience
> 6. Design
>
> Go to other sections of this report:
>
> *Establishing a Purpose*—Before you embark on this voyage . . .
> *Implementing Links*—In order to make your site easy to use . . .
> *Handling Visuals and Text*—These seem to spice up your website, but . . .
> *Interactive Websites*—People do not just want to see a webpage, they want to . . .

List of links includes all sections, except planning.

ESTABLISHING A PURPOSE FOR YOUR WEBSITE

Before you decide what you are going to put up on your webpage, it is important that you define what purpose your website intends to serve. A lot of people have a webpage just for the sake of having one. Their homepage is just kind of there for others to look at. There is no way to interact with the webpage. Webpages should have the purpose clearly stated on them, so the reader knows what they are getting into.

Go to other sections of this report:

Planning Is the Key—Planning is essential in order for you to . . .
Implementing Links—In order to make your site easy to use . . .
Handling Visuals and Text—These seem to spice up your website, but . . .
Interactive Websites—People do not just want to see a webpage, they want to . . .

Wording of links repeats keywords in titles of other sections.

IMPLEMENTING LINKS IN YOUR WEBPAGE

People need to be able to navigate their way around your website once they are in it. They should be able to move back, forward, and to other sites if they want to. The more links a webpage has, the more freedom the reader has to pick and choose whatever it is he or she wants to read.

One of the biggest complaints I've heard from Web users is the fact that they often feel "trapped" inside of webpages, with no way out except to use the "Back" key. However, one must decide carefully the number of links to include, and which ones to exclude as well.

Go to other sections of this report:

Planning Is the Key—Planning is essential in order for you to . . .
Establishing a Purpose—Before you embark on this voyage . . .
Handling Visuals and Text—These seem to spice up your website, but . . .
Interactive Websites—People do not just want to see a webpage, they want to . . .

HANDLING VISUALS AND TEXT

Much like the traditional medium of written communication, webpages, too, benefit from visual aids. Visuals tend to catch the eye of the reader and take the place of text as well. However, when placing visuals on the Web, they must be planned just as carefully as they are when they are placed on paper. The same technical writing rules apply to visuals on the Web.

(continued)

Example 13.1

(continued)

Text needs to be thought about, too. Especially on the Web, short "chunks" are necessary to keep the reader's interest. Because a computer screen seems to be smaller than what we would actually see on a regular sheet of paper, readers seemed to be turned off by large blocks of text on the Web. They are forced to keep scrolling down in order to get everything. Also, the bolding of heads and increasing their font sizes makes these items dominant over all other text, as they should be.

Go to other sections of this report:

Planning Is the Key—Planning is essential in order for you to . . .
Establishing a Purpose—Before you embark on this voyage . . .
Implementing Links—In order to make your site easy to use . . .
Interactive Websites—People do not just want to see a webpage, they want to . . .

INTERACTIVE WEBSITES

It is nice to have a webpage that allows readers to interact with the webpage somehow. Some examples of this may include a webpage that allows readers to e-mail the authors with questions or comments. Including your e-mail address on your webpage allows others to give you some input and constructive criticism about your website.

Another way a webpage could be interactive is through links that allow you to order something or inquire about something. Many companies today have their catalogs on the Web, and people can order things directly from the Internet. It is amazing what kind of feedback you can receive, or how your sales can increase, if you make your website interactive.

Go to other sections of this report:

Planning Is the Key—Planning is essential in order for you to . . .
Establishing a Purpose—Before you embark on this voyage . . .
Implementing Links—In order to make your site easy to use . . .
Handling Visuals and Text—These seem to spice up your website, but . . .

Example 13.2

Informational
Web Report,
Using Anchor
Link Strategy

Links that jump
reader to appropriate
section.

IMRD: RESEARCH REPORT

Introduction / Method / Results / Discussion

Introduction

With the advanced use of the electronic job-search, it is becoming difficult to ignore the increasingly important role of the on-line résumé. Since computers are becoming increasingly user-friendly, even those with relatively little computer experience are becoming familiar with this technique. One eminent problem, however, is the question of how users can get the information they need to the end they desire. This page will focus on how to translate a current résumé into one that is readily accessible for use in this type of a search. We aim to overcome the problem that there are limitless ways for an employer to request such résumés: as an attachment, as text format, as HTML, and still others limiting the characters per line to 80. A solution is a résumé done in HTML, since anyone with Web access can get it open — in most cases, if the reader is getting your e-mail, he or she also has a browser installed. The only problem that remains is that of the fonts you choose — the user may have different default fonts set. Add a line "Best read in xxx font" — is there another way?   **Top**

"Top" links return reader to links at beginning of document.

Method

I went on to create an electronic résumé, to be either included or posted on the Web. To do this, I saved my résumé, which was previously in Word format, as HTML. I inserted horizontal lines, as well as targets and anchors. Then I reopened the copy of my résumé that had been saved in Word and saved it this time in .txt format. I then changed the page layout from a contemporary design to a more standardized paragraph form.   **Top**

Results

Attempt	Result
Word to HTML	When I changed my résumé from Word to HTML format, the result was that I lost much of the formatting I had done, because HTML does not support it. To force my résumé to look essentially the same, I inserted tables. Moreover, the targets and anchors were inserted to help eliminate the problem of not being able to view all of the information on the screen, as you would if the résumé were in front of you. The horizontal lines further helped to separate each section of information for the reader's understanding.
Word to text	When I changed my résumé from Word to .txt format, the result was that I lost much of the formatting I had done. In this case, I had even fewer options and ended up deciding on a different layout for my résumé altogether, simplifying it greatly. This, however, includes the issue of having 80 characters per line, as it was easy to change the page layout so that, despite any line specification, it would match. **Top**

(continued)

Example 13.2

(continued)

Discussion

Because specifications vary from employer to employer, there are limitless ways of sending a résumé electronically. Hence, there is no perfect or universal résumé—a relatively frustrating result of this process. However, it is useful at this time to have a copy of your résumé in multiple formats so that they are readily available despite the request—particularly in .txt format, as this can be attached, sent as an e-mail, and easily translated to HTML. A solution to this problem is, just as a paper résumé has standards, to have standards for the electronic résumé. This will take time and the acceptance of the résumé in this medium. The best way to be safe right now is to save your résumé as text.

Top

Example 13.3

Instructions on a
Webpage

DEVELOPING ELECTRONIC RÉSUMÉS

Developing an electronic résumé is a useful technique, particularly when doing electronic job searches. It gets your credentials across to the reader, as well as allowing the freedom to move around efficiently. These instructions will lead you step by step through the process of creating an electronic résumé.

RECOMMENDATIONS

Links lead to more
detailed discussions.

- *It is recommended that you create your résumé before beginning this process.* Typically, résumés require a great deal of information, and the beginner will find it simpler to understand the contents of these instructions without being concerned with this information. Traditional résumé sections

ASSUMPTIONS

These instructions assume that the user

- Has a basic knowledge of how to operate a personal computer.
- Has a basic knowledge of how to operate basic features of a word processor program.
- Has not previously created an HTML-formatted résumé.
- Has a basic concept of a résumé.

Operational assumptions:

- The user is currently operating in Microsoft Word.
- The user has previously saved a copy of the résumé in Word format.

DIRECTIONS

1. Open your résumé as you would normally.
    - Check to assure that you have it saved in Word format.

Links lead to more
detailed discussions.

2. Save your résumé as an HTML document.
3. As a result of the previous step, *your document will lose all formatting* that was in place in the Word document. To get the results of formatting in your new HTML document,
   a. Insert a table.
   b. Insert horizontal lines to visually separate section bodies of the résumé.
4. Insert a horizontal menu between your name and the first section of your résumé. This menu should be horizontal, with the main sections of your résumé and index items. See example.
5. Create internal hyperlinks from each index item in the menu to the respective section body within the résumé.
   - This will let the user view the sections of your résumé that they wish to see.
6. Other formatting to enhance your résumé:
   a. Horizontal menus at the end of each résumé section (example)
      - Add internal hyperlinks for usability.
7. Add a Letter of Application.
   a. Save as a separate HTML document.
   b. Insert a hyperlink at the bottom of each page
      - So that the reader can move back and forth between your Résumé and Letter of Application.
8. Open a copy of a Web browser. Open your document and view it to assure that it is functioning properly.
9. *Your resume is now ready!* You can now
   - Send it as an attachment to anyone with a browser that can view it.
   - Contact your local Web administ rator to post it on your website.
   - Do an electronic job search and post your résumé on the employers' sites.

Example 13.4

Instruction Set

This short document uses the "scroll" strategy.

DOWNLOADING AND SAVING IMAGES OFF THE WEB

1. Open an Internet browser program like Microsoft Internet Explorer or Netscape Navigator, and "surf" the Internet until you find an image that you would like to download.
2. Using the mouse, move the pointer arrow until it is on the image that you wish to download, and click on the right mouse button. A pop-up menu will appear next to the image you have selected.

(continued)

Example 13.4

(continued)

The pop-up menu that appears will let you choose several options:

Save Picture lets you save the picture as a .gif file to your disk or the computer's hard drive.

Set as Wallpaper lets you use the picture as the "wallpaper" or background on your computer's desktop screen.

Copy lets you copy the picture into another program or document (e.g., Microsoft Word) using the program's edit and paste features.

Add to Favorites lets you bookmark the webpage that the image is on into the "favorites" section of your Internet browser software.

Properties will tell you the name of the picture file, the address of the webpage on which it is found, the type of file the picture is (e.g., .gif), and the size of the picture file.

3. Click on Save Picture. A Save Picture dialogue window will open that looks like this (Figure 1):

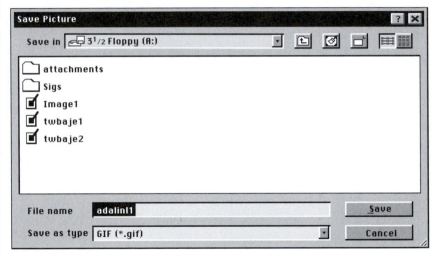

Figure 1

Highlighted terms duplicate wording and look of terms in Figure 1.

Save in allows you to choose the disk drive that you want the picture to be saved on.

File name allows you to type in the name you wish the picture to be saved as.

Save as type lets you choose what type of file you wish the picture to be saved as. Image or picture files are usually saved as .gif files or .jpg files. Notice that the .gif file type popped up in the "Save as type" box as a default.

4. Fill in the appropriate information in the "Save in," "File name," and "Save as type" dialogue boxes and then click Save.

Crop an Image / Resize an Image / Change Image Resolution

Globalization/Localization and Websites

As Internet use grows worldwide, so does the number of non-English-speakers who use it. According to International Data Corporation (IDC), by 2005 non-English speakers will represent 70 percent of all Web users ("Globalization"). In order to reach a wider audience and/or client base, websites written in English must be able to be translated for use in non-English-speaking countries. Preparing a website for localization in other countries, languages, and cultures presents certain challenges.

One of the most common problems in website translation is the expansion and contraction of text. European languages often take up as much as 30 percent more space than English on a webpage. For example, Italian, Czech, German, and Greek will all expand to over 100 percent of English text, whereas Arabic, Hebrew, and Hindi will contract to less than 100 percent (Hoft). The languages that rely on characters (e.g., Japanese, Chinese, Korean—referred to as "double-byte" languages) can take up more or less space. It is difficult to predict exactly how a double-byte translation will affect your screen design until the actual translation is in place. This in turn interferes with spacing in the text and effective positioning of visual aids. Keep this contraction and expansion in mind when designing your screens, composing content, setting table properties, and creating graphics. Give extra space around items such as control buttons, menu trees, and dialog boxes (Macromedia).

When creating tables on your websites, base the properties on the longest translation that will be used on your site. If you use percentages rather than pixels, you can set your tables to 100 percent width. This will allow the table and the text to expand according to the user's browser—the text and the graphics will appear the same to users viewing it in French as it will to those using the English-language version (Macromedia).

The symbols you choose are critical also. Avoid culturally specific symbols. Graphics with multiple meanings, pictures of body parts, religious symbols (e.g., a cross), or even symbols such as a stop sign may cause confusion, misunderstanding, and may even be construed as offensive ("Globalization"). Both the American A-Okay sign (thumb and forefinger forming a circle) and the thumbs-up sign are considered obscene gestures in parts of Europe, Middle Eastern countries, Brazil, and Australia. A website that utilized these symbols would signal to users in those areas that you hadn't done any research on their culture and customs.

Be sure to budget time for adjustments, even if you think you've optimized the site for localization.

For further reference:

Nancy Hoft Consulting at <www.world-ready.com> is a training and consulting firm that specializes in effective techniques for communicating with multilingual and multicultural audiences. Look here for their advice on writing and designing for a global community and for links to other helpful sites.

(continued)

(continued)

The Macromedia DreamWeaver Support Center at <www.macromedia.com/support/dreamweaver/manage/localization_design/localization_design03.html> has a great example, based on a French chocolate manufacturer's website, of text expansion and contraction and other issues with website design.

Works Cited

"Globalization, Internationalization, Localization: An Overview." 2001–2004. Globalization.com. 7 Feb. 2004 <www.globalization.com/index.cfm ?MycatID=1&MysubCatID=1&pageID=1322>.

Hoft, Nancy. "Writing and Designing for an International Audience." 5 Apr. 2002. 7 Feb. 2004 <www.world-ready.com/stcorlando.htm>.

Macromedia Dreamweaver Support Center. "Accounting for text expansion and contraction." 27 Aug. 2001. 7 Feb. 2004. <www.macromedia.com/ support/dreamweaver/manage/localization_design/localization_design03 .html>.

Exercises

These exercises assume that they will occur in a computer lab where it is possible to project a site onto many screens or one large one. In that situation, small groups of two or three and oral reporting seem most effective. However, if individual work and written reports work better in the local situation, use that approach.

▶ You Create

1. Create a simple webpage that includes a title, text about yourself, and at least one visual aid.

2. Using the page you created in Exercise 1, create two other versions of it. Keep the content the same, but change the design.

3. Create a series of paragraphs that presents the same information in promotional, concise, and scannable text.

4. Create or download an image. Present it on a webpage in three different sizes. In groups of two or three, discuss how you achieved the differences and the effect of the differences. Report your findings orally to the class.

5. Using the page you created in Exercise 1, make several different backgrounds. In groups of three or four, review the effectiveness of the background (Is it distracting? Does it obscure the text?), and demonstrate both a good and bad version to the class.

▶ You Analyze/Group

6. Go to any website. In groups of two or three, assess the role the reader is asked to assume. Orally report your findings to the class. Alternative: Write a brief analytical report in which you identify the role and present support for your conclusion.

7. Go to any website. In groups of two or three, assess the style of the text. Is it promotional, concise, or scannable? Present your findings orally to the class. Or, as in Exercise 6, write a brief analysis.

8. Go to any website. In groups of two or three, assess the use of visuals at the site. Review for clarity, length of time to load, physical placement on the site. Present an oral report to the class, or write an information analysis.

9. Go to any website. In groups of two or three, assess the template. Are types of information effectively grouped? Is it easy to figure out where the links will take you? Report orally or in writing as your instructor requires.

10. In groups of two or three, critique any of the Examples (pp. 342–348) or use one of the samples in the website that accompanies this text. Judge them in terms of style, screen design, and audience role. Explain where you think the examples are strong and where they could be improved.

Writing Assignment

Create an informational website; if possible, load it onto the Web so that others may review it. Determine a purpose and an audience for the site. Create a homepage and documents that carry out the purpose. The site should have at least three nodes. Before you create the site, fill out the planning sheet on page 340. Your instructor will place you in a "review group"; set up a schedule with the other group members so that they can review your site for effectiveness at several points in your process. To review the site, use the points in the section "Testing" (pp. 336–340) or use the Worksheet for Evaluating a Website (pp. 341–342).

▶ Web Exercise

Review two or three websites of major corporations in order to determine how they use the elements of format. Review the homepage, but also review pages that are several layers "in" (e.g., Our Products/Cameras/UltraCompacts) in the Web; typically, pages further "in" look more like printed pages. Write a brief analytical or IMRD report discussing your results.

Works Cited and Consulted

Brooks, Randy M. "Principles for Effective Hypermedia Design." *Technical Communication* 40.3 (August 1993): 422–428.

Coney, Mary, and Michael Steehouder. "Role Playing on the Web: Guidelines for Designing and Evaluating Personas Online." *Technical Communication* 47.3 (August 2000): 327–340.

December, John. "An Information Development Methodology for the World Wide Web." *Technical Communication* 43.4 (November 1996): 369–376.

Farkas, David K., and Jean B. Farkas. "Guidelines for Designing Web Navigation." *Technical Communication* 47.3 (August 2000): 341–358.

Gallagher, Susan. "Your First Web Page." *intercom* 44.4 (May 1997): 13–15.

Grice, Roger A., and Lenore S. Ridgway. "Presenting Technical Information in Hypermedia Format: Benefits and Pitfalls." *Technical Communication Quarterly* 4.1 (Winter 1995): 35–46.

Horton, William. *Designing and Writing Online Documentation: Hypermedia for Self-Supporting Products.* 2nd ed. New York: Wiley, 1994.

Hunt, Kevin. "Establishing a Presence on the World Wide Web: A Rhetorical Approach." *Technical Communication* 43.4 (November 1996): 376–387.

Nielson, Jakob. "How Users Read the Web." 1 Oct. 1997. 24 April 2004 <www.useit.com/alertbox/981129.html>.

Pearrow, Mark. *Web Site Usability Handbook.* Rockland, MA: Charles River Media, 2000.

Spyridakis, Jan. "Guidelines for Authoring Comprehensible Web Pages and Evaluating Their Success." *Technical Communication* 47.3 (August 2000): 359–382.

Tatters, Wes. *Teach Yourself Netscape Web Publishing in a Week.* Indianapolis: Samsnet, 1996.

Wilkinson, Theresa A. "Web Site Planning." *intercom* 44.10 (December 1997): 14–15.

Williams, Thomas R. "Guidelines for Designing and Evaluating the Display of Information on the Web." *Technical Communications* 47.3 (August 2000): 383–396.

Focus on HTML

HTML (*Hypertext Markup Language*) is the invisible structure of the Web. Viewers can see a Web document because the browser (e.g., Netscape Communicator or Internet Explorer) "reads" an HTML document and displays the results on the screen. Actually, HTML is a code, a series of typed orders placed in the document. For instance, to make a word appear boldfaced on the screen, the writer places a "start bold" () and an "end bold" () command in the HTML document:

I want you to read this book.

The browser displays the sentence

I want you to **read** this book.

HTML code exists for everything that makes a document have a particular appearance on screen. If the item appears on screen, it appears because the code told the browser to display it. Codes exist for paragraphing, fonts, font sizes, color, tables, and all other aspects of a document. Codes tell which visual aid should appear in a particular place in a document. Figures 1 and 2 show the HTML code for a simple document and the document as it displays on a browser.

```
<html>
<head>
<title>
Sample Display Techniques Illustrated          browser
</title></head>                                 title
<body>
<b><h3>Sample Display Techniques Illustrated    title
</b></h3> by Dan Riordan

<p>These pages illustrate several techniques    text
for displaying information. I have illustrated
ways to use lists, the align command, the
anchor command, and escape links.

<p><b>List</b> I like to teach, especially
<ul>
<li>in groups and                               list
<li>using technology.</ul>

<p><b>Align Center</b> My wife and I often
visit a cabin up north. Here is what the cabin  visual
looks like: <center><img src="twrida1.gif"      aid
width=100 height=167></center>

<p><b>Anchor</b> This device allows you to
"link" inside a document. I illustrate the
device by letting you read a series of
<a href="twrida5.htm">letters</a> my family's
immigrants wrote in the 1850s.
<p><b>Escape Links</b> Here are "escape links"
to sites connected to this one:
<br><b><i><small>Return to <br><a
href="http://www.uwstout.edu/english/riordan/   "escape
techwrit/techwrit.htm">Technical Writing         links"
</a>|</b>
<b><i> <a href="http//www.uwstout.edu/english/
projects.htm">English Department Student
Projects</a>|</b>
<b><i> <a href="http//www.uwstout.edu/english/
english.htm">English Department</a></p></b></i>
</small>

</body>
</html>
```

Figure 1
HTML Code for a Web Document

(continued)

(continued)

The classic way to develop a site is to create material using an ASCII text editor like Notepad (DOS) or SimpleText (Mac). The method for creating it is easy.

- Open a file in one of these two programs.
- Type in certain HTML commands.
- Type in your text.

But typing in code is time consuming and susceptible to errors. If, for instance, one of the brackets (>) is omitted in the boldface code, the word will not appear as boldfaced. As a result, most Web authors use a Web authoring program that creates the code as the writer designs the webpage on screen. Many such programs exist. Some of the most frequently used are Front Page, DreamWeaver, AdobeGoLive, and Netscape Composer. Instructions on using such programs are beyond the scope of this book; however, many good instruction books are available, and all the programs have help menus and training tutorials. The best advice is to begin to practice with the programs to learn their features and to develop enough proficiency so that you can achieve the effects that you visualize.

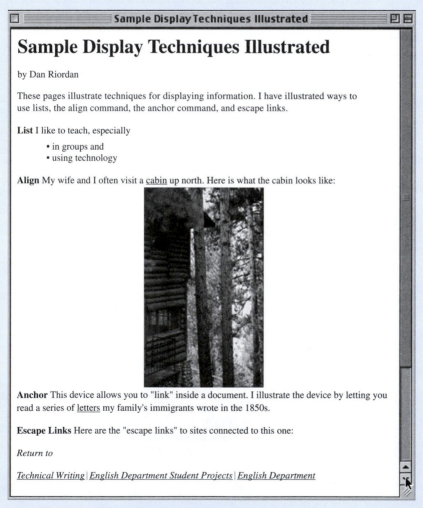

Figure 2
Browser Display of the Code in Figure 1

Chapter 14

Formal Reports

Chapter 14
In a Nutshell

Formal format presents documents in a way that makes them seem more "official." Often the format is used with longer (10 or more pages) documents, or else in documents that establish policy, make important proposals, or present the results of significant research.

Formal format requires a title page, a table of contents, a summary, and an introduction, in that order.

The *title page* gives an overview of the report—title, author, date, report number if required, and report recipient if required. Place all these items, separated by white space, at the left margin of the page.

The *table of contents* lists all the main sections and subsections of the report and the page on which each one begins.

The *summary*—often called "executive summary" and sometimes "abstract"—presents the report in brief. The standard method is to write the summary as a "proportional reduction"; each section of the summary has the same main point and the relative length as the original section. After your readers finish the summary, they should know your conclusions and your reasons.

The *introduction* contains all the usual introductory topics but gives each of them a head—background, scope, purpose, method, and recommendations.

Formal reports are those presented in a special way to emphasize the importance of their contents. Writers often use formal reports to present recommendations or results of research. Other reasons for using a formal approach are length (over 10 pages), breadth of circulation, perceived importance to the community, and company policy. Although a formal report looks very different from an informal report, the contents can be exactly the same. The difference is in the changed perception caused by the formal presentation. This chapter explains the elements of formal reports and discusses devices for the front, body, and end material.

The Elements of a Formal Report

To produce a formal report, the writer uses several elements that orient readers to the report's topics and organization. Those elements unique to the formal report are the front material and the method of presenting the body. Other elements—appendixes, reference sections, introductions, conclusions, and recommendations—are often associated with the formal report but do not necessarily make the report formal; they could also appear in an informal report.

The formal front material includes the title page, the table of contents, and the list of illustrations. Almost all formal reports contain a summary at the front, and many also have a letter of transmittal. The body is often presented in "chapters," each major section starting at the top of a new page.

Because these reports often present recommendations, they have two organizational patterns: traditional and administrative (Freeman and Bacon; ANSI). The traditional pattern leads the reader through the data to the conclusion (Freeman and Bacon). Thus conclusions and recommendations appear at the end of the report. The administrative pattern presents readers with the information they need to perform their role in the company, so conclusions and recommendations appear early in this report.

Traditional	Administrative
Title page	Title page
Table of contents	Table of contents
List of illustrations	List of illustrations
Summary or abstract	Summary or abstract
Introduction	Introduction
Discussion—Body Sections	Conclusions
Conclusions	Recommendations/Rationale
Recommendations/Rationale	Discussion—Body Sections
References	References
Appendixes	Appendixes

Front Material

. . . .

Transmittal Correspondence

Transmittal correspondence is a memo or letter that directs the report to someone. A memo is used to transmit an internal, or in-house, report. An external, or firm-to-firm, report requires a letter. (See Chapter 19 for a sample letter.) In either form, the information remains the same. The correspondence contains

▶ The title of the report.
▶ A statement of when it was requested.
▶ A very general statement of the report's purpose and scope.
▶ An explanation of problems encountered (for example, some unavailable data).
▶ An acknowledgment of those who were particularly helpful in assembling the report.

Sample Memo of Transmittal

Date:	May 1, 2005
To:	Ms. Elena Solomonova, Vice-President, Administrative Affairs
From:	Rachel A. Jacobson, Human Resources Director
Subject:	Proposal for the Spousal Employment Assistance Program

Title of report
Cause of writing

Attached is my report "Proposal for the Implementation of a Spousal Employment Assistance Program," which you requested after our March 15 meeting.

Purpose of report

The report presents a solution to the problems identified by our large number of new hires. In brief, those new hires all had spouses who had

Statement of request

to leave careers to move to Rochester. This proposal recommends initiating a spousal employment assistance program to deal with relocation problems.

Praise of coworkers

Compiled by the Human Resources staff, this report owes a significant debt to the employees and their spouses who agreed to be interviewed as part of its preparation.

Title Page

Well-done title pages (see Figure 14.1, p. 358) give a quick overview of the report, while at the same time making a favorable impression on the reader. Some firms have standard title pages just as they have letterhead stationery for business letters. Here are some guidelines for writing a title page:

Figure 14.1

Title Page for a
Formal Report

> PROPOSAL FOR THE IMPLEMENTATION
> OF A SPOUSAL EMPLOYMENT
> ASSISTANCE PROGRAM
>
>
> By
> Rachel A. Jacobson
> Director, Human Resources
>
>
> May 1, 2005
>
>
> Corporate Proposal
> HRD 01-01-2005
>
>
> Prepared for
> Elena Solomonova
> Vice-President, Administrative Affairs

▶ Place all the elements at the left margin (ANSI). (Center all the elements if local policy insists.)
▶ Name the contents of the report in the title.
▶ Use a 2-inch left margin.
▶ Use either all caps or initial caps and lowercase letters; use boldface when appropriate. Do not use "glitzy" typefaces, such as outlined or cursive fonts.
▶ Include the writer's name and title or department, the date, the recipient's name and title or department, and a report number (if appropriate).

Table of Contents

A table of contents lists the sections of the report and the pages on which they start (see Figure 14.2). Thus it previews the report's organization, depth, and emphasis. Readers with special interests often glance at the table of contents, examine the abstract or summary, and turn to a particular section of the report. Here are some guidelines for writing a table of contents:

▶ Title the page *Table of Contents.*
▶ Present the name of each section in the same wording and format as it appears in the text. If a section title is all caps in the text, place it in all caps in the table of contents.
▶ Do not underline in the table of contents; the lines are so powerful that they overwhelm the words.

Figure 14.2

Table of Contents
for an
Administrative
Report

TABLE OF CONTENTS

▶ Do not use "page" or "p." before the page numbers.
▶ Use only the page number on which the section starts.
▶ Set margins so that page numbers align on the right.
▶ Present no more than three levels of heads; two is usually best.
▶ Use *leaders,* a series of dots, to connect words to page numbers.

List of Illustrations

Illustrations include both tables and figures. The list of illustrations (Figure 14.3) gives the number, title, and page of each visual aid in the report. Here are guidelines for preparing a list of illustrations:

▶ Use the title *List of Illustrations* if it contains both figures and tables; list figures first, then tables.
▶ If the list contains only figures or only tables, call it *List of Figures* or *List of Tables.*
▶ List the number, title, and page of each visual aid.
▶ Place the list on the most convenient page. If possible, put it on the same page as the table of contents.

Figure 14.3

List of
Illustrations for a
Formal Report

LIST OF ILLUSTRATIONS

Summary or Abstract

A summary or an abstract (or executive summary) is a miniature version of the report. (See Chapter 8 for a full discussion of summaries and abstracts.)

In the summary, present the main points and basic details of the entire report. After reading a summary, the reader should know

▶ The report's purpose and the problem it addresses.
▶ The conclusions.
▶ The major facts on which the conclusions are based.
▶ The recommendations.

Because the summary "covers" many of the functions of an introduction, recent practice has been to substitute the summary for all or most of the introductory material, placing the conclusions and recommendations last. Used often in shorter (10- to 15-page) reports, this method eliminates the sense of overrepetition that is sometimes present when a writer uses the entire array of introductory elements.

Follow these guidelines to summarize your formal report:

▶ Concentrate this information into as few words as possible—one page at most.
▶ Write the summary *after* you have written the rest of the report. (If you write it first, you might be tempted to explain background rather than summarize the contents.)
▶ Avoid technical terminology (most readers who depend on a summary do not have in-depth technical knowledge).

SUMMARY

Recommendation given first Background	This report recommends that the company implement a spousal assistance program. Swift expansion of the company has brought many new employees to us, most of whom had spouses who left professional careers. Because no assistance program exists, our employees and their spouses have found themselves involved in costly, time-consuming, and stressful situations that in several instances have affected productivity on the job.
Basic conclusions Benefits	A spousal assistance program will provide services that include home- and neighborhood-finding assistance, medical practitioner referrals, and employment-seeking assistance. Advantages include increased employee morale, increased job satisfaction, and greater company loyalty.
Cost Implementation	Cost is approximately $54,000/year. The major benefit is productivity of the management staff. The program will take approximately six months to implement and will require hiring one spousal employment assistance counselor.

Introduction

The introduction orients the reader to the report's organization and contents. Formal introductions help readers by describing purpose, scope, procedure, and background. Statements of purpose, scope, procedure, and background orient readers to the report's overall context.

To give readers the gist of the report right away, many writers now place the conclusions/recommendation right after the introduction. Recently, writers have begun to combine the summary and the introductory sections to cut down on repetition. Example 14.1 (pp. 368–371) illustrates this approach.

Purpose Statement

State the *purpose* in one or two sentences. Follow these guidelines:

- State the purpose clearly. Use one of two forms: "The purpose of this report is to present the results of the investigation" or "This report presents the results of my investigation."
- Use the *present* tense.
- Name the alternatives if necessary. (In the purpose statement in the example below, the author names the problem [lack of a spousal assistance program] and the alternatives that she investigated.)

Scope Statement

A *scope statement* reveals the topics covered in a report. Follow these guidelines:

- In feasibility and recommendation reports, name the criteria; include statements explaining the rank order and source of the criteria.
- In other kinds of reports, identify the main sections, or topics, of the report.
- Specify the boundaries or limits of your investigation.

Procedure Statement

The *procedure statement*—also called the *methodology statement*—names the process followed in investigating the topic of the report. This statement establishes a writer's credibility by showing that he or she took all the proper steps. For some complex projects, a methodology section appears after the introduction and replaces this statement. Follow these guidelines:

- Explain all actions you took: the people you interviewed, the research you performed, the sources you consulted.
- Write this statement in the *past tense*.

▶ Select heads for each of the subsections. Heads help create manageable chunks, but too many of them on a page look busy. Base your decision on the importance of the statements to the audience.

Brief Problem (or Background) Statement

In this statement, which you can call either the *problem* or *background statement,* your goal is to help the readers understand—and agree with—your solution because they view the problem as you do. You also may need to provide background, especially for secondary or distant readers. Explain the origin of the problem, who initiated action on the problem, and why the writer was chosen. Follow these guidelines:

▶ Give basic facts about the problem.
▶ Specify the causes or origin of the problem.
▶ Explain the significance of the problem (short term and long term) by showing how new facts contradict old ways.
▶ Name the source of your involvement.

In the following example, the problem statement succinctly identifies the basic facts (relocating problems), the cause (out-of-state hires), the significance (decline in productivity), and the source (complaints to Human Resources). Here are the purpose, scope, procedure, and background statements of the proposal for Spousal Employment Assistance:

INTRODUCTION

Purpose

Two-part purpose: to present and to recommend

This proposal presents the results of the Human Resources Department's investigation of spousal employment assistance programs and recommends that XYZ Corp. implement such a program.

Scope

Lists topics covered in the report

This report details the problems caused by the lack of a spousal employment assistance program. It then considers the concerns of establishing such a program here at XYZ. These concerns include a detailed description of the services offered by such an office, the resources necessary to accomplish the task, and an analysis of advantages, costs, and benefits. An implementation schedule is included.

Procedure

Enough information given to establish credibility

The Human Resources Department gathered all the information for this report. We interviewed all 10 people (8 women and 2 men) hired within the past 12 months and 6 spouses (4 men and 2 women). We gathered information from professional articles on the subject. The human resources office provided all the salary and benefits figures. We also interviewed the director of a similar program operating in Arizona and a management training consultant from McCrumble University.

Problem

Background (cause)

Basic facts

In the past year, XYZ has expanded swiftly, and this expansion will occur throughout the near future. In the past year, 10 new management positions were created and filled. Seven of these people moved here from out of state. Several of these people approached the Human Resources Department for assistance with the problems involved in relocating.

Source of impetus to solve problem

Possible solution

Some of these problems were severe enough that some decline in productivity was noted and was also brought to the attention of Human Resources. Four of the managers left, citing stress as a major reason. That turnover further affected productivity. A spousal employment assistance office is one common way to handle such concerns and offset the potential bad effects of high turnover.

Lengthy Problem (or Background) Statements

Some reports explain both the problem and its context in a longer statement called either *Problem* or *Background*. A *background statement* provides context for the problem and the report. In it you can often combine background and problem in one statement.

Some situations require a lengthier treatment of the context of the report. In that case, the background section replaces the brief problem statement. Often this longer statement is placed first in the introduction, but practices vary. Place it where it best helps your readers.

To write an effective background statement, follow these guidelines:

▸ Explain the general problem.
▸ Explain what has gone wrong.
▸ Give exact facts.
▸ Indicate the significance of the problem.
▸ Specify who is involved and in what capacity.
▸ Tell why you received the assignment.

BACKGROUND

General problem

Management increases have brought many new persons into the XYZ team in the past year. This increase in personnel, while reflecting an excellent trend in a difficult market, has had a marked down side. The new personnel have all experienced significant levels of stress and some slide

Data on what is
wrong

in productivity as a result of the move. All 10 of the recent hires had spouses who left professional career positions to relocate in Rochester. These people have experienced considerable difficulty finding career opportunities in our smaller urban region, and all the families have reported a certain amount of stress related to everything from finding a home to finding dentists. Four of these managers subsequently left our employ, citing stress as the major reason to leave. These departures caused us to undertake costly, time-consuming personnel searches.

Significance

Why the author
received the
assignment

After interviews revealed the existence of such stress, the Executive Committee of Administrative Affairs discussed the issue at length and authorized Human Resources to carry out this study. The Director of Human Resources chaired a committee composed of herself, one manager who did not leave, and a specialist on budget. HR staff conducted the data gathering.

Conclusions and Recommendations/Rationale

Writers may place these two sections at the beginning of the report or at the end. Choose the beginning if you want to give readers the main points first and if you want to give them a perspective from which to read the data in the report. Choose the end if you want to emphasize the logical flow of the report, leading up to the conclusion. In many formal reports, you present only conclusions because you are not making a recommendation.

Conclusions

The conclusions section emphasizes the report's most significant data and ideas. Base all conclusions only on material presented in the body. Follow these guidelines:

▶ Relate each conclusion to specific data. Don't write conclusions about material you have not discussed in the text.
▶ Use concise, numbered conclusions.
▶ Keep commentary brief.
▶ Add inclusive page numbers to indicate where to find the discussion of the conclusions.

CONCLUSIONS

This investigation has led to the following conclusions. (The page numbers in parentheses indicate where supporting discussion may be found.)

Conclusions
presented in same
order as in text

1. The stresses experienced by the new hires are significant and are expected to continue as the company expands (6).

2. Stress is not related to job difficulties but instead is related more to difficulties other family members are experiencing as a result of the relocation (6).
3. Professionals exist who are able to staff such programs (7).
4. The program will result in increased employee morale, increased job satisfaction, and greater company loyalty (9).
5. A program could begin for a cost of $54,000 (10).
6. The major benefits of the program will be increased productivity of the management staff and decreased turmoil created by frequent turnover (11).
7. A program would take six months to initiate (13).

Recommendations/Rationale

If the conclusions are clear, the main recommendation is obvious. The main recommendation usually fulfills the purpose of the report, but do not hesitate to make further recommendations. Not all formal reports make a recommendation.

In the rationale, explain your recommendation by showing how the "mix" of the criteria supports your conclusions. Follow these guidelines:

▶ Number each recommendation.
▶ Make the solution to the problem the first recommendation.
▶ If the rationale section is brief, add it to the appropriate recommendation.
▶ If the rationale section is long, make it a separate section.

RECOMMENDATIONS

Solution to the basic problem

1. XYZ should implement a spousal employment assistance program. This program is feasible and should eliminate much of the stress that has caused some of the personal anxiety and productivity decreases we have felt with the recent expansion.

Other recommendations on implementation

2. The Executive Committee should authorize Human Resources to begin the procedure of writing position guidelines and hiring an SEA counselor.

The Body of the Formal Report

The body of the formal report, like any other report, fills the needs of the reader. Issues of planning and design, covered in other chapters, all apply here. You can use any of the column formats displayed in Chapter 6 for laying out pages. Special concerns in formal reports are paginating and indicating chapter divisions.

Paginating

Be consistent and complete. Follow these guidelines:

▶ Assign a number to each piece of paper in the report, regardless of whether the number actually appears on the page.
▶ Assign a page number to each full-page table or figure.
▶ Place the numbers in the upper right corner of the page with no punctuation, or center them at the bottom of the page either with no punctuation or with a hyphen on each side (-2-).
▶ Consider the title page as page 1. Do not number the title page. Most word processing systems allow you to delete the number from the title page.
▶ In very long reports, use lowercase roman numerals (i, ii, iii) for all the pages before the text of the discussion. In this case, count the title page as page i, but do not put the i on the page. On the next page, place a ii.
▶ Paginate the appendix as discussed in "End Material" (below).
▶ Use headers or footers (phrases in the top and bottom margins) to identify the topic of a page or section.

Indicating Chapter Divisions

To make the report "more formal," begin each new major section at the top of a page (see Example 14.2, which starts on p. 372).

End Material

The end material (glossary and list of symbols, references, and appendixes) is placed after the body of the report.

Glossary and List of Symbols

Traditionally, reports have included glossaries and lists of symbols. However, such lists tend to be difficult to use. Highly technical terminology and symbols should not appear in the body of a report that is aimed at a general or multiple audience. Place such material in the appendix. When you must use technical terms in the body of the report, define them immediately; informed readers can simply skip over the definitions. Treat the glossary as an appendix. If you need a glossary, follow these guidelines:

▶ Place each term at the left margin, and start the definition at a tab (2 or 3 spaces) farther to the right. Start all lines of the definition at this tab.
▶ Alphabetize the terms.

References

The list of references (included when the report contains information from other sources) is discussed along with citation methods in Appendix B.

Appendix

The appendix contains information of a subordinate, supplementary, or highly technical nature that you do not want to place in the body of the report. Follow these guidelines:

▶ Refer to each appendix item at the appropriate place in the body of the report.
▶ Number illustrations in the appendix in the sequence begun in the body of the report.
▶ For short reports, continue page numbers in sequence from the last page of the body.
▶ For long reports, use a separate pagination system. Because the appendixes are often identified as Appendix A, Appendix B, and so on, number the pages starting with the appropriate letter: A-1, A-2, B-1, B-2.

Worksheet for Preparing a Formal Report

☐ **Determine the audience for this report.**
Who is the primary audience and who the secondary? How much does the audience understand about the origins and progress of this project? How will they use this report? Will it be the basis for a decision?

☐ **Plan the visual aids that will convey the basic information of your report.**

☐ **Construct those visual aids.**
Follow the guidelines in Chapter 7.

☐ **Prepare a style sheet for up to four levels of heads and for margins, page numbers, and captions to visual aids.**

☐ **Decide whether each new section should start at the top of a new page.**

☐ **Create a title page.**

☐ **Prepare the table of contents.**
How many levels of heads will you include? (Two is usual.) Will you use periods for leaders?

(continued)

(continued)

☐ **Prepare the list of illustrations.**
Present figures first, then tables.

☐ **Determine the order of statements (purpose, scope, procedure, and so forth) in the introduction.**
In particular, where will you place the problem and background statements? in the introduction? in a section in the body?

☐ **Prepare a glossary if you use key terms unfamiliar to the audience.**

☐ **List conclusions.**

☐ **List recommendations, with most important first.**

☐ **Write the rationale to explain how the mix of conclusions supports the recommendations.**

☐ **Write the summary.**

☐ **Prepare appendixes of technical material.**
Use an appendix if the primary audience is nontechnical or if you have extensive tabular or support material.

Examples

Example 14.1 is the body of the report whose introduction is explained in this chapter. Example 14.2 is a brief formal report.

Example 14.1

Formal Report Body

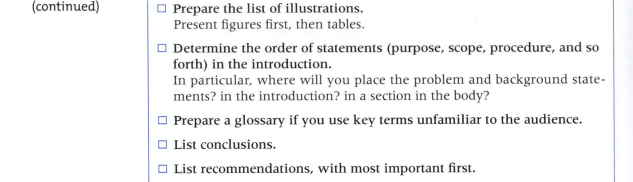

DISCUSSION

In this section, I will describe spousal employment assistance (SEA), discuss the advantages and benefits of it, and develop a time schedule for the implementation of it.

NATURE OF THE PROBLEM

Many complex issues arise when relocating a dual-career family. Issues such as a new home, a new mortgage, two new jobs in the family, a new and reliable child care service—to name only a few. These issues, if not dealt with in an efficient manner, can create tremendous stress in the new employee—stress that dramatically affects productivity on the job.

Productivity and protection of our company's human resources investment are the key issues we are dealing with in this program. The intention is that the more quickly the employee can be productive and settled in a new area, the less costly it will be for our company.

DESCRIPTION OF THE PROGRAM

I am proposing a separate office within the company for the SEA program. It would be staffed by a consultant who would research and develop the following areas:

- Home-finding counseling
- Neighborhood finding
- Mortgage counseling
- Spouse and family counseling
- Spouse employment assistance
- Child care referrals
- School counseling
- Cost-of-living differences
- Doctor and dentist referrals

All counseling services would be handled by our SEA office employee except for formal employment assistance, which would be contracted with a third-party employment firm. A third-party firm can provide the advantage of objectivity as well as a proper level of current employment information.

ADVANTAGES OF THE PROGRAM

The program is a service provided by us, and paid for by us, that is for the sole purpose of assisting the new employee. The advantages are increased employee morale, increased job satisfaction, and greater company loyalty. The employee feels that the company is concerned with the problems he or she is facing in the relocation process. The assistance the employee receives makes the move easier, so adjusting to the new job is quicker. The result is a more productive employee.

WHAT ARE THE COSTS VERSUS THE BENEFITS?

Costs

The comprehensive program will cost the company approximately $54,500 per year. As illustrated in Table 1, this includes $2,500 for research and

(continued)

Example 14.1

(continued)

development, $27,000 for the SEA consultant, and $25,000 for the third-party employment firm (10 contracts at $2,500 each). This figure doesn't include the cost of office completion, which would run about $1,100 to finish the first-floor office space (room 120), which isn't currently occupied.

Also in Table 1, I have estimated the dollar amount our company invests yearly on new relocating managers. $54,500 is a drop in the bucket when you realize that we spend at least $290,000 yearly on new hires alone.

Table 1
Cost/Employee Investment Comparison

	Estimated Yearly Cost of Program*	Estimated Value of Human Resource Investment
Research/development	$ 2,500	New relocating managers
SEA consultant	27,000	Approx. 10 @ $29,000 each
		$290,000
Employment firm contracts		
10 @ approx. $2,500 each	25,000	
TOTAL	$ 54,500	TOTAL $290,000

*Doesn't include the one-time cost of office completion (about $1,100).

Benefits

The benefit to our company is the increase in productivity of the management staff. The cost to our company shouldn't be considered a luxury or frill expense, but a way to protect and enhance the company's human resources investment. The yearly cost of the program ($54,500) compared with the estimated yearly cost of new employees who would use it ($290,000) shows that the expense is far outweighed by the investment we've made in new management hires

WHAT ABOUT IMPLEMENTATION?

Implementation time is estimated at six to seven months depending on when the SEA consultant is hired. This is because, after a three-month hiring and selection period, the new consultant would be given three months to begin the research and development of the program. After these three months, research would continue, but client consultation would also begin (refer to Figure 1).

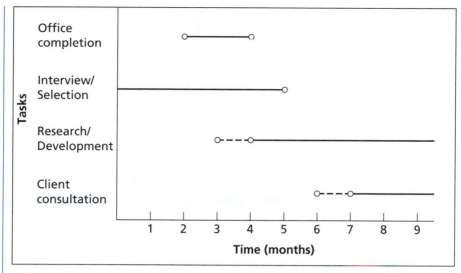

Figure 1
Schedule for Program Implementation

Example 14.2

Formal Report

FEASIBILITY OF FINDING PROGRAM
LANGUAGE CODE ON THE INTERNET

By
Chad Seichter

May 8, 2003

Prepared for
Kim O'Neil
Program Director

TABLE OF CONTENTS

(continued)

Example 14.2

(continued)

ABSTRACT

This report determines whether or not it is feasible for Applied Mathematics majors to find program language code on the Internet. To make this judgment, I looked for two separate programming languages: Java and Smalltalk. The pages I looked at were evaluated according to how current they were, if the source was credible, and if there actually was code on them. I found that I was easily able to find code for both of these languages and therefore have concluded that the Internet is a feasible source of information for my major.

INTRODUCTION

Background

Goal   The goal of this project was to decide if the Internet is a feasible source of information for Applied Math majors. I researched to see if I could find source code for Java and Smalltalk. Because of my research, I have come to the conclusion that the Internet is a good source for information for Applied Math majors. The purpose of this report is to explain my research and tell how I came to this decision.

Rationale   For my study I chose two different programming languages to try to find information on. The two languages were Java and Smalltalk. I picked these because writing computer programs is a large part of the Applied Math program. I chose the Java language because it is an up-and-coming language that is very popular right now. I figured it would be easy to find information on this. On the other hand, Smalltalk is a lot less known and used in the present day. I didn't know how much I would find on a language that is not very popular. I figured that choosing languages on the opposite ends of the popularity spectrum would give me a good look at not only these two languages but also all the other ones.

Method

Choosing Topics   My topics of study were whether or not I could find source code for programming languages on the Internet. I focused my search on two separate languages. One was a popular language, Java, and one less popular, Smalltalk. I figured that these two languages would give me a good overview about finding code for all of the different computer languages there are.

Determining Feasibility   I had to research my topics in order to figure out if the Internet was a feasible source of information. I used two separate search engines to do this: WebCrawler and Dogpile. Both of these search engines seemed to give me good information on my topics. Once I did the search, I evaluated the websites that came up and based my evaluation on my criteria (see "Explanation of Criteria" section below). In order to conclude that the Internet was useful for me, I needed to find websites that met most of the criteria for each topic.

Explanation of Criteria   I used the following criteria to come up with a conclusion:

1. Current   For the Java language, I chose that the pages had to have been updated sometime within the last six months because many changes are still being done to this language. For the Smalltalk language, I decided that anything within the last two years was good because it is an older language with fewer revisions still taking place.
2. Credible   In order for the source to be credible, it couldn't be someone's personal webpage. I also tried out the code myself to see if the page was credible. If it didn't work for me, then I considered it noncredible.
3. Was there code?   To meet this criterion, the webpage had to have actual examples of source code on it that could be used by others.

Conclusion

After doing my searches using WebCrawler and Dogpile, I am able to conclude that the Internet is a feasible source of information for Applied Math majors. I was able to find numerous webpages that met my three criteria for both the Java language and the Smalltalk language.

DISCUSSION

Finding Java Code on the Web

Introduction   I am an Applied Math major and am trying to figure out if the Internet and my major are interrelated. The purpose of this search is to find code that I could use to assist me in making computer programs. The language of code I would like to find is Java code. Specifically, I want to find Java applets that can be inserted into my programs. To help answer my question, I used the WebCrawler search engine. I checked the first five sites and evaluated them according to my criteria. If I could find at least five sites that had code, I could come to the conclusion that the Web is useful for finding Java code.

(continued)

Example 14.2

(continued)

Findings   Using the WebCrawler search engine, I typed in the keywords "java applets" and then looked at the sites that came up. Over 45,000 sites came up using this search, so I started looking at them starting at the beginning. I found that all of the first five sites met my criteria. Table 1 below shows my results using my criteria.

Table 1
Java Code Sites on the Internet

	Current	Credible	Were There Applets?
http://java.sun.com/starter.html	Yes	Yes	No, but had links to some
http://www.javasoft.com/applets/index.html	Yes	Yes	Yes
http://javapplets.com/	Yes	Yes	Yes
http://www.conveyor.com/conveyor-java.html	Yes	Yes	Yes
http://javaboutique.internet.com/	Yes	Yes	Yes

Conclusion   After doing this search and going through these top five sites, I have concluded that the Internet is very helpful for me to find applets for Java. All these pages appeared to be very recent. Most of them had a copyright in the year 2000, so I know they would still be useful now. I wanted to check to see if the code was correct, so I moved some of the applets off these webpages onto one that I had created, and they all worked in my code as well. Therefore I knew that these sites were credible.

Java has become such a big language to program that there appears to be an almost unlimited amount of code on the Web for it. I think that the sites that I checked out were probably above-average sites, because all the code I used worked correctly. I'm sure that there are some sites that aren't as good as these and could have bad code. If I was ever to use any of these pages, I would make sure I understood the code and made sure it was an efficient way to do the task. At worst, there are many examples to give you an idea of how to create an applet, so this search was very helpful to me.

Finding Smalltalk Code on the Web

Introduction   I am an Applied Math major and am trying to figure out if the Internet and my major are interrelated. The purpose of this search is to find code that I could use to assist me in making computer programs. The language of code I would like to find is Smalltalk. I chose this language because it isn't very popular, so I figure if I can find this, then I should be able to find almost any other language. To help answer my question, I used the Dogpile search engine. I checked the first five sites and evaluated them according to my criteria. If I could find at least five sites that had code, I could come to the conclusion that the Web is useful for finding Java code.

Findings Using the Dogpile search engine, I typed in the keywords "Smalltalk example code" and then looked at the sites that came up. There weren't a whole lot of sites, but the ones that came up usually had good information and links to other good pages. Table 2 below shows my results from the five websites I looked at.

Table 2
Smalltalk Code Sites on the Internet

	Current	Credible	Code on the Page?
www.site.gmu.edu	1995, but information was still useful	Yes	Yes
www.mk.dmu.ac.uk	No date given	Yes	Yes
www.ics.hawaii.edu	Yes	Yes	Yes
www.phaidros.com	Yes	Yes	Yes
www.objectconnect.com	No date given	Yes	Yes

Conclusion After doing this search and going through these top five sites, I have concluded that the Internet is helpful for me to find Smalltalk code. Some of the pages weren't very current or didn't have a date, but because there haven't been too many changes to this language recently, that isn't a big issue. Most of them were within the last couple of years, so that was sufficient. I wanted to check to see if the code was correct, so I used some of the code in other programs I have written in the past, and it all seemed to work. Therefore, I knew that these sites were credible. Also, because Smalltalk isn't a popular language to program in, I feel that because I can use the Internet as a tool for this language, I can probably also use it for almost any other. Besides having code, I found several other helpful tools while using the Web that related to the programming language that could help me in writing my own programs.

Exercises

▶ You Create

1. For Exercise 10, Chapter 6 (pp. 163–164), create one or all of the following: a title page, table of contents, summary, conclusion, and recommendation.

2. For Exercise 10, Chapter 6 (pp. 163–164), create a page layout.

▶ You Revise

3. Redo this table of contents to make it more readable.

Introduction	2–3	chefmate	6
LIST OF FIGURES	2	conclusion	page 5
recommendation	4	table 1 cost	page 6
discussion	5–12	cost	7–8
width of front panel	5	hitachi 7	
hitachi 5			

4. Edit the following selection, which is taken from the discussion section of a formal report. Add at least two levels of heads, and construct one appropriate visual aid. Write a one- or two-sentence summary of the section.

> The cost of renting a space will not exceed $150. Our budget allows $140 per month investment at this time for money available from renting space. The cost of renting a space is $175 per month for 100 square feet of space at Midtown Antique Mall. This exceeds the criteria by a total of $25 per month. The cost for renting 100 square feet of space at Antique Emporium is $100 per month, which is well within our criteria based on our budget. Antique Emporium is the only alternative that meets this criterion. The length of the contract cannot exceed 6 months because this is what we have established as a reasonable trial period for the business. Within this time, we will be able to calculate average net profit (with a turnover time no longer than 3 months) and determine if it is worth the time invested in the business. We will also be able to determine if we may want to continue the business as it is or on a larger scale by renting more floor space. The Midtown Antique Mall requires a 6-month contract. The Antique Emporium requires an initial 6-month contract which continues on a month-to-month basis after the contract is fulfilled. Both locations fulfill the contract length desired in the criteria. The possibility of continuing monthly at the Antique Emporium is an attractive option compared with renewing contracts bi-yearly.

▶ Group

5. In groups of two to four, discuss whether the conclusions and recommendations in Examples 15.2 or 15.3 (pp. 397–407) should appear at the beginning. Be prepared to give an oral report to the class. If your instructor requires, rewrite the introductory material so it follows the pattern of the human resources proposal (pp. 361–362) Alternate: Rewrite the human resources introduction by using the executive summary method.

6. Your instructor will hand out a sample report from the *Instructor's Manual.* In groups of two to three, edit it into a formal report. Change introductory material as necessary.

7. If you are working on formal format elements, bring a draft of them to class. In groups of two or three, evaluate each other's material. Use the guidelines (for Title Page, Table of Contents, Summary, Introduction, Conclusion, Recommendation) in this chapter as your criteria. Rewrite your material as necessary.

Writing Assignments

1. Create a formal report that fulfills a recommendation, feasibility, proposal, or research assignment, as given in other chapters of this book.

    a. Create a template for your formal report. Review Chapter 6, pages 153–155, and Chapter 17, pages 447–449.

    b. Choose an introductory combination and write it.

    c. Write the conclusions and recommendations/rationale sections.

    d. Divide into pairs. Read each other's draft from the point of view of a manager. Assess whether you get all the essential information quickly. If not, suggest ways to clarify the material.

2. Write a learning report for the writing assignment you just completed. See Chapter 5, Writing Assignment 7, pages 134–135, for details of the assignment.

Web Exercise

Review two or three websites of major corporations in order to determine how they use the elements of format. Review the homepage, but also review pages that are several layers into the site (e.g., Our Products/Cameras/Ultra-Compacts); typically, pages further "in" look more like hard copy. Write a brief analytical or IMRD report discussing your results.

Works Cited

American National Standards Institute (ANSI). *Guidelines for Format and Production of Scientific and Technical Reports*. ANSI 239. 18-1974. New York: ANSI, 1974.

Freeman, Lawrence H., and Terry R. Bacon. *Writing Style Guide*. Rev. ed. GM1620. Warren, MI: General Motors, 1991.

15

Recommendation and Feasibility Reports

Chapter 15
In a Nutshell

Feasibility studies and recommendations present a position based on credible criteria and facts. *Feasibility studies* use criteria to investigate an item in order to tell the reader whether or not to accept the item. *Recommendations* use criteria to compare item A to item B in order to tell the reader which one to choose. To decide whether or not to air condition your house is a feasibility issue; to decide which air conditioning system to purchase is a recommendation issue.

Report strategy. In the introduction, *set the context:* tell the background of the situation, explain the methods you used to collect data, and state why you chose these criteria. In the body, *deal with one criterion per section.* A helpful outline for a section is

▶ Brief introduction to set the scene

▶ Discussion of data, often subdivided by alternative

▶ A helpful visual aid

▶ A brief, clear conclusion

Based on criteria. Criteria are the framework through which you and the reader look at the subject.

▶ Select topics that an expert would use to judge the situation. (For the air conditioner, a criterion is cost.)

▶ Select a standard, to limit the criterion. (The limitation is "the system may not cost more than $6000.")

▶ Apply the criteria. (Look at the sales materials of two reputable systems.)

▶ Present the data and conclusion clearly. Report the appropriate facts from your investigation, create a useful visual aid, and use heads and chunks to guide the reader through the subsections.

Professionals in all areas make recommendations. Someone must investigate alternatives and say "choose A" or "choose B." The "A" or "B" can be anything: which type of investment to make, which machine to purchase, whether to make a part or buy it, whether to have a sale, or whether to relocate a department. The decision maker makes a recommendation based on *criteria:* standards against which the alternatives are judged.

For professionals, these choices often take the form of *recommendation reports* or *feasibility reports.* Although both present a solution after alternatives have been investigated, the two reports are slightly different. Recommendation reports indicate a choice between two or more clear alternatives: this distributor or that distributor, this brand of computer or that brand of computer (Markel). Feasibility reports investigate one option and decide whether it should be pursued. Should the client start a health club? Should the company form a captive insurance company? Should the company develop this prototype? (Alexander and Potter; Angelo; Bradford). This chapter explains how to plan and write both types of reports.

Planning the Recommendation Report

In planning a recommendation, you must consider the audience, choose criteria for making your recommendations, use visual aids, and select a format and an organizational principle.

Consider the Audience

In general, many different people with varying degrees of knowledge (a multiple audience) read these reports. A recommendation almost always travels up the organizational hierarchy to a group—a committee or board—that makes the decision. These people may or may not know much about the topic or the criteria used as the basis for the recommendation. Usually, however, most readers will know a lot about at least one aspect of the report—the part that affects them or their department. They will read the report from their own point of view. The human resources manager will look closely at how the recommendation affects workers, the safety manager will judge the effect on safety, and so on. All readers will be concerned about cost. To satisfy such readers, the writer must present a report that enables them all to find and glean the information they need.

Choose Criteria

To make data meaningful, analyze or evaluate them according to criteria. Selecting logical criteria is crucial to the entire recommendation report because you will make your recommendation on the basis of those criteria and because your choice of the "right" criteria establishes your credibility.

The Three Elements of a Criterion

A *criterion* has three elements: a name, a standard, and a rank (Holcombe and Stein). The *name* of the criterion, such as "cost," identifies some area relevant to the situation. The *standard* is a statement that establishes the limit of the criterion—for instance, "not to exceed $500.00." The standard heavily influences the final decision. Consider two very different standards that are possible for cost:

1. The cost of the water heater will not exceed $500.

2. The cheapest water heater will be purchased.

If the second standard is in effect, the writer cannot recommend the more expensive machine even if it has more desirable features.

The *rank* of the criterion is its weight in the decision relative to the other criteria. "Cost" is often first, but it might be last, depending on the situation.

Discovering Criteria

Criteria vary according to the type of problem. In some situations, a group or individual will have set up all the criteria in rank order. In that case, you show how the relevant data for the various alternatives measure up to these criteria.

When criteria have not been set up, you need to discover them by using your professional expertise and the information you have about needs and alternatives in the situation. One helpful way of collecting relevant data is to investigate appropriate categories: technical, management/maintenance, and financial criteria (Markel).

Technical criteria apply to operating characteristics such as the necessary heat and humidity levels in an air-moving system. *Management/maintenance criteria* deal with concerns of day-to-day operation, such as how long it will take to install a new air-moving system. *Financial criteria* deal with cost and budget. How much money is available, and how big a system will it purchase?

Applying Criteria

Suppose you were to investigate which of two jointers to place in a high school woods lab. To make the decision, you need to find the relevant data and create the relevant standards. To find relevant data, answer questions derived from the three categories.

▶ Technical—Does the jointer have appropriate fence size? table length? cutting capacity?

▶ Financial—How much does each jointer cost? How much do optional features cost? How much money is available? What is the standard?

▶ Management/Maintenance—Which one is safer? Will we need to reconfigure the lab or its electrical service? Will the jointers be available by the start of school in August?

To create standards, you must formulate statements that turn these questions into bases for judgment. You derive these standards from your experience, from an expert authority (such as another teacher), or from policy. For instance, because you know from your own experience the length of your typical stock, your standard will read "Must be able to handle up to 52 inches." Another teacher who has worked with these machines can tell you which features must be present for safety. School policy dictates how you should phrase the cost standard.

Use Visual Aids

Although you might use many kinds of visuals—maps of demographic statistics, drawings of key features, flow charts for procedures—you will usually use tables and graphs. With these visuals you can present complicated information easily (such as costs or a comparison of features). For many sections in your report, you will construct the table or figure first and then write the section to explain the data in it. Visual aids help overcome the problem of multiple audiences. Consider using a visual with each section in your report.

In the following example, the author first collected the data, then made the visual aids, and *then* wrote the section. Note that the table combines data from several criteria; this technique avoids many small, one-line tables. (The entire report appears as Example 15.1 at the end of the chapter.)

Fence Size

The fence serves as a guide for planing face and edge surfaces. The size of the fence, width and length, is directly related to cutting efficiency. The fence size of the jointer currently in operation is 3″ × 28″. In purchasing a new jointer the fence size should be increased for improved accuracy and squaring efficiency.

- Delta DJ-20. As Table 1 shows, the Delta fence size is 5″ × 36″. These dimensions represent a 2″ × 8″ increase, which will result in more efficient operations.
- Powermatic-60. The Powermatic fence is 4″ × 34$\frac{1}{2}$″, a 1″ × 6$\frac{1}{2}$″ increase over that of the existing fence (see Table 1).

Conclusion.   Both machines exceed the fence size criterion of 3″ × 28″. Delta DJ-20 has the greatest increase, 2″ × 8″, and will result in greater squaring accuracy and longitudinal control when jointing edge surfaces.

Table 1

8″ Jointer Capabilities Comparison

Criteria	Standard	Delta DJ-20	Powermatic-60
Fence size	Minimum of 3″ wide × 28″ long	5″ × 36″	4″ × 34½″
Table length	Minimum of 52″	76½″	64″
Cutting capacity	Minimum depth of ⅜″	⅝″	½″
Cost	Not to exceed $2300	$2128.00	$2092.00

Select a Format and an Organizational Principle

As you plan your report, you must select a format, an organizational principle for the entire report, and an organizational principle for each section.

Select a Format

Your choice of format depends on the situation. If the audience is a small group that is familiar with the situation, an informal report will probably do. If your audience is more distant from you and the situation, a formal format is preferable. The informal format is explained in Chapter 12, the formal format in Chapter 14.

In addition to selecting format type, create a style sheet of heads and margins. Review Chapter 6 and the examples in Chapters 12 and 14. Your style sheet should help your audience find what they need to do their job.

Organize the Discussion by Criteria

Organize the discussion section according to criteria, with each criterion receiving a major heading. Review Examples 15.1 and 15.2; each major section is the discussion of one criterion. Your goal is to present comparable data that readers can evaluate easily.

Organize Each Section Logically

Each section deals with one criterion and evaluates the alternatives in terms of that criterion. Each of these sections should contain three parts: an introduction, a body, and a conclusion. In the introduction, define the criterion and discuss its standard, rank, and source, if necessary. (If you discuss the standard, rank, and source somewhere else in the report, perhaps in the introduction, do not repeat that information.) In the body, explain the relevant facts about each alternative in terms of the criterion; in the conclusion, state the judgment you have made as a result of applying this criterion to the facts. You will find a sample section on page 389.

Drafting the Recommendation Report

As you draft the recommendation report, carefully develop the introduction, conclusions, recommendations/rationale, and discussion sections.

Introduction

After you have gathered and interpreted the data, develop an introduction that orients the readers to the problem and to the organization of the report. Your goal is to make readers confident enough to accept your recommendation. In recommendation reports, as in all reports, you can mix the elements of the introduction in many ways. Always include a purpose statement and add the other statements as needed by the audience. Four common elements in the introduction are

▶ Statement of purpose.
▶ Explanation of method of investigation.
▶ Statement of scope.
▶ Explanation of the problem.

Purpose

Begin a recommendation report with a straightforward statement, such as "The purpose of this report is . . . " or, more simply, "This report recommends. . . . " You can generally cover the purpose, which is to choose an alternative, in one sentence.

Method of Gathering Information

State your method of gathering information. As explained in Chapter 5, the four major methods of gathering data are observing, testing, interviewing, and reading. Stating your methodology not only gives credit where it is due but also lends authority to your data and thus to your report.

In the introduction, a general statement of your model of investigation is generally sufficient: "Using lab and catalog resources here at the university and after discussion with other Industrial Arts teachers in this area, I have narrowed my choices to two: Delta Model DJ-20 and Powermatic Model 60."

Scope

In the scope statement, cite the criteria you used to judge the data. You can explain their source or their rankings here, especially if the same reasons apply to all of them. Name the criteria in the order in which they appear in your report. If you have not included a particular criterion because data are unavailable or unreliable, acknowledge this omission in the section on scope so that your readers will know you have not overlooked that criterion. Here is an example:

Each machine has been evaluated using the following criteria, in descending order of importance:

1. Fence size
2. Table length
3. Cutting capacity
4. Cost

Background

In the background, discuss the problem, the situation, or both. To explain the problem, you must define its nature and significance: "Considering that the machine has been under continuous student use for 27 years and has reduced accuracy because of the small table and fence size, I indicated I would contact you regarding a new jointer." Depending on the audience's familiarity with the situation, you may have to elaborate, explaining the causes of various effects (Why does it have reduced accuracy? Just what is the relationship of the table and the fence? What *are* tables and fences?).

To explain the situation, you may need to outline the history of the project, indicate who assigned you to write the report, or identify your position in the corporation or organization.

The following informal introduction effectively orients the reader to the problem and to the method of investigating it.

Sample Informal Introduction

Memo head

Date: December 2, 2006
To: Joseph P. White, Superintendent of Schools
From: David Ayers
Subject: Purchase recommendation for 8" jointer

Situation and background

Recently, Jim DeLallo and I discussed at length the serious problems he was having in operating the jointer at the high school. Considering that the machine has been under continuous student use for 27 years and has reduced accuracy because of the small table and fence size, I indicated I would contact you regarding a new jointer. You asked that I forward 2007–2008 budget requests by December 15.

Cause of writing

Method

Therefore, I have prepared this recommendation report for choosing a new 8" jointer. Using lab and catalog resources here at the university and after discussion with other Industrial Arts teachers in this area, I have narrowed my choices to two: Delta Model DJ-20 and Powermatic Model 60. Each machine has been evaluated using the following criteria, in descending order of importance:

Scope

1. Fence size
2. Table length

3. Cutting capacity
4. Cost

Preview

The remainder of the report will compare both machines to the criteria.

Conclusions

Your conclusions section should summarize the most significant information about each criterion covered in the report. One or two sentences about each criterion are usually enough to prepare the reader for your recommendation. Writers of recommendation and feasibility reports almost always place these sections in the front of the report. Remember, readers want the essential information quickly.

All elements in the criteria have been met. The slightly higher cost of the Delta, $36.00, is more than offset by increased efficiency and capacity as noted below:

1. A larger fence size—for better control of stock when squaring
2. A larger table size—resulting in more efficient planing
3. A greater depth of cutting capacity—for improved softwood removal and increased rabbeting capacity

Recommendations/Rationale Section

The recommendation resolves the problem that occasioned the report. For short reports like the samples presented here, one to four sentences should suffice. For complex reports involving many aspects of a problem, a longer paragraph (or even several paragraphs) may be necessary.

Recommendation

It is recommended that the district budget for capital purchase of the Delta Model DJ-20 8″ Jointer in 2007–2008. Selection of the Delta jointer is a departure from the practice of purchasing Powermatic equipment for the woodworking shop. It is my feeling that the Delta Jointer is best suited

Possible negative factor explained

for the current and future needs of the woodworking program. Service and repair will not be a problem in changing equipment manufacturers, since N. H. Bragg services both lines of equipment.

Discussion Section

As previously noted, you should organize the discussion section by criteria, from most to least important. Each criterion should have an introduction, a body discussing each alternative, and a conclusion. Here is part of the discussion section from the recommendation report on ink-jet printers.

DISCUSSION

A. Impressions Per Hour

Definition For a press to be economically efficient it should be able to print a minimum of 8000 impressions per hour. Most of the presses on the market today are capable of this speed, but some produce at even faster speeds.

Comparison

Press	Impressions Per Hour
AB Dick 9840	10,000
Multigraphic 1860	8000

As you can see, both of the proposed presses are rated as acceptable according to the criterion set in this area. As the comparison shows, the AB Dick 9840 is capable of printing 2000 more impressions per hour than the Multigraphic 1860. With impressions per hour being the most important criterion, the AB Dick would be the best choice.

Planning the Feasibility Report

Feasibility reports investigate whether to undertake a project. They "size up a project before it is undertaken and identify those projects that are not worth pursuing" (Ramige 48). The project can be anything: place a golf course at a particular site, start a capital campaign drive, or accept a proposal to install milling machines. The scope of these reports varies widely, from analyses of projects costing millions of dollars to informal reviews of in-house proposals. Your goal is to investigate all relevant factors to determine whether any one factor will prevent the project from continuing. Basically you ask, "Can we perform the project?" and you provide the rationale for answering yes or no. Follow the same steps as for planning recommendation reports. In addition, consider the following guidelines:

▸ Consider the audience.
▸ Determine the criteria.
▸ Determine the standards.
▸ Structure by criteria.

Consider the Audience

Generally, the audience is familiar with the situation in broad outline. Your job is to give specific information. They know, for example, that in any project a certain time frame is allowed for cost recovery, but they do not know how much time this project needs. Your goal is to make them confident enough of you and the situation to accept your decision.

Determine the Criteria

Criteria are established either by a management committee or by "prevailing practice." Either a group directs investigators to consider criteria such as cost and competition level, or "prevailing practice"—the way knowledgeable experts investigate this type of proposed activity—sets the topics. For instance, cost recovery is always considered in the evaluation of a capital investment project.

If you have to discover the topics yourself, as you often do with small projects, use the three categories described on page 383—technical, management/maintenance, and financial criteria. The criteria you choose will affect the audience's sense of your credibility.

Determine the Standards

To determine standards is to state the limits of the criteria. If the topic is reimbursement for acceptable expenses, you must determine the standards to use to judge whether the stated expenses fall within the acceptable limit. These standards require expert advice unless they exist as policy. If the policy is that a new machine purchase must show a return in investment of 20 percent, and if the machine under consideration will return 22 percent, buying the machine is feasible.

Structure by Criteria

The discussion section of a feasibility report is structured by criteria. The reimbursement report could include sections on allowable growth, time of recovery of investment, and disposal costs.

Writing the Feasibility Report

· ·

To write the feasibility report, choose a format and write the introduction and the body.

Choose a Format

The situation helps you determine whether to use a formal or an informal format for your feasibility studies. As a rule of thumb, use the formal format for a lengthy report intended for a group of clients. The informal format is suitable for a brief report intended to determine the feasibility of an internal suggestion.

Write the Introduction and Body

In the introduction, present appropriate background, conclusions, and recommendations. Treat this introduction the same as a recommendation introduction. In the discussion, present the details for each topic. As in the recommendation report, you should present the topic, the standard, relevant details, and your conclusion. Organize the material in the discussion section from most to least important. As with all reports, use appropriate visual aids, including tables, graphs, and even maps, to enhance your readers' comprehension.

The following section from an informal internal feasibility report presents all four discussion elements succinctly:

Sample Feasibility Section

WHOLESALE COST

Introduction

This section examines the wholesale costs associated with producing the Heaven 'n Nature Sticker Christmas Card Kit. The standard set for this criterion was, "The wholesale cost of the product should not exceed $6.00 per set."

Analysis

A list of potential suppliers was provided by Illuminated Ink. These sources were then reviewed by searching through wholesale supply catalogs, visiting local vendors, and searching on-line sources. After an acceptable supplier had been located for each component of the kit, the total price of the kit was calculated and compared to the standard set for this criterion. See Table 1.

Table 1

Breakdown of Wholesale Component Costs

Component	Description	Qty	Supplier	Cost/ Unit
Card base	Cranberry Red, Forest Green, and Midnight Blue cardstock	3	Picture This (1)	$0.54
Card insert	white 20# bond, green ink	3	Picture This (1)	$0.12
Winter friends stickers	Frances Meyer	3	Picture This (1)	$2.64
Snowflake stickers	silver, gold, & white	24	Picture This (1)	$0.48
Brads	gold minibrads	6	Picture This (1)	$0.24

(continued)

Table 1 (continued)

Component	Description	Qty	Supplier	Cost/Unit
Cotton ball	small, white	3	Wal-Mart (2)	$0.03
Envelope	white fiber A2 envelope	3	Impact Images (3)	$0.72
Soft fold box	6″ × 9″	1	Impact Images (3)	$0.28
Instruction sheet	white 20# bond paper, black ink	1	Illuminated Ink (4)	$0.04
Mailing labels	1″ × 2⅝″, white	8	Sam's Club (5)	$0.04
			Total Cost Per Kit	**$5.13**

CONCLUSION

This research reveals that it is possible for Illuminated Ink to produce this kit for $5.13 per kit, a price that is lower than the standard required.

Several brief informal feasibility reports appear in the examples and exercises of this chapter. In addition, you can find a wide range of examples by searching Google with the keyword phrases "feasibility report" or "feasibility study."

Worksheet for Preparing a Recommendation/Feasibility Report

☐ **Analyze the audience.**
Who will receive this report?
Who will authorize the recommendation in this report?
How much do they know about the topic?
What is your purpose in writing to them? How will they use the report?
What will make you credible in their estimation?

☐ **Name the two alternatives or name the course of action that you must decide whether to take.**

☐ **Determine criteria.**
Ask technical, management/maintenance, and financial questions.

☐ **For each criterion, provide a name, a standard, and a rank.**

☐ **Rank the criteria.**

☐ **Prepare background for the report.**
Who requested the recommendation report? Name the purpose of the report. Name the method of investigation. Name the scope. Explain the problem. What is the basic opposition (such as need for profit versus declining sales)?
What are the causes or effects of the facts in the problem?

☐ Select a format—formal or informal.

☐ Prepare a style sheet including treatment of margins, headings, page numbers, and visual aid captions.

☐ Select or prepare visual aids that illustrate the basic data for each criterion.

☐ Select an organizational pattern for each section, such as introduction, alternative A, alternative B, visual aid, conclusion.

Worksheet for Evaluating Your Report

☐ Evaluate the introduction.
  • Does the introduction give you the gist of the report?
  • Does the introduction give you the context (situation, criteria, reason for writing) of the report?
  • Do you know the recommendation after reading 5 to 10 lines?

☐ Evaluate the criteria.
  • Do they seem appropriate?
  • Are all the appropriate ones included? If not, which should be added?
  • Can you find a statement of the standard for each one? If no, which ones?
  • Can you really evaluate the data on the statement of standard?
  • Do you understand why each criterion is part of the discussion?
  • Do you understand the rank of each criterion?

☐ Evaluate the discussion.
  • Is the standard given so you can evaluate?
  • Are there enough data so you can evaluate?
  • Do you agree with the evaluation?
  • Do you understand where the data came from?

☐ Evaluate the visual aid and the paper's format.
  • Are the two levels of heads different enough? See pages 142–144.
  • Is the discussion called "discussion"?
  • Does a visual appear in each spot where one would help communicate the point?
  • Are any of the visuals more or less useless; that is, they really do not interact with any points in the text?
  • Is the visual clearly titled and numbered?
  • Is the visual on the same page as the text that describes it?
  • Does the text tell you what to see in the visual?

Worksheet for Evaluating a Peer's Report

☐ **Interview a peer. Ask these questions:**
   1. Why did you include each sentence in the introduction? (Your partner should explain the reason for each one.)
   2. Why did you use the head format you used?
   3. Why did you choose each criterion?
   4. Why did you write the first sentence you wrote in each criterion section?
   5. Why did you organize each section the way you did? Do you think a reader would like to read it this way?
   6. What one point have you made with the visual aid? Why did you construct it the way you did and place it where you did?
   7. If you had to send this paper to someone who paid you money regularly for doing a good job, would you? If not, what would you do differently? Why don't you do that for the final paper?
   8. Are you happy with the level of writing in this paper? Do you think these sentences are appropriately professional, the kind of thing you could bring forward as support for your promotion? If not, how will you fix them? If you try to fix them, do you know what you're doing?

Examples

Examples 15.1–15.3 illustrate informal recommendation and feasibility reports. For other recommendation examples, see "Brief Analytical Reports," Chapter 12, pages 297–301. For another feasibility example, see "Feasibility of Finding Program Language Code on the Internet," Chapter 14, Example 14.2, pages 372–377.

Example 15.1

Informal Recommendation Report

Date: December 2, 2006
To: Joseph P. White, Superintendent of Schools
From: David Ayers
Subject: Purchase recommendation for 8″ jointer

Recently, Jim DeLallo and I discussed at length the serious problems he was having in operating the jointer at the high school. Considering that the machine has been under continuous student use for 27 years and has reduced accuracy because of the small table and fence size, I indicated I would contact you regarding a new jointer. You asked that I forward 2006–2007 budget requests by December 15.

Therefore, I have prepared this recommendation report for choosing a new 8″ jointer. Based on lab and catalog resources and after discussion with other Industrial Arts teachers in this area, I have narrowed my choices to two: Delta Model DJ-20 and Powermatic Model 60. Each machine has been evaluated using the following criteria, in descending order of importance:

1. Fence size
2. Table length
3. Cutting capacity
4. Cost

RECOMMENDATION

It is recommended that the district budget for capital purchase of the Delta Model DJ-20 8″ jointer in 2006–2007. Selection of the Delta jointer is a departure from the practice of purchasing Powermatic equipment for the woodworking shop. It is my feeling that the Delta jointer is best suited for the current and future needs of the woodworking program. Service and repair will not be a problem in changing equipment manufacturers, since N. H. Bragg services both lines of equipment.

All elements in the criteria have been met. The slightly higher cost of the Delta, $36.00, is more than offset by increased efficiency and capacity, as noted below.

1. A larger fence size—for better control of stock when squaring
2. A larger table size—resulting in more efficient planing
3. A greater depth of cutting capacity—for improved softwood removal and increased rabbeting capacity

The remainder of the report will compare both machines to the criteria.

CRITERIA

Fence Size

The fence serves as a guide for planing face and edge surfaces. The size of the fence, width and length, is directly related to cutting efficiency. The fence size of the jointer currently in operation is 3″ × 28″. In purchasing a new jointer the fence size should be increased for improved accuracy and squaring efficiency.

- Delta DJ-20. As Table 1 shows, the Delta fence size is 5″ × 36″. This represents a 2″ × 8″ increase, which will result in more efficient operations.
- Powermatic-60. The Powermatic fence is 4″ × 34½″, a 1″ × 6½″ increase over that of the existing fence (see Table 1).

(continued)

Example 15.1

(continued)

Conclusion Both machines exceed the fence size criterion of 3″ × 28″. Delta DJ-20 has the greatest increase, 2″ × 8″, and will result in greater squaring accuracy and longitudinal control when jointing edge surfaces.

Table 1

8″ Jointer Capabilities Comparison

Criteria	Standard	Delta DJ-20	Powermatic-60
Fence size	Minimum of 3″ wide × 28″ long	5″ × 36″	4″ × 34½″
Table length	Minimum of 52″	76½″	64″
Cutting capacity	Minimum depth of ⅜″	⅝″	½″
Cost	Not to exceed $2300	$2128.00	$2092.00

Table Length

In-feed and out-feed tables are combined and referred to as table length. Increased table length improves accuracy when jointing and provides greater stability when planing face surfaces. On the existing machine, the table length is 52″. When planing and jointing stock over 40″, it is difficult to maintain accuracy. To realize improved handling and accuracy on a new jointer, the table length should be above 52″.

- Delta DJ-20. As Table 1 shows, the table length of the Delta is 76½″, a 24½″ increase. This increased size will allow for greater efficiency when planing stock to approximately 60″.
- Powermatic-60. As Table 1 shows, the table length of this jointer exceeds the minimum length by 12″. Improved planing can be increased to approximately 50″.

Conclusion Both jointers exceed the 52″ minimum table length size. The significant increase in the Delta jointer table length will offer improved planing and jointer accuracy and increased handling capacity.

Cutting Capacity (Depth of Cut)

Jointer cutting capacity is determined by the maximum depth of cut. This depth of cut is created when the in-feed table is lowered. For production work with softwoods and edge rabbeting, a large depth of cut is desired. The existing jointer has a ⅜″ maximum depth of cut. This is a limiting factor when doing softwood production work and constructing edge rabbets over ⅜″. When purchasing a new machine, the depth of cut should be at least ⅜″.

- Delta DJ-20. As Table 1 shows, the depth of cut on this machine is $\frac{5}{8}$", $\frac{1}{4}$" above the minimum standard. This will be an important feature when edge rabbeting and doing softwood production work.
- Powermatic-60. As Table 1 shows, the depth of cut for this machine is $\frac{1}{2}$", a $\frac{1}{8}$" increase above the minimum standard.

Conclusion Both machines exceed the $\frac{3}{8}$" minimum criterion set. The Delta jointer has the greatest depth of cut, $\frac{5}{8}$", which will allow for greater soft-wood removal and maximum rabbeting.

Cost

The jointer is a capital equipment item and the cost cannot be department budgeted if in excess of $2300, unless prior approval is granted by the secondary committee. Costs (including shipping, stand, and three-phase conversion) are

- Delta DJ-20: $2128.00
- Powermatic-60: $2092.00

Conclusion Both machines meet the fourth criterion. The Powermatic is slightly lower in cost, but does not have all the capacity and features of the Delta model. The additional cost of the Delta jointer ($6.80 on a 20-year depreciation schedule) is more than offset by the increase in table and fence size and improved cutting depth.

Example 15.2

Informal Recommendation Report

Date:	December 3, 2006
To:	Steve Zubek, President/Widgit Printing
From:	Mark Jezierski, Production Manager
Subject:	Which of the proposed printing presses we should purchase

I am writing in response to your recent request to determine which of the proposed printing presses, the Multigraphic 1860 or the AB Dick 9840, would best satisfy our needs. The age and inefficiency of the current equipment used in production have prompted this report. I have thoroughly researched these machines through trade journals and with other printers who currently use these machines to determine which of these machines best fit the criteria set by upper management.

My recommendation to you is that the Widget Printing Company purchase the AB Dick 9840. The criteria I have used to determine the feasibility of this purchase according to their rank and importance consist of:

1. Machine must be able to print a minimum of 8000 impressions per hour.
2. The press must be able to print an area image up to 11.5" × 17.25".

(continued)

Example 15.2

(continued)

3. The press must be able to print a paper size of 12″ × 17.5″.
4. The cost to purchase the new machine must be under $25,000.

DISCUSSION

A. Impressions Per Hour

Definition For a press to be economically efficient it should be able to print a minimum of 8000 impressions per hour. Most of the presses on the market today are capable of this speed, but some produce at even faster speeds.

Comparison

Press	Impressions Per Hour
AB Dick 9840	10,000
Multigraphic 1860	8000

As you can see, both of the proposed presses are rated as acceptable according to the criterion set in this area. As the comparison shows, the AB Dick 9840 is capable of printing 2000 more impressions per hour than the Multigraphic 1860. With impressions per hour being the most important criterion, the AB Dick would be the best choice.

B. Maximum Image Area

Definition The press purchased must be able to print an image up to 11.5″ × 17.25″. This image area will allow the printing of 8.5″ × 11″ and 11″ × 17″ jobs that have from one side to four sides that bleed.

Comparison

Press	Maximum Image Area
Multigraphic 1860	13.19″ × 17.50″
AB Dick 9840	12.50″ × 17.25″

Again you can see both of the presses meet the suggested criterion of 11.5″ × 17.25″. However, the Multigraphic 1860 is capable of the largest maximum image area (13.19″ × 17.50″). This gives a considerable excess amount of image area to be used to print jobs that require larger image areas than the specified maximum of the criterion. The AB Dick 9840 is capable of an image area of 12.5″ × 17.25″, which is acceptable according to the standards we have set. When considering only image area, the Multigraphic 1860 would be the best choice.

C. Paper Size

Definition The press purchased must be able to print a paper size of 12″ × 17.5″. This paper size will allow the printing of 8.5″ × 11″ and 11″ × 17″ jobs that have from one side to four sides that bleed.

Comparison

Press	Maximum Paper Area
Multigraphic 1860	15.00″ × 18.00″
AB Dick 9840	13.50″ × 17.25″

Again, as you can see, both of the proposed presses meet the criterion established for this area. The Multigraphic 1860 is capable of printing on the largest size paper. It can print on paper up to 15″ × 18″, which by far exceeds the paper size requirement needed by Widget Printing. This capability, however, will allow us a greater flexibility in projects that require larger than specified paper sizes. The AB Dick 9840 also exceeds the established criterion, although by a smaller margin than the Multigraphic 1860, but it will also allow us a degree of flexibility in the undertaking of larger projects. When considering only paper size, the Multigraphic 1860 would be the best choice.

D. Price

Definition The price of the press purchased is not a major criterion. The capabilities of the press are more important factors. However, a price range has been set and cannot be overlooked when deciding on which press to purchase. We have established a price ceiling of $25,000.

Comparison

Press	Price
AB Dick 9840	$17,395
Multigraphic 1860	$20,000

Once again both of the proposed machines meet the established criterion. Both of these presses are well within the price range set. Based on only price as a criterion, the AB Dick would be the best choice.

RECOMMENDATION

After comparing the AB Dick 9840 and the Multigraphic 1860 presses, I feel the AB Dick 9840 is the press that the Widget Printing Company should purchase. The AB Dick meets all of the established criteria and has the largest capabilities in the area of impressions per hour, which is the most important criterion established. It also is the best press per dollar of purchase available. Therefore I feel that it is in our best interest to purchase the AB Dick 9840.

Example 15.3

Informal
Feasibility Report

NEW PRODUCT DEVELOPMENT FEASIBILITY REPORT

Heaven 'n Nature Sticker Christmas Card Kit

October 28, 2005

Frances Butek

Product Development Specialist

Prepared for
Illuminated Ink
15825 - 160th Ave.
Bloomer, WI 54724

REPORT ABSTRACT

This feasibility report addresses the question, "Should Illuminated Ink produce Heaven 'n Nature Sticker Christmas Card kits?" The criteria used to answer this question are wholesale cost, component availability, and product desirability. Our conclusion is that Illuminated Ink should go forward with production plans.

INTRODUCTION

Question

This feasibility report addresses and answers the question, "Should Illuminated Ink produce Heaven 'n Nature Sticker Christmas Card kits?"

Purpose

The purpose of this report is to provide the co-owners of Illuminated Ink with information on if they should invest their time, effort, and resources into the development of a particular new product. Illuminated Ink is a small, faith-based business specializing in developing, designing, and marketing educational toys, games, crafts, and curriculum.

Product Description

The Heaven 'n Nature Sticker Christmas Card Kit is a product designed to encourage a child to create six unique Christmas cards that display a beautiful winter scene of realistic woodland animals decorating a majestic Christmas tree. The interior of the card is preprinted with the greeting, "Joy to the World, the Lord is come! Let every heart prepare Him room, while Heaven and Nature sing the wonders of His love!"

CRITERIA AND STANDARDS

Three criteria were considered to make an informed decision. The standards set for each criteria were determined by the co-owners of Illuminated Ink. These criteria and standards were:

1. **Wholesale Cost.** The wholesale cost of the product should not exceed $6.00 per set.
2. **Component Availability.** All of the components of the kit must be received within 10 days of ordering.
3. **Product Desirability.** The product must be deemed desirable by 75 percent or more of Illuminated Ink's customers.

(continued)

Example 15.3

(continued)

METHOD

Two lists were supplied by Illuminated Ink. These lists contained the names and contact information for potential customers and wholesale suppliers. The lists were reviewed, customers and suppliers were contacted, and an analysis performed on the information provided. The results of the analysis were compared to the criterion standards, and a decision was made.

CONCLUSION

The results of all three criteria were positive. First, the final wholesale cost of the components was determined to be $5.13 per set. Second, the maximum amount of time Illuminated Ink must wait to receive the kit components was established to be 10 days. And third, the kit was rated as highly desirable by Illuminated Ink's customers. Thus, having met all three criteria standards, our conclusion to this feasibility question is a resounding "YES! Illuminated Ink should produce this product."

WHOLESALE COST

Introduction

This section examines the wholesale costs associated with producing the Heaven 'n Nature Sticker Christmas Card kit. The standard set for this criterion was, "The wholesale cost of the product should not exceed $6.00 per set."

Analysis

A list of potential suppliers was provided by Illuminated Ink. These sources were then reviewed by searching through wholesale supply catalogs, visiting local vendors, and searching on-line sources. After an acceptable supplier had been located for each component of the kit, the total price of the kit was calculated and compared to the standard set for this criterion. See Table 1.

CONCLUSION

This research reveals that it is possible for Illuminated Ink to produce this kit for $5.13 per kit, a price that is lower than the standard required.

COMPONENT AVAILABILITY

Introduction

This section examines the availability of the components required for the Heaven 'n Nature Sticker Christmas Card kit. The standard set for this criterion was, "All of the components of the kit must be received within ten days of ordering."

Analysis

A list of potential suppliers was provided by Illuminated Ink. These sources were then reviewed by searching through wholesale supply catalogs, visiting local vendors, and searching on-line sources. After an acceptable supplier had been located for each component of the kit, the amount of time required for delivery was noted. See Table 2.

Table 1
Breakdown of Wholesale Component Costs

Component	Description	Qty	Supplier	Cost / Unit
Card base	Cranberry Red, Forest Green, and Midnight Blue cardstock	3	Picture This (1)	$0.54
Card insert	white 20# bond, green ink	3	Picture This (1)	$0.12
Winter friends stickers	Frances Meyer	3	Picture This (1)	$2.64
Snowflake stickers	silver, gold, & white	24	Picture This (1)	$0.48
Brads	gold minibrads	6	Picture This (1)	$0.24
Cotton ball	small, white	3	Wal-Mart (2)	$0.03
Envelope	white fiber A2 envelope	3	Impact Images (3)	$0.72
Soft fold box	6" × 9"	1	Impact Images (3)	$0.28
Instruction sheet	white 20# bond paper, black ink	1	Illuminated Ink (4)	$0.04
Mailing labels	1" × 2⅝", white	8	Sam's Club (5)	$0.04
			Total Cost Per Kit	**$5.13**

Table 2
Breakdown of Component Supply Times

Component	Supplier	Description	Supply Time
Card base	Picture This (1)	Cranberry Red, Forest Green, and Midnight Blue cardstock	3 days
Card insert	Picture This (1)	white 20# bond, green ink	2 days
Winter friends stickers	Picture This (1)	Frances Meyer	7 days
Snowflake stickers	Picture This (1)	silver, gold, & white	3 days
Brads	Picture This (1)	gold minibrads	10 days
Cotton ball	Wal-Mart (2)	small, white	2 days
Envelope	Impact Images (3)	white fiber A2 envelope	4 days
soft fold box	Impact Images (3)	6" × 9"	4 days
instruction sheet	Illuminated Ink (4)	white 20# bond paper, black ink	2 days
mailing labels	Sam's Club (5)	1" × 2⅝", white	2 days
		Maximum Supply Time	**10 days**

(continued)

Example 15.3

(continued)

Conclusion

This research reveals that it is possible for Illuminated Ink to obtain the components for this kit within 10 days, a time frame that falls within the standard required.

PRODUCT DESIRABILITY

Introduction

This section examines the desirability of the Heaven 'n Nature Christmas Card kit. The standard set for this criterion was, "The product must be deemed desirable by 75 percent or more of Illuminated Ink's customers."

Analysis

A list of customers was provided by Illuminated Ink. These customers were sent an e-mail that requested that they fill out an on-line Product Desirability Survey (6) that was posted on the Illuminated Ink website. This survey displayed an illustration of the product and provided a detailed description of the kit and its contents. The survey then asked viewers if they found the product appealing, and what price range they felt was most appropriate for the product. The survey responses were recorded in a table and the results analyzed. See Table 3.

Table 3
Breakdown of Desirability Survey Responses

Question: Do you find this product appealing?	Yes	No
Total Responses	176	16

Conclusion

Of the 192 responses received, 92 percent of Illuminated Ink's customers found this product desirable. This finding is considerably higher than the standard that was set, and indicates that Illuminated Ink should strongly consider producing this product.

CHRISTMAS CARD KIT SURVEY

Card Description

A beautiful winter scene of realistic woodland animals decorating a majestic Christmas tree is created by a child from a selection of stickers and then applied to a white background. A Midnight Blue cardstock base with a hinged church window is folded in half, and the forest scene is placed "in" the window. The interior of the card is preprinted with the greeting, "Joy to the World, the Lord is come! Let every heart prepare Him room, while Heaven and Nature sing the wonders of His love!" The interior and exterior of the card are then decorated with additional snowflake and woodland animal stickers. Finally, two tiny gold brads are attached to the window frame as handles.

Kit Contents
(All materials needed to make six cards.)

6 die-cut card exteriors (2 each of Midnight Blue, Forest Green, and Cranberry Red)
6 white, interior sheets
6 white fiber, acid-free-paper envelopes
12 gold brads
Stickers (silver, gold, and white snowflakes, woodland animals)
3 cotton balls
Illustrated instructions

(continued)

Example 15.3

(continued)

Our Questions

Please answer the questions below. Your answers will not be used for any purpose other than to help us determine whether or not we should develop this product.

Please Note

We are confident that the design of this card kit will be well received by our customers. What we are concerned about is the cost involved in producing it. We are very cost conscious about our products and strive to make our products affordable to all. Please give special consideration to the question regarding the price you would be willing to pay for this kit, as this will most likely be the factor that ultimately determines whether or not we are able to produce this product.

First Name _____

Last Name _____

Email Address _____

As described above, does this product idea appeal to you?
○   Yes   ○   No

Would you be inclined to purchase such a product if it were available?
○   Yes   ○   No   ○   Maybe

What is the maximum price range that you would be willing to pay for this type of product?
○ $4.00 - $5.99   ○ $6.00 - $7.99   ○ $8.00 - $9.99
○ $10.00 - $11.99   ○ $12.00 - $13.99   ○ $14.00 - $15.99

Your Advice

We value your advice! Please use the space below to share with us any comments/suggestions/concerns that you feel might help us improve this product.

1. _____

2. _____

3. _____

4. _____

5. _____

Thank You for your help!

| Submit | | Clear |

REFERENCES

(1) Picture This
6000 Hwy 93
Eau Claire, WI
Price Quote Date: October 28, 2005

(2) Wal-Mart
3915 Gateway Drive
Eau Claire, WI
Visited On: October 28, 2005

(3) Impact Images
4919 Windplay Drive Suite 7
El Dorado Hills, CA 95762
Price Quote Date: October 23, 2005

(4) Illuminated Ink
15825 – 160th Ave.
Bloomer, WI 54724
Price Quote Date: October 23, 2005

(5) Sam's Club
4001 Gateway Dr.
Eau Claire, WI 54701
Price Quote Date: October 28, 2005

(6) Christmas Card Kit Survey
<www.illuminatedink.com/christmas_card_kit_survey.htm>
Posted On: October 28, 2005

Exercises

▶ Group

1. In groups of two to four, analyze the community attitudes that are addressed by the authors of Examples 15.1–15.3 or of the examples in Exercises 2 and 3 below. What factors have the writers obviously tried to accommodate? What kind of memo is expected? What length? Do they desire to prove conclusively that the material is accurate? Or is there an informal understanding that only a few words are necessary?

    Alternate: In the groups, role-play the sender and receiver of the reports. Receivers interview the senders to decide whether to implement the recommendation.

▶ You Analyze

2. Analyze this sample for organization, format, depth of detail, and persuasiveness. If necessary, rewrite the memo to eliminate your criticisms. Create the visual aid that the author mentions at the end of the report. Alternate: Rewrite the memo as a much "crisper," less chatty document. Alternate: Construct a table that summarizes the data in the report.

    The purpose of this report is to determine from which insurance company I should purchase liability insurance for my 2001 Chevrolet Cavalier. Data for this report were gathered from personal interviews with agents representing their companies. After comparing different companies, I narrowed my choice to decide which one I should buy. I evaluated Ever Safe and Urban Insurance using the following criteria, which are ranked in importance:

    1. Cost—Could annual insurance of liability be less than $250?
    2. Payments—Could it be paid semiannually?
    3. Service—Is the agent easily accessible?

    After this evaluation, I concluded that Ever Safe was the best company to purchase my liability insurance. First, this insurance company costs $245, which is less than the $250 limit that I proposed to spend. Second, it can be paid semiannually. And, third, Ever Safe offers toll-free claim service 24 hours a day.

    Ever Safe costs $245 a year, with Urban Insurance costing $240 a year, which both met my required criteria of purchasing liability insurance for under $250. Urban Insurance is $5 less; however, Ever Safe does have other options that are worth the extra money in means of purchasing.

    Ever Safe and Urban Insurance both offer semiannual payments. In terms of this aspect, they are both weighted the same.

Ever Safe offers toll-free claim service 24 hours a day. Urban Insurance is long distance with limited working hours. They are available after working hours but only through an answering machine that will record your message for the agent to get in touch with you on the following day.

In the decision of an insurance company, it is plain to see that Ever Safe meets the requirements of my criteria and that Urban Insurance does not. Urban Insurance is cheaper, allows semiannual payments, but does not fulfill the service that I was looking for. For the extra dollars of payment, the service in Ever Safe is worth it.

3. Analyze this section for organization, format, depth of detail, and persuasiveness. Rewrite the memo, if necessary, to eliminate your criticisms. Create a visual aid that the author mentions at the end of the report. Summarize the data that support the recommendation. Alternate: Rewrite the memo as a much "crisper," less chatty document. Alternate: Construct a table or figure that effectively summarizes the data in the report.

COST

INTRODUCTION

Since the 49'R Pulling Team does not have any sponsors, they can only spend money on parts that are necessary and feasible. It was proposed that a new supercharger should not cost more than $5000 after the trade-in of the 49'R Pulling Teams' current supercharger. Just the purchase of the supercharger minus the estimated value of the current supercharger would have been fine. But there would have been a cost associated with the equipment criterion that would push the total cost too high.

RESEARCH

The 49'R Pulling Team wanted to trade in their existing supercharger, so a sales representative would have to be contacted to negotiate a price. Since the 49'R Pulling Team has been in this sport for many years, they provided some information about who to contact about purchasing a new supercharger. The one person they have done business with is John Knox, who is with Sassy Engines in New Hampshire. It was estimated the value of the existing supercharger would be $1500.

Research from the Internet provided only a few companies that offer the style and size supercharger that the 49'R Pulling Team was looking for. The companies included Littlefield, SSI, SCS, Kobelco, and Kuhl. The cost for a new Kobelco 14-71 hi-helix supercharger was $5800. For a SSI 14-71 hi-helix supercharger, the outright cost was $5800 dollars. The cost for a Kuhl 14-71 hi-helix retro-fit supercharger was $5250. Those prices did not include any money from a trade-in or sale.

So far, all three superchargers were under $5000 and this criterion would have had a positive recommendation. But after looking into the equipment criterion and finding out there would be an additional $3500 cost, all three superchargers exceed the $5000 limit. The reason for the equipment cost can be found under the equipment link.

www.kuhlsuperchargers.com/cat_p02.htm
www.kocoa.com/2005_pricing_schedules.htm
www.sassyengines.com/Blowerdriveparts.html

CONCLUSION

This table gives the results that were accumulated and the cost after a trade-in or possible sale estimated at $1500 and with the $3500 equipment cost.

Supercharger Brand	Cost for New	Cost After Trade-In	Final Cost
Kobelco	$5800	$4300	$7800
Kuhl	$5250	$3750	$7250
SSI	$5800	$4300	$7800

After the estimated value of the existing supercharger was subtracted from the cost of a new 14-71 supercharger from three different companies and the addition of the equipment cost, all three were more than the $5000 that was set in this criterion. Under the restrictions of this criterion, the purchase of a new supercharger would not be recommended to the 49'R Pulling Team.

▶ You Revise

4. Rewrite this brief section. Create a table that illustrates the data.

INTRODUCTION

The University of Wisconsin–Stout has been submitted to a tremendous amount of budget cuts. The athletic department has been granted an estimated budget of $250,000 for updates of current facilities. For a new floor installation to be feasible, the total cost must not reach over $250,000.

FIGURES

According to Connors Flooring, the total installation cost of a maple sports floor is $81,300 with over 38 years of life expectancy. This is based on a 10,000-square-foot floor.

The Maple Floor Manufacturers Association has concluded that wood floors also require regular cleaning, sanding, lines repainted, and floor refinishing approximately every three years. This is a cost of approximately

$8,000.00 per three years, equaling $2,666.00 p/year. With a 38 year life expectancy, the total for cleaning, sanding, painting, etc. = $101,308.

CONCLUSION

The total cost of the floor is under $200,000, which meets the first criteria of not exceeding $250,000.

5. Rewrite this text for a more professional tone. Evaluate the table for effectiveness; if necessary create a new table.

HAS TO BE UNDER $75

INTRODUCTION

One of the biggest things for our team is going to be the fact of money and how much these sweat suits will be costing. First off none of us have much money to begin with and we also have limited funds for our sport as is. I feel that if you can set a limit that the whole team can agree is reasonable while at the same time getting a quality product. I feel that this is actually one of the most important criterion.

RESEARCH

First of all to get the price set at $75 was very easy, at practice one day when we were discussing the sweat suits, all I had to say was that price and everyone agreed right on the spot. The next part of the research was the tough part. The three places I ended up going to look for this price was eastbay.com, askjeeves.com, and to Fleet Feet. The research that I did turned out pretty helpful for this project although a lot more work then I had intended.

RESULTS

When I first started looking at sweat suits on eastbay, it was a big disappointment, the selection was really good and they had quite a bit of stuff but I knew that there was no way that anyone on our team could afford those. All the suits were at least $100 or more and I won't even mention some of the prices that were listed. When I went to askjeeves, I ran into basically the same problem. All that it was really giving me were links to really expensive name brand products. Although name brands are very dependable there is always another no name brand that can be just as durable for a cheaper price, so pushed on in my journey. When I went into Fleet Feet I just had a feeling that they would at least help me find something if they didn't have it there with them. Sure enough they gave me a catalog along with a website that I could visit. When I went to the website which is hollowayusa.com, I found what I was looking for right away. When I saw it, it was just too good to be true, the pants were listed at

$35.90 and the top is listed at $37.90. Which is a grand total of $73.80. You can see the different prices I ended up finding on the different sites located in Table 1 below.

Table 1

Different Price Options

Brand	Top	Bottom	Total
Nike	$55.00	$60.00	$115.00
Holloway	$37.90	$35.90	$73.80
Addidas	$52.50	$50.25	$102.75

CONCLUSION

When I saw the sweats and then saw the price of them I knew that they had a good shot of getting accepted. This price is a perfect price and a price that everyone on the team had agreed on before I even started looking for the suits. When I let the team know the price they were all very excited.

▶ You Create

6. As your instructor requires, perform the following exercises in conjunction with one of this chapter's Writing Assignments.

   a. Perform the actions required by the worksheet.

   b. Write a discussion section. Construct a visual aid that depicts data for each criterion. Write an introduction for the section: define the criterion and tell its significance, rank, and source. Point out the relevant data for each section. Write a one-sentence conclusion. Word it positively. (Say X is cheaper than Y, not Y is more expensive than X.)

   c. Write an introduction that orients the reader to the situation and to your recommendation. Choose one of the several methods shown in this chapter and Chapter 12.

   d. In groups of two or three, review each other's problem statements and the criteria derived from them. Make suggestions for improvement.

   e. In groups of two or three, read a body section from each other's reports. Assess whether it presents the data that support the conclusion.

   f. In groups of two or three, compare conclusions to the recommendations. Do the conclusions support the recommendations?

   g. In groups of two or three, assess each other's introductions. Do they contain enough information to orient the reader to the situation and the recommendation?

h. In groups of two or three, read the near-final reports for consistency of format. Are all the heads at the appropriate level? Are all the heads really informative? Is the style sheet applied consistently? Does it help make the contents easy to group? Do the visual aids effectively communicate key points?

Writing Assignments

1. Assume that you are working for a local firm and have been asked to evaluate two kinds, brands, or models of equipment. Select a limited topic (for instance, two specific models of 10-inch table saws, the Black and Decker model 123 and the Craftsman model ABC), and evaluate the alternatives in detail. Write a report recommending that one of the alternatives be purchased to solve a problem. Be sure to explain the problem. Both alternatives should be workable; your report must recommend the one that will work better.

    Gather data about the alternatives just as you would when working in industry—from sales literature, dealers, your own experience, and the experience of others who have worked with the equipment. Select a maximum of four criteria by which to judge the alternatives and use a minimum of one visual aid in the report. Aim your report at someone not familiar with the equipment. Fill out the worksheet in this chapter, and perform the parts of Exercise 6 that your instructor requires.

2. Assume that you are working for a local firm that wants to expand to a site within 50 miles. Pick an actual site in your area. Then write a feasibility report on the site. Devise criteria based on the situation. Do all the research necessary to discover land values, transportation systems, governing agencies, costs, and any other relevant factors. Your instructor will provide you with guidance about how to deal with the local authorities and how to discover the facts about these topics. Use this chapter's worksheet, and perform those parts of Exercise 6 that your instructor requires.

3. Assume that you have been asked to decide on the feasibility of a proposed course of action. Name and describe the proposal. Then establish the relevant criteria to determine feasibility. Apply the criteria and write an informal report. Use this chapter's worksheet, and perform those parts of Exercise 6 that your instructor requires.

4. Find a firm or an agency in your locale that has a problem that it will allow you to solve. Research the problem, and present the solution in a report. The report may be either formal or informal, recommendation or feasibility. Your instructor will help you schedule the project. This project should not be an exercise in format and organization, but a solution that people need in order to perform well on their jobs. Use this chapter's worksheet, and perform those parts of Exercise 6 that your instructor requires.

5. Assume that your manager wants to create a webpage. Investigate the situation, and write a report explaining the feasibility of creating and maintaining a website.

6. Write a learning report for the writing assignment you just completed. See Chapter 5, Writing Assignment 7, pages 134–135, for details of the assignment.

Web Exercises

1. Assume that your manager wants to create a webpage. Investigate the situation and write a report explaining the feasibility of creating and maintaining a website.

2. Write a report on whether or not the Web is a feasible source of information that you can use to perform your duties as a professional in your field. For instance, is the Web a more feasible source than hard copy of OSHA regulations or ASTM standards?

Works Cited

Alexander, Heather, and Ben Potter. "Case Study: The Use of Formal Specification and Rapid Prototyping to Establish Product Feasibility." *Information and Software Technology* 29.7 (1987): 388–394.

Angelo, Rocco M. *Understanding Feasibility Studies.* East Lansing, MI: Educational Institute of the American Hotel and Motel Association, 1985.

Bradford, Michael. "Four Types of Feasibility Studies Can Be Used." *Business Insurance* (19 June 1989): 16.

Holcombe, Marya W., and Judith K. Stein. *Writing for Decision Makers: Memos and Reports with a Competitive Edge.* Belmont, CA: Lifelong, 1981.

Markel, Mike. "Criteria Development and the Myth of Objectivity." *The Technical Writing Teacher* 18.1 (1991): 37–47.

Ramige, Robert K. "Packaging Equipment, Twelve Steps for Project Management." *IOPP Technical Journal* X.3 (1992): 47–50.

Chapter 16 Proposals

Chapter 16
In a Nutshell

The goal of a proposal is to persuade readers to accept a course of action as an acceptable way to solve a problem or fill a need. Internal proposals show that the situation is bad and your way will clearly make it better. External proposals show that your way is the best.

Basic proposal issues. Four issues for you to discuss convincingly in a proposal are

▶ The *problem*—how some fact negatively affects positive expectations (high absenteeism on manufacturing line 1 is causing a failure to meet production goals) and that you know the cause (workers are calling in sick because of sore backs).

▶ The *solution*—actions that will neutralize the cause (eliminate bending by reconfiguring the work tables and automating one material transfer point).

▶ The *benefits* of the solution—what desirable outcome each person or group in the situation will obtain.

▶ The *implementation*—who will do it and how, how long it will take.

Develop credibility. To accept your solution, your readers must feel you are credible. Your methods must be clear and sound—an expert's assessment of the situation. Your analyses of the problem, the cause, the benefits, how long it will take, the cost, etc., must show a reasoned regard of each concern, one that will not cause surprises later on.

Basic guidelines. Follow these guidelines:

▶ Use a top-down strategy.

▶ Describe the situation and use visual aids.

▶ Provide content in the introduction.

▶ Provide a summary that clearly states the proposed solution.

A proposal persuades its readers to accept the writer's idea. There are two kinds of proposals: external and internal. In an *external proposal,* one firm responds to a request—from another firm or the government—for a solution to a problem. Ranging from lengthy (100 pages or more) to short (4 or 5 pages), these documents secure contracts for firms. In an *internal proposal,* the writer urges someone else in the company to accept an idea or to fund equipment purchases or research.

The External Proposal

A firm writes external proposals to win contracts for work. Government agencies and large and small corporations issue a *request for proposal (RFP),* which explains the project and lists its specifications precisely. For example, a major aircraft company, such as British Airlines, often sends RFPs to several large firms to solicit proposals for a specific type of equipment—say, a guidance system. The RFP contains extremely detailed and comprehensive specifications, stating standards for minute technical items and specifying the content, format, and deadline for the proposals.

The companies that receive the RFP write proposals to show how they will develop the project. A team assembles a document demonstrating that the company has the technical know-how, managerial expertise, and budget to develop the project.

After receiving all the proposals, the firm that requested them turns them over to a team of evaluators, some of whom helped write the original specifications. The evaluators rate the proposals, judging the technical, management, and cost sections in order to select the best overall proposal (Bacon).

Not all proposals are written to obtain commercial contracts. Proposals are also commonly written by state and local governments, public agencies, education, and industry. University professors often write proposals, bringing millions of dollars to campuses to support research in fields as varied as food spoilage and genetic research.

Discussion of a lengthy, 50- to 200-page proposal is beyond the scope of this book; it is a subject for an entire course. But brief external proposals are very common. They require the same planning and contain the same elements as a lengthy proposal. The following sections illustrate the planning and elements of a brief external proposal.

Planning the External Proposal

To write an external proposal, you must consider your audience, research the situation, use visual aids, and follow the usual form for this type of document.

Consider the Audience

The audience for an external proposal consists of potential customers. These customers know that they have a need, and they have a general idea of how to fill that need. Usually they will have expressed their problem to you in a written statement (an RFP) or in an interview. Generally, a committee decides whether to accept your proposal. Assess their technical awareness and write in such a way that not only do they understand your proposal, but they also have confidence in it and in you. To write to them effectively, you should follow these guidelines:

▶ Address each need they have expressed.
▶ Explain in clear terms how your proposal fills their needs.
▶ Explain the relevance of technical data.

For instance, if you want to sell a computer system to a nonprofit arts organization, you cannot just drop terms for computer parts—say, 2 gigs of RAM—and expect them to know what that means. You need to explain the data so that the people who make the decision to commit their money will feel comfortable doing so.

Research the Situation

To write the proposal effectively, understand your customer's needs as well as the features of your own product or service. Your goal is to show how the features will fill the needs. Discover this by interviewing the customer or by reading their printed material. Showing that you understand the situation and have taken proper research steps enhances your credibility.

Writers devise different ways to develop their research. To relate needs and features, many writers compile a two-column table like this:

Need	Feature That Meets Need
Director must be able to access latest financial data and public relations data.	Available in content management database.
Director must be able to access data at any time.	Director needs personal digital assistant with wireless capabilities.
Secretary enters data, but not continuously.	Secretary needs wireless access to workstation.
Secretary does accounting.	Secretary can use Accountant Inc. 2.1c and Excel.
Artist enters data.	Artist needs wireless access to workstation.

Need	Feature That Meets Need
Artist does desktop publishing.	Artist needs InDesign, Photoshop, Illustrator, and six-color laser printer.
$15,000 maximum.	Airport station, 2 laptop computers, 1 pda, software for artwork, word processing, accounting, and desktop publishing.

Once you establish the client's needs, you can easily point out a reasonable way to meet them.

Use Visual Aids

Many types of visual aids may be appropriate to your proposal. Tables might summarize costs and technical features. Maps (or layouts), for instance, might show where you will install the workstation and the electrical lines in the office complex. Illustrations of the product with callouts can point out special features. Remember that your goal is to convince the decision makers that your way is the best; good visuals are direct and dramatic, drawing your client into the document.

Writing the External Proposal

To write an external proposal, follow the usual form for writing proposals. The four main parts of a proposal are an executive summary and the technical, managerial, and financial sections.

The Executive Summary

The executive summary contains information designed to convince executives that the proposers should receive the contract. In short external proposals, this section should be reduced in proportion to the body (see Chapter 8). It should succinctly present the contents of the technical, managerial, and financial sections. Generally write this section last.

The Technical Section

A proposal's technical section begins by stating the problem to be solved. The proposers must clearly demonstrate that they understand what the customer expects. The proposal should describe its approach to solving the problem and present a preliminary design for the product, if one is needed. Sometimes the firm offers alternative methods for solving the problem and invites the proposal writer to select one. In the computer network example, the proposal might explain three different configurations that fulfill needs slightly differently but still stay within the $15,000 maximum cost.

The Management Section

The management section describes the personnel who will work directly on the project. The proposal explains the expertise of the people responsible for the project. In a short proposal, this section usually explains qualifications of personnel, the firm's success with similar projects, and its willingness to service the product, provide technical assistance, and train employees. This section also includes a schedule for the project, sometimes including deadlines for each phase.

The Financial Section

The financial section provides a breakdown of the costs for every item in the proposal. This section varies in depth. Often a brief introduction and table may be sufficient, but if you need to explain the source or significance of certain figures, do so.

The Internal Proposal

The internal proposal persuades someone to accept an idea—usually to change something, or to fund something, or both. Covering a wide range of subjects, internal proposals may request new pieces of lab equipment, defend major capital expenditures, or recommend revised production control standards. The rest of this chapter explains the internal proposal's audiences, visual aids, and design.

Planning the Internal Proposal

The goal of a proposal is to convince the person or group in authority to allow the writer to implement his or her idea. To achieve this goal, the writer must consider the audience, use visual aids, organize the proposal well, and design an appropriate format.

Consider the Audience

The audience profile for a proposal focuses on the audience's involvement, their knowledge, and their authority.

Ethics and Proposals

Proposals are an attempt to persuade an audience to approve whatever it is that is being proposed. Whether the proposal is internal or external, solicited or unsolicited, it is a kind of contract between the technical writer (or company) and the audience. Because proposals often deal with time and money, your trustworthiness and accountability are at stake. Consider your audience's needs and write sympathetically and knowledgeably for them. The ethical writer considers the audience's requirements, not what he or she can get out of the situation.

How Involved Is the Audience?

In most cases, readers of a proposal either have assigned the proposal and are aware of the problem or have not assigned the proposal and are unaware of the problem. For example, suppose a problem develops with a particular assembly line. The production engineer in charge might assign a subordinate to investigate the situation and recommend a solution. In this assigned proposal, the writer does not have to establish that a problem exists, but he or she does have to show how the proposal will solve the problem.

More often, however, the audience does not assign the proposal. For instance, a manager could become aware that a new arrangement of her floor space could create better sales potential. If she decides to propose a rearrangement, she must first convince her audience—her supervisor—that a problem exists. Only then can she go on to offer a convincing solution.

How Knowledgeable Is the Audience?

The audience may or may not know the concepts and facts involved in either the problem or the solution. Estimate your audience's level of knowledge. If the audience is less knowledgeable, take care to define terms, give background, and use common examples or analogies.

How Much Authority Does the Audience Have?

The audience may or may not be able to order the implementation of your proposed solution. A manager might assign the writer to investigate problems with the material flow of a particular product line, but the manager will probably have to take the proposal to a higher authority before it is approved. So the writer must bear in mind that several readers may see and approve (or reject) the proposal.

Use Visual Aids

Because the proposal is likely to have multiple audiences, visual aids are important. Visuals can support any part of the proposal—the description of the problem, the solution, the implementation, and the benefits. In addition to the tables and graphs described in Chapter 7, Gantt charts (see Chapter 17) and diagrams can be very helpful.

Gantt Charts

As described in Chapter 7, Gantt charts visually depict a schedule of implementation. A Gantt chart has an X axis and a Y axis. The horizontal axis displays time periods; the vertical axis, individual processes. Lines inside the chart show when a process starts and stops. By glancing at the chart, the reader can see the project's entire schedule. Figure 16.1 is an example of a Gantt chart.

Figure 16.1

Gantt Chart

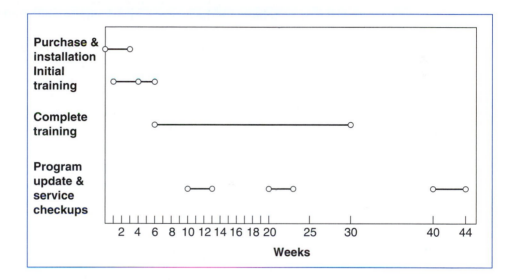

Diagrams

Many kinds of diagrams, such as flow charts, block diagrams, organization charts, and decision trees, can enhance a proposal. Layouts, for instance, are effective for proposals that suggest rearranging space.

Organize the Proposal

The writer should organize the proposal around four questions:

1. What is the problem?

2. What is the solution?

3. Can the solution be implemented?

4. Should the solution be implemented?

What Is the Problem?

Describing the problem is a key part of many proposals. You must establish three things about the problem:

- ▶ The data
- ▶ The significance
- ▶ The cause

The *data* are the actual facts that a person can perceive. The *significance* is the way the facts fail to meet the standard you hope to maintain. To explain the significance of the problem, you show that the current situation negatively affects productivity or puts you in an undesirable position. The *cause* is the problem itself. If you can eliminate the cause, you will eliminate the negative effects. Of course, almost every researcher soon discovers that there are chains of causes. You must carry your analysis back to the most reasonable cause. If the problem is ultimately the personality of the CEO, you might want to stop the chain before you say that. To be credible, you must show that you have investigated the problem thoroughly by talking to the right people, looking at the right records, making the right inspection, showing the appropriate data, or whatever. In the following section from a proposal, the writers describe a problem:

CONFUSING PARKING SIGNS

Significance

Cause

Data

Significance

Table 1 shows a big jump in the number of parking tickets given out in 2006–2007, an increase of over 3000 tickets. We feel that the increase occurred because of the inadequate parking lot signs. The current signs are old, plain, vague, and not very sensible. They only state that a permit is required, and one often does not know what kind of permit is needed. The signs don't specify whether they are for faculty, students, or commuters. In addition, the current signs are only 12 inches by 18–24 inches and can be overlooked if people are unaware of them.

In our survey of some West Central University students, we found that many students who received tickets either did not know that they could not park in the specific lot, were unsure of which lot they were able to park in, or did not see any specific signs suggesting that they could not park there.

Table 1

Tickets Given Out per 2500 Parking Stalls at West Central University

Year	No. of Tickets
2004–2005	13,202
2005–2006	13,764
2006–2007	16,867

What Is the Solution?

To present an effective solution, explain how it will eliminate the cause, thus eliminating whatever is out of step with the standard you hope to maintain. If the problem is causing an undesirable condition, the solution must show how that condition can be eliminated. If the old signage for parking lots gives insufficient information, explain how the solution gives better information. A helpful approach is to analyze the solution in terms of its impact on the technical, management/maintenance, and financial aspects of the situation.

NEW SIGNS FOR ENTRANCE

Solution named

Our solution is to create new permanent signs to be installed at the entrance of each parking lot. The new signs (in their entirety) will measure 3 feet by 4 feet so they will be visible to anyone entering the lot. Each sign will include the name of the lot; a letter to designate if the lot is for students (S), faculty (F), or administration (A); a color code for the particular permit needed; and the time and the days that the lot is monitored. The signs will not only present the proper information but will also look nice, making the campus more appealing. See Figure 1 for the proposed design.

Details show how the solution solves the problem.

Benefits

Figure 1
Entrance Sign

Can the Solution Be Implemented?

The writer must show that all the systems involved in the proposal can be put into effect. To make this clear to the audience, you would explain

- The cost
- The effect on personnel
- The schedule for implementing the changes

This section may be difficult to write because it is hard to tell exactly what the audience needs to know.

IMPLEMENTATION

Agents involved in implementation

The businesses we suggest that you deal with are Fulweil Structures, CE Signs, and University Grounds Services. The reason for choosing these businesses is that you will please the community of Menomonie by doing business in town and these businesses have good prices for a project like this. Also, these companies can provide services over the summer.

Schedule

Implementation of the new signs will take approximately one summer. A suggested schedule is

1. Order signs from Fulweil.   1 week
2. Fulweil constructs signs.   1 week
3. CE paints signs.   1 week
4. Grounds crew erects signs.   2 weeks

Schedule explained

If you compare this schedule to the estimates below, you will see that we have built in some time for delays. The project can be easily finished in a month. We suggest June because it has the fewest students for the most weeks; our second suggestion is August, but then you will have to finish by about the 20th or risk much confusion when school starts on the 25th.

Cost

Cost background

Below is a list of supplies and approximate costs from Fulweil Structures and CE Signs. The total project cost is $13,892.16. Fulweil Structures asked us to inform you that these prices are not binding quotes.

Table 2

List of Supplies and Approximate Costs for New Entrance Signs

Table presents all cost figures

Fulweil Structures (each sign)		
6′ × 2″ × 2″ solid bar aluminum (2 in quantity)	$118.91	
3′4″ × 3′4″ aluminum sheet (1 in quantity)	64.93	
3 hours of labor at $25/hour	75.00	
Total cost		$ 258.84
CE Signs (each sign)		
3′ × 3′ Reflective Scotchguard	$ 50.00	
10–15 letters painted	125.00	
2½ hours labor at $34/hour	85.00	

Total cost		$ 260.00
Total cost of each sign		518.84
Total cost of 24 signs		12,452.16
Projected cost of erecting signs		
2 hours/sign @ 25.00/hr (24 signs)	$1200.00	
Materials/sign @ 10.00 (24 signs)	240.00	
Total cost of erecting signs		1440.00
Total cost of project (24 signs)		$13,892.16

Should the Solution Be Implemented?

Just because you can implement the solution does not mean that you should. To convince someone that you should be allowed to implement your solution, you must demonstrate that the solution has benefits that make it desirable, that it meets the established criteria in the situation, or both.

THE BENEFITS OF THIS PROJECT

List of people who
benefit

Discussion of each
area of benefit

The benefits of the signs will be felt by you, the students, the faculty, and the administration. You will see the number of appeals decline because the restrictions will be clearly visible, saving much bookwork and time for appeals. You will also answer fewer phone calls from persons needing to know where to park and you will write fewer tickets, thus saving much processing time.

    The students, faculty, and administration will be happier because they will know exactly where and when they can and cannot park. Students will not receive as many parking tickets and will save money. Faculty and administration will also benefit by not having students park in their reserved parking spots (or at least not as often).

Design the Proposal

To design a proposal, select an appropriate format, either *formal* or *informal*. A formal proposal has a title page, table of contents, and summary (see Chapter 14). An informal proposal can be a memo report or some kind of preprinted form (see Chapter 12). The format depends on company policy and on the distance the proposal must travel in the hierarchy. Usually the shorter the distance, the more informal the format. Also, the less significant the proposal, the more informal the format. For instance, you would not send an elaborately formatted proposal to your immediate superior to suggest a $50 solution to a layout problem in a workspace.

Writing the Internal Proposal

Use the Introduction to Orient the Reader

The introduction to a proposal demands careful thought because it must orient the reader to the writer, the problem, and the solution. The introduction can contain one paragraph or several. You should clarify the following important points:

- ◗ Why is the writer writing? Is the proposal assigned or unsolicited?
- ◗ Why is the writer credible?
- ◗ What is the problem?
- ◗ What is the background of the problem?
- ◗ What is the significance of the problem?
- ◗ What is the solution?
- ◗ What are the parts of the report?

An effective way to provide all these points is in a two-part introduction that includes a context-setting paragraph and a summary. The context-setting paragraph usually explains the purpose of the proposal and, if necessary, gives evidence of the writer's credibility. The summary is a one-to-one miniaturization of the body. (Be careful not to make the summary a background; background belongs in a separate section.) If the body contains sections on the solution, benefits, cost, implementation, and rejected alternatives, the summary should cover the same points.

A sample introduction follows.

DATE: April 8, 2006
TO: Jennifer Williamson
FROM: Steve Vinz
             Mike Vivoda
             Michele Welsh
             Marya Wilson
SUBJECT: Installing new parking lot signs

Reason for writing: sets context

Parking on campus has been a topic of many discussions here at West Central University and one of much concern. The topics on parking include what lots students are able to park in, when students can park in the lots, and the availability of parking on campus. We believe that students do not know exactly when and where they can park in the campus lots because of the vague and confusing signs.

Summary

We feel that the school should post at each entrance new, more informative, and more readable signs containing all the rules and regulations. These signs would say exactly who can and cannot park in the lot, the times when the lots are patrolled, and what type of permit is needed. The project could be completed in 5 weeks and would cost $13,892.16. Major

Preview of sections   benefits include fewer administrative hassles and happier university community members. This memo will first discuss the problem, then the solution, implementation, and the benefits.

Use the Discussion to Convince Your Audience

The discussion section contains all the detailed information that you must present to convince the audience. A common approach functions this way:

The problem

> ▶ Explanation of the problem
> ▶ Causes of the problem

The solution

> ▶ Details of the solution
> ▶ Benefits of the solution
> ▶ Ways in which the solution satisfies criteria

The context

> ▶ Schedule for implementing the solution
> ▶ Personnel involved
> ▶ Solutions rejected

In each section, present the material clearly, introduce visual aids whenever possible, and use headings and subheadings to enhance page layout.

Which sections to use depends on the situation. Sometimes you need an elaborate implementation section; sometimes you don't. Sometimes you should discuss causes, sometimes not. If the audience needs the information in the section, include it; otherwise, don't.

The section above (pp. 422–425) illustrates one approach to the body. Other examples appear in the examples.

Worksheet for Preparing a Proposal

□ **Determine the audience for the proposal.**
Will one person or group receive this proposal?
Will the primary audience decide on the recommendations in this proposal?
How much do they know about the topic?
What information do you need to present in order to be credible?

(continued)

(continued)

☐ **Prepare background.**
Why did the proposal project come into existence?

☐ **Select a format—formal or informal.**

☐ **Prepare a style sheet of margins, headings, page numbers, and visual aid captions.**

External Proposal

☐ **Write a statement of the customer's needs.**

☐ **Prepare a two-column list (pp. 417–418) of the customer's needs and the ways your proposal meets those needs.**

☐ **Present your features in terms of the customer's needs, using the customer's terminology.**

☐ **Clearly explain the financial details.**

☐ **Explain in detail why your company has the expertise to do the job.**

☐ **Prepare a schedule for implementing. Assess any inconveniences implementation may cause.**

Internal Proposal

☐ **Define the problem.**
Tell the basic standard that you must uphold (we must make a profit). Cite the data that indicate that the standard is not being upheld (we lost $5 million last quarter). Explain the data's causes (we lost three large sales to competitors) and significance (we cannot sustain this level of loss for another year).

☐ **Construct a visual aid that illustrates the problem or the solution.**
Write a paragraph that explains this visual aid.

☐ **List all the parameters within which your proposal must stay.**
Examples include cost restrictions, personnel restrictions (can you hire more people?), and space restrictions.

☐ **Outline your methodology for investigating the situation.**

☐ **Prepare a list of the dimensions of the problem, and show how your proposed solution eliminates each item.**
(This list is the basis for your benefits section.)

☐ **Write the solution section.**
Explain the solution in enough detail so that a reader can fully understand what it entails in terms of technical aspects, management/maintenance, and finances. Also clearly show how it eliminates the causes of the undesirable condition.

☐ **Construct the benefits section.**
Clearly relate each benefit to some aspect of the problem. A benefit eliminates causes of the problem (the bottleneck is eliminated) or causes the solution to affect something else positively (worker morale rises).

☐ **Prepare a schedule for implementation.**
Assess any inconveniences.

☐ **List rejected alternatives, and in one sentence tell why you rejected them.**

Worksheet for Evaluating a Proposal

☐ **Answer these questions about your paper or a peer's. You should be able to answer "yes" to all of the following questions. If you receive a "no" answer, you must revise that section.**
a. Is the problem clear?
b. Is the solution clear?
c. Do you understand (and believe) the benefits?
d. Does the implementation schedule deal with all aspects of the situation?
e. Does the introduction give you the basics of the problem, the solution, and the situation?
f. Is the style sheet applied consistently? Does it help make the contents clear?
g. Do the visual aids communicate key ideas effectively?

Examples

Examples 16.1, 16.2, and 16.3 illustrate three different methods of handling internal proposals.

Example 16.1

Internal Proposal

Date:	November 7, 2007
To:	George Schmidt, Chief Engineer
From:	Greg Pittsch, Assistant Engineer
Subject:	Unnecessary shearing from joint welds

After talking to you on the phone last week, I mentioned that the Block Corporation is having difficulties with shearing on their engine mount supports. I contacted Mr. Jackson, a research expert, who said the stress

(continued)

Example 16.1

(continued)

from the weight of the engine causes the weld to shear. The shearing then causes the motor to collapse onto the engine mount supports. He advised me to purchase a higher-tensile-strength weld. The new weld I propose will reduce the defect rate from 10% to 0%. This memo includes the following information: weld shearing, weld constraints, and shearing solution.

WELD SHEARING

Unnecessary weld shearing of the engine mount supports has been a problem for the Block Corporation since 2004. The company is suffering a 10% defective rate on every 100 engine mounts welded.

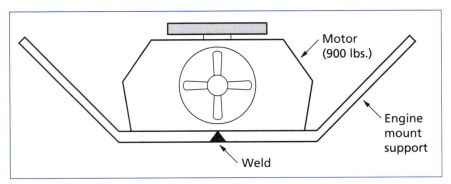

Figure 1
Engine Mount Weld

As seen in Figure 1, the weld must hold together when 900 lbs. of force are applied to the motor mount supports. A quality weld with a high tensile strength should withstand temperature fluctuation without shearing.

WELD CONSTRAINT

The Block Corporation listed the following constraints for implementing a new weld:

1. Material costs must increase by less than .01¢ per engine mount support welded.
2. Welding machines must not exceed 240 volts.
3. Current welding machines have to be used.
4. Each electrical outlet has to have a separate transformer.

SHEARING SOLUTION

The solution to the company's problem is to implement a higher-tensile-strength weld. The weld is projected to increase material and electrical costs, but is not expected to exceed the company's 1% budget increase for the 2008 fiscal year.

Cost

New welding wire with a higher tensile strength will increase 2¢ for every 100 yards of wire. All engine mount welds require 3 yards of wire to secure a solid weld. The overall cost increase per engine mount welded will be only .006¢.

Voltage

There will be an increase in the amount of electricity used in the new welding process. The welding machines will be required to switch from 120 to 240 volts.

Use of Current Machines

The welders will use the same welding machines as in the past. The welding machines are compatible with the new welds and do not need to be replaced.

Separate Transformers

An electrical hookup from 120 to 240 volts will be needed at each electrical outlet. A transformer will be required at each individual box to ensure an increase in voltage flow.

Example 16.2

Internal Proposal

Date: February 14, 2006
To: Irene Gorman
From: Chris Lindblad
Subject: Replacing Voltage Buss Bars

INTRODUCTION

Right now, the voltage buss bars on the C90 modules are not ohmed until after all of the option chips have been bonded to the circuit board. If a voltage buss-to-ground short is found after the bonding process, the short must be located, which takes an average of over 10 hours.

RECOMMENDATION

Based on time savings, cost, space available on the bonders, training involved, and savings to the company, I recommend installing a Fluke model 73 multimeter at each bonding station and to have operators ohm the buss bars after each option is bonded to the circuit board.

(continued)

Example 16.2

(continued)

TIME SAVINGS WILL RESULT

Installing a Fluke model 73 multimeter at each bonding station would have a positive effect on the time spent on locating voltage buss-to-ground shorts, as shown in Table 1. It also shows that ohming the voltage buss bars after each bond would increase the bonding process time but would result in a time savings of 7.1 hours per module.

Table 1

Time Savings with Ohming Capabilities Installed at Bonding Stations

Time	Without Using a Fluke 73 (in hours)	Using a Fluke 73 (in hours)
Average time required to bond all options on circuit board	49.5	51.9
Average time required to locate buss-to-ground shorts	10.0	0.5
Total time required	59.5	52.4
Time savings	—	7.1

WITHIN ALLOWED BUDGET

Currently the budget allows for $5000.00 in bonder improvements. The cost of equipping each bonding machine with a Fluke model 73 multimeter and associated test leads would amount to $350.00. The total cost of equipping the four option bonders would be $1400.00. This cost is well within the allowed budget and would also allow any future bonders to be installed with this equipment.

THE SPACE IS AVAILABLE

The space required for the installation of the ohming equipment is minimal, and it can easily be installed at the base of the bonder at the buss bar end of the module without any loss of mobility of the bonding head. It would also be within easy reach of the operator and cause no safety hazards. Also, no special power requirements are necessary because the Fluke 73 operates on an internal battery source.

TRAINING IS MINIMAL

The training required by the operator to learn how to use the Fluke 73 could be handled by the company's training department, which already has training in place for its use. Only one hour of class time is required with three hours on-the-job training to become proficient in its use.

SAVINGS ARE SIGNIFICANT

By installing a Fluke model 73 multimeter at each of the option bonding stations, the company could save a considerable amount of money. The current cost of troubleshooting a voltage buss-to-ground short is $130.00. This cost, with ohming equipment installed, would drop to around $13.00 per voltage buss-to-ground short, with a savings of $117.00.

CONCLUSION

If you require any further information or documentation on my recommendation, please contact me at the module test department.

Example 16.3

Internal Proposal

REPLACEMENT OF PRESENT SINGLE-PHASE VENTILATION MOTOR WITH A NEW THREE-PHASE INDUCTION MOTOR

INTRODUCTION

The purpose of this report is to inform you of the inadequacies of the present ventilation system, and the benefits of replacing the current motor. In this report I will first give you a quick summary of the proposal, followed by the necessary background information required. I will then discuss in detail the following: the problems with the present system, the proposed solution to correct the problem, the implementation of the new system, the rationale behind the decision, followed by the conclusion.

Summary of Proposal

The problem in the ventilation system came to light during a routine inspection. I noticed the following problems with the present system:

1. Insufficient air flow at the southern end of plant
2. Current motor wastes too much electricity

The combination of these two problems creates both an unsafe and an inefficient system. Fortunately, the solution is quite simple and inexpensive. To correct the problem, the present single-phase motor in the system must be removed and replaced with a new three-phase induction motor. This new motor will not only correct the problems of the present system, it will also produce the following benefits:

1. Longer life
2. Decreased power factor
3. Expandability
4. Minimal downtime at installation

(continued)

Example 16.3

(continued)

Background Information

One must know the difference between a single-phase and a three-phase motor. A single-phase motor runs on only one electrical phase, but requires additional starting circuitry. Three-phase motors, on the other hand, require all three electrical phases, but do not need any starting circuitry. It is also important to know that the amount of air flow in a system is measured in cubic yards per hour.

DISCUSSION

This section covers problems with the present system, a proposed solution, implementation of the new system, rationale, and benefits of the system.

Problems with the Present System

Air Flow   The main problem with the present ventilation system is that it is unable to produce enough air flow to the southern end of the plant. During my inspection, I took various measurements of air flow throughout the plant using an air flow meter. I noticed that the southern end is receiving only 1800 cubic yards ventilation an hour. OSHA standards require that 2000 cubic yards must be replaced every hour. If this situation is not corrected, we may be endangering the well-being of our employees, not to mention being slapped with a possible fine from OSHA. After closer examination of the ventilation system, I discovered that the only thing wrong with the system is the motor driving the fan. The motor is old and worn out, and therefore unable to produce the necessary air flow.

Electrical Consumption   The other problem with the present system is its abnormally high power-consumption, which I discovered while taking measurements on the present motor with a digital VOM meter. With these measurements, I calculated that the motor is running at only 50% peak efficiency. An average three-phase induction motor runs at approximately 90% peak efficiency. Over the course of a year, the company loses about $900 from the inefficiency of the present motor.

Proposed Solution

After careful analysis of all information, I have come to the following conclusion: replace the present single-phase motor in the ventilation system with a new three-phase induction motor. A new three-phase motor will not only increase air flow, it will also do it more efficiently.

Air Flow   If a three-phase induction motor were installed in place of the single-phase motor, it would increase air flow by almost 20%. I calculated this by using the torque and speed characteristics of a three-phase motor.

This would boost the air flow to the southern end of the plant from 1800 cubic yards to over 2100 cubic yards per hour. This is well within OSHA standards.

Electrical Consumption   One of the biggest advantages of a three-phase motor over a single-phase motor is the efficiency. An average new three-phase motor can run at up to 90% efficiency. An average single-phase motor of the same horsepower could achieve only 80% efficiency at best. The more efficient a device, the less expensive it is to run. See Figure 1.

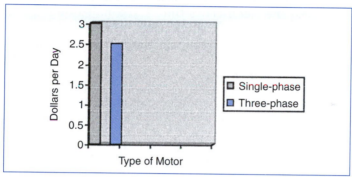

Figure 1
Electrical Running Cost of a Single-Phase Motor vs. a Three-Phase Motor

As you can see from this figure, a three-phase motor requires much less electricity to run per day than the single-phase motor. The reason a three-phase is more efficient than the single-phase motor is that it needs no additional starting circuitry. The addition of this starting circuitry in a single-phase motor is what robs it of maximum efficiency.

Implementation of the New System

The installation of a new three-phase induction motor should pose no problems. In this section, I will concentrate on the main aspects of installation: cost, time, and inconvenience.

Cost   The overall cost of replacing and installing the new motor should not exceed $500. The motor and control box together are $300. The wiring must be done by a certified electrician and overall labor cost should not exceed $150. The remaining $50 will buy new motor mountings, brackets, and wire. A three-phase junction box is within 20 feet of the ventilation system and should pose no installation difficulties for the electrician.

(continued)

Example 16.3

(continued)

Time   The installation time from start to finish should be no more than five hours. It will take one hour for us to remove the old single-phase motor. Installing the three-phase motor should take no more than an hour and a half. The remaining hour would be used for cleanup work and initial start-up of the system. I received all of these time and cost figures from a certified electrician.

Inconveniences   The new three-phase motor could be installed with only a few slight inconveniences, the most obvious of which is the shutdown of the ventilation system. This cannot be done during working hours, so it will have to be done on a Saturday. The labor costs I have stated earlier reflect the electrician's time-and-a-half rate imposed by working on the weekend. There is also the minor inconvenience of having someone here that Saturday to let the electrician into the building.

Rationale/Benefits

A three-phase induction motor in a ventilation system will provide three main benefits: longer life, decreased power factor, and expandability.

Longer Life   If a three-phase induction motor were installed into the ventilation system, it would provide much longer life than an equivalent single-phase motor. This point is made clear in Figure 2.

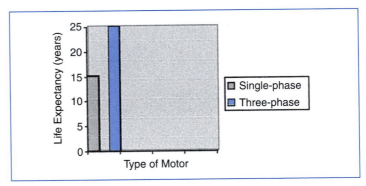

Figure 2
Expected Life of a Single-Phase Motor vs. a Three-Phase Motor

As you will notice from Figure 2, a three-phase motor will last much longer than will a single-phase motor doing the same task. The reason for this is the simplicity of operation of a three-phase motor; the single-phase requires starting circuitry, which has a tendency to break down more quickly.

Decreased Power Factor   Power factor is a confusing factor involved with the use of any inductive device, including motors. Power factor, if left unchecked, alters the electricity we receive. Although both single-phase and three-phase motors have some amount of power factor associated with them, a three-phase motor has less. A three-phase motor will reduce the amount we are charged for power factor.

Expandability   Another advantage of a three-phase motor is the expandability we would receive in our ventilation system. With the increase in air flow, we could easily add on to the ventilation system.

CONCLUSION

We cannot afford to let this problem continue. A three-phase motor in the ventilation system will best suit our needs both now and in the future.

Exercises

▶ You Create

1. Create a visual aid that demonstrates that a problem exists.
2. With the visual aid from Exercise 1, write a paragraph that includes the data, the significance, and the cause of the problem, and write a second paragraph that suggests a way to eliminate the problem.
3. Make a Gantt chart of a series of implementation actions. Write a paragraph that explains the actions.
4. Create two different page designs for the proposal about parking signs on campus (pp. 422–425).

▶ You Revise

5. Rewrite the following paragraphs. The writer is a recreation area supervisor who has discovered the problem; the reader is the finance director of a school district. Shorten the document. Make the tone less personal. Make a new section if necessary. Adopt the table or create new visual aids. If your instructor requires, also add an introduction and a summary.

DISCUSSION OF TRENDS

I have data that establish trends in the building's use (see Table 1). These data show peak adult and student use during the winter months. When school is out (June–August), we have more students and children using the building. Our slow months are in the Spring (April and May) and in the Fall (September and October). These trends coincide with what we know to be true about revenue loss. I have a more difficult time controlling the adult and student population using the building during the winter months. This results in a higher (25%) revenue loss for these months. On the other end, the children and students using the building during the summer are easier to control. This results in a lower (10%) revenue loss.

Table 1

Building Use for Open Recreation

	Adults	Students (Grades 7–12)	Children (Grade 6 & Under)	Total
January	621	583	412	1616
February	645	571	407	1623
March	597	545	393	1535
April	428	372	279	1070
May	210	330	239	779
June	365	701	587	1653
July	276	823	650	1749
August	327	859	718	1904
September	189	268	225	682
October	226	314	275	815
November	398	292	412	1102
December	589	494	384	1467

PROBLEM

The problem of revenue loss really involves two issues. The loss of revenue leads directly to a secondary issue, which is loss of control. When I cannot control the people entering the building, we lose revenue. When these people assume they can get in free, they also assume I cannot control their actions thereafter.

As a supervisor I have many duties. During open rec. I am expected to be at the office window collecting fees. This is all well and fine *if* I could stay there the entire time! Unfortunately, I must occasionally check activities in the weight room, the fieldhouse, the pool, and the locker rooms. At these times I am out of the office and cannot control people from just walking in. Even answering the phone causes problems. I must cross the office to the desk, and then I lose direct eye contact to the front entrance.

I really have no good explanation for why I have more problems with the adult-student users during the winter months as compared to the student-children users in the summer. All I know is that when these winter "bucket shooters" start pouring in for open rec., control goes right out the window. The only answer is a barrier to contain them in the lobby area until they have paid.

SOLUTION

The solution is a barrier that extends from the entrance door into the lobby, to the office wall. This is a length (open space) of about 14 feet. I suggest a chain as a temporary solution. Attached at the entrance door frame, it should extend 5½ feet to a stationary post (nonpermanent support), feed through an opening at the top of this post, and continue on another 5½ feet to another post. The remaining 3 feet to the office wall will be the entrance area. This will be chained off as well, and passage will be allowed only after paying the open rec. fee. A sign that reads "DO NOT ENTER" should be attached to the chain at the entrance area by the office. When I am out of the office, people may think twice and remain in the lobby until I return.

The cost in hardware for this barrier will be minimal, and I suggest it only as a temporary measure. I would like to establish the effectiveness of a simple barrier before considering a more permanent structure. There will always be some who ignore the barrier. There is never a perfect solution.

▶ You Analyze

6. Analyze Examples 16.1 and 16.2. Follow the instructions for Exercise 8. Alternate: If your instructor requires, rewrite and redesign one of the examples.

▶ Group

7. In groups of two to four, discuss one of the proposals given in this chapter. What do you like? dislike? Would you agree to implement the solution? Report your results to the class.

8. In groups of three or four, analyze Example 16.3. Prepare a memo to your class that pinpoints its weaknesses and strengths. Focus on depth of detail, appropriateness for audience, and unnecessarily included items.

9. In groups of three or four, write a proposal using the details given for a nonprofit organization's need for a computer system (pp. 417–418).

Writing Assignments

For each of the following assignments, first perform the activities required by the worksheet (pp. 427–429).

1. Write a proposal in which you suggest a solution to a problem. Topics for the assignment could include a problem that you have worked on (and perhaps solved) at a job or a problem that has arisen on campus, perhaps involving a

student organization, or at your workplace. Explain the problem and the solution. Show how the solution meets established criteria or how it eliminates the causes of the problem. Explain cost and implementation. If necessary, describe the personnel who will carry out the proposal. Explain why you rejected other solutions. Use at least two visual aids in your text. Your instructor will assign either an informal or a formal format. Fill out the worksheet from this chapter, and perform the exercises that your instructor requires.

Your instructor may make this a group assignment. If so, follow the instructions for developing a writing team (Chapter 2), and then analyze your situation and assign duties and deadlines.

2. In groups of three or four, write a simple request for a proposal (RFP). Ask for a common item that other people in your class could write about. (If you've all taken a class in computerized statistics, for example, ask for a statistics software program.) Try to find a real need in your current situation. Interview affected people (such as the statistics instructor) to find out what they need. Then trade your RFP with that of another group. Your group will write a proposal for the RFP you receive. Your instructor will help you with the day-to-day scheduling of this assignment.

3. Write a learning report for the writing assignment you just completed. See Chapter 5, Writing Assignment 7, pages 134–135, for details of the assignment.

Web Exercise

Write a proposal suggesting that you create a website for a campus club or a company division (including a "special interest" site, such as for the company yoga club). Explain how you will do it, why you are credible, the cost, the benefits to the company, and the schedule for production.

Work Cited

Bacon, Terry. "Selling the Sizzle, Not the Steak: Writing Customer-Oriented Proposals." *Proceedings of the First National Conference on Effective Communication Skills for Technical Professionals.* Greenville, SC: Continuing Engineering Education, Clemson University, November 15–16, 1988.

Chapter 17 · User Manuals

Chapter 17
In a Nutshell

A manual should be written and designed so that readers are comfortable enough with the machine or object to confidently interact with it. Effective manuals teach readers that machines are objects that require humans to use and control them. Your readers can achieve this position as you help them relate to the machine.

Supply context. Help them see the machine from the designer's point of view. What does this machine or this part do, and why, and what kinds of concerns does that function imply? Once readers get the big picture, they will usually try to use the item for its intended purpose.

Explain what the parts do. List all the visible parts, and explain what they cause, how to stop or undo what they cause, what other parts work in sequence with them.

Explain how to perform the sequences. Think of readers as users or doers. What actions will they perform? Think of common ones like turning the machine on and off. Spend time working on the machine yourself so you can clearly explain how to work it.

Use visual logic. One major section should discuss each of the three areas mentioned above. Divide each section into as many subsections as needed. Use heads and white space so readers can easily find sections and subsections. Use clear text and visual aids so readers figure out how to do the actions confidently.

Develop credibility. Give brief introductions that tell the end goal of a series of steps; give warnings before you explain the step; state the results of actions or give clear visual aids so that readers can decide if they are progressing logically through the steps.

C ompanies sell not only their products but also knowledge of how to use those products properly. This knowledge is contained in manuals. Both the manufacturer and the buyer want a manual that will allow users safely and successfully to assemble, operate, maintain, and repair the product.

Very complex mechanisms have separate multivolume manuals for different procedures such as installation and operation. The most common kind of manual, however, is the user's manual, which accompanies almost every product.

User's manuals have two basic sections: descriptions of the functions of the parts and sets of instructions for performing the machine's various processes. In addition, the manual gives information on theory of operation, warranty, specifications, parts lists, and locations of dealers to contact for advice on parts. This chapter explains how to plan and write an operator's manual.

Planning the Manual

Your goal is for the manual to help readers make your product work. To plan effectively, determine your purpose, consider the audience, schedule the review process, discover sequences, analyze the steps, analyze the parts, select visual aids, and format the pages.

Determine Your Purpose

The purpose of a manual is to enable its readers to perform certain actions. But manuals cannot include every detail about any system or machine. Decide which topics your readers will need, or can deal with. For example, you would choose to explain simple send and receive commands for e-mail beginners, but not complicated directory searching.

Decide the level of detail. Will you provide a sketchy outline, or will you "hand-hold," giving lots of background and explanation? To see the results of a decision to "hand-hold," follow the "Background Sound" instructions in Example 11.2, pages 278–279. Making these key decisions will focus your sense of purpose, allowing you to make the other planning decisions detailed in this chapter.

Consider the Audience

Who is your audience? Create an audience profile. Characterize your readers and their situation so that you can include text, visuals, and page design that give them the easiest access to the product. First, determine how much they know about general terms and concepts. Readers who are learning their first word processing program know nothing about "save," "cut," "paste," "open," "close," and "print." Readers learning their fifth program, however, already understand these basic word processing concepts. Early in the planning process, make a list of all the words the readers must understand.

Second, consider your goal for your readers. What should they be able to do as a result of reading the manual? A common answer, of course, is to be able to operate the product, but what are those key abilities they *must* have to do so? Those abilities will help you decide what sections to include and how to write them.

Third, consider how your readers will read the manual. Both beginning and expert audiences usually are "active learners." They do not want to read; instead, they want to accomplish something relevant quickly (Redish). When they do read, they do not read the manual like a story, first page to last. Instead, they go directly to the section they need. To accommodate these active learners who differ widely in knowledge and experience, use format devices—such as heads and tables of contents—that make information accessible and easy to find. This type of thinking will help you with the layout decisions you must make later and will help you decide what information to include in the text.

Fourth, consider where the audience will use the manual. This knowledge will help you with page design. For instance, manuals used in poor lighting might need big pages and typefaces, whereas manuals used in constricted spaces or enclosed in small packages need small pages and typefaces.

Fifth, consider your audience's emotional state. For various reasons, many, if not most, users do not like, or even trust, manuals (Cooper). Further, users are often fearful, hassled, or both. Your goal is to both allay their fears and develop their confidence. The presentation of your manual—its sequence and format—and of your identity as a trustworthy guide will develop a positive relationship.

Determine a Schedule

Early in your planning process, set up a schedule of the entire project. Typically, a manual project includes not just you, the writer, but also other people who will review it for various types of accuracy—technical, legal, and design. In industrial situations, this person might be the engineer who designed the machine. If you write for a client, it will be the client or some group designated by the client.

Think of each draft as a cycle. You write, and then someone reviews, and as a result of their review you rewrite or redesign. At the outset of the project, set dates for each of these reviews and decide who will be part of the review team. In addition, agree with your reviewers on when you expect them to return the draft and on what types of comments they are to make. You can handle the actual schedule in several ways, perhaps write in the actions you will perform during various weeks on a calendar. Or you could make a Gantt chart.

Suppose your tasks are to interview an engineer, create a design, write a draft, have a reader's review, write a second draft, have a second review, and print the manual. Suppose also that your schedule allows you 8 weeks. Your Gantt chart might look like Figure 17.1 (p. 444).

Figure 17.1

Gantt Chart

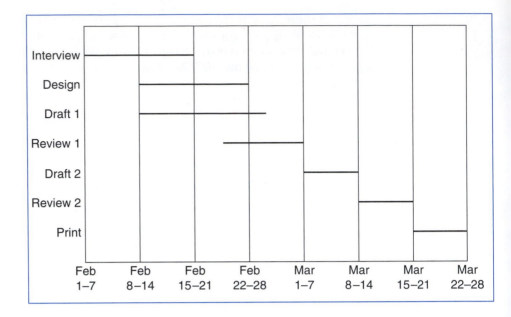

Discover Sequences

Discovering all the sequences means that you learn what the product does and what people do as they use it (Cohen and Cunningham). To learn what the product does, learn the product so thoroughly that you are expert enough to talk to an engineer about it. Because this process takes a good deal of time, you need to plan the steps you will take to gain all this knowledge. Schedule times to use the product. Talk to knowledgeable people—either users or designers, or both. Your goal is to learn all the procedures the product can perform, all the ways it performs them, and all the steps users take as they interact with the product.

For example, the writer of a manual for a piston filler, a machine that inserts liquids into bottles, must grasp how the machine causes the bottle to reach the filling point and how the machine injects the liquid into the bottle. Gaining this knowledge requires observing the machine in action, interviewing engineers, and assembling and disassembling sections.

But the writer must also know what people do to make the machine work. The most practical way to gain this knowledge is to practice with the product. These acts become the basis for the sections in the procedures section. As you practice, make flow charts and decision trees. In your flow charts, list each action and show how it fits into a sequence with other actions (see Figure 17.2).

The sequences your manual must teach the user typically include

▶ How to assemble it.
▶ How to start it.
▶ How to stop it.
▶ How to load it.

Figure 17.2

Flow Chart

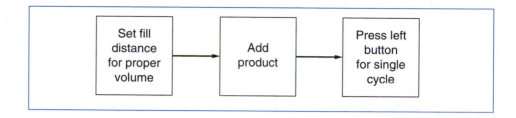

▶ How it produces its end product.
▶ How each part contributes to producing the end product.
▶ How to adjust parts for effective performance.
▶ How to change it to perform slightly different tasks.

Analyze the Steps

To analyze the steps in each sequence means to name each individual action that a user performs. This analysis is exactly the same as that for writing a set of instructions (review Chapter 11). In brief, determine both the end goal and the starting point of the sequence, and then provide all the intermediary steps to guide the users from start to finish. Try constructing a decision tree. Make a flow chart for the entire sequence, and then convert the chart into a decision tree.

For an example of such a conversion, compare Figures 17.2 and 17.3. In these steps, taken from a piston filler manual, the writer wants to explain how to insert a specified amount of liquid into a bottle. Figure 17.2 shows the flow chart; Figure 17.3 (p. 446) shows a decision tree based on the flow chart.

Here is the text developed from the two figures:

1. Set the fill distance for the proper volume.
   a. Check specifications for bottle volumes (p. 10).
   b. To determine this distance, find out the diameter of your piston.
   c. Go to the volume chart on p. 11.
   d. Find the piston diameter in the left column.
   e. Read across to the volume you need.
   f. Read up to determine the length you need.
   g. Adjust the distance from A to B (Figure 6 [not shown]) to the length you need.
2. Add the product to the hopper. If you are unsure of the product type, see specifications (p. 11).
3. Press the left button (A on Figure 6 [not shown]) for single cycle.

Analyze the Parts

To analyze the parts, list each important part and explain what it does. Then convert these notes into a sentence. If you look at a few common user manuals, say, for a DVD player, you will always find this section in the front of the manual. A helpful method is to make a three-part row for each part. Name the

Figure 17.3

Decision Tree
Based on Flow
Chart

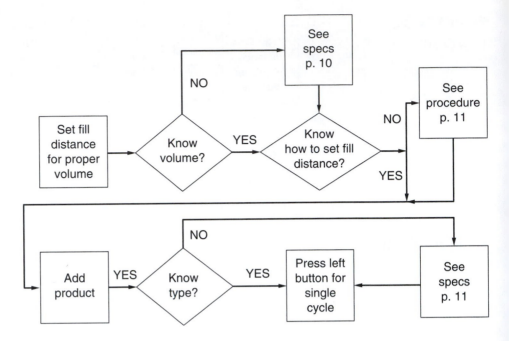

part, write the appropriate verb, and write the effect of the verb. Then turn that list into a comprehensible sentence. Here is the list for a stop button:

Name of Part	*Verb*	*Effect*
stop button	stops/ends	all functions stop

Here is the sentence for the button:

> The red emergency stop button immediately stops all functions of the machine.

Select Visual Aids

Visual aids—photographs, drawings, flow charts, and troubleshooting charts—all help the reader learn about the product. In recent years, with the advent of desktop publishing and many graphics software and hardware programs, the use of visual aids has proliferated. Including many visuals is now the norm. Many manuals have at least one visual aid per page; many provide one per step.

Your goal is to create a text-visual interaction that conveys knowledge both visually and textually. Consider this aspect of your planning carefully. If you can use a visual aid to eliminate text, do so. Notice one key use of visual aids in manuals. Visuals give permission. Although many visual aids are logically unnecessary because the text and the product supply all the knowledge, they are still useful. Consider, for instance, Figure 5.3 on page 454 of the video camera manual or the phone visuals on pages 460–463 of the telephone manual.

Neither of these is strictly needed because the user could read the text and look at the machine and see what is described in the text. But the visual reassures the readers that they are "in the right place." Use visuals liberally in this manner. Your readers will appreciate it.

You must decide whether each step needs a visual aid. Most manual writers now repeat visual aids. As a result, the reader does not have to flip back and forth through pages. To plan the visual image needed to illustrate a step, decide which image to include and from which angle users will view it. If they will see the part from the front, present a picture of it from the front. Use a storyboard (Riney), such as the one shown in Figure 17.4, to plan the visual aid. Storyboards are discussed in Chapter 18.

Figure 17.4

Storyboard

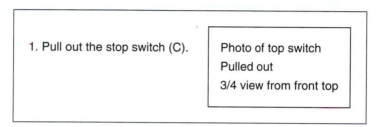

Format the Pages

The pages of a manual must be designed to be easy to read. Create a style sheet with a visual logic (see Chapters 6 and 7) that associates a particular look or space with a particular kind of information (all figure captions italic, all page numbers in the upper outside corner, and all notes in a different typeface). You must also design a page that moves readers from left to right and top to bottom. (Review Chapter 6 for format decisions.) This process is more complex than you might think, so carefully consider your options. You might review several consumer manuals that accompany software products or common home appliances.

To produce effectively laid out pages, use a grid and a template. A *grid* is a group of imaginary lines that divide a page into rectangles (see Figure 17.5, p. 448). Designers use a grid to ensure that similar elements appear on pages in the same relative position and proportion. One common grid divides the page into two unequal columns. Writers place text in the left column and visual aids in the right column, as shown on page 448.

A *template* is an arrangement of all the elements that will appear on each page, including page numbers, headers, footers, rules, blocks of text, headings, and visual aids. Figure 17.6 (p. 449) is a template of a page. The arrows indicate all the spots at which the author made a deliberate format decision. Create a tentative template before you have gone very far with your writing because your visual logic is part of your overall strategy (see p. 449) and will influence your word choice dramatically.

Figure 17.5

Page Grids

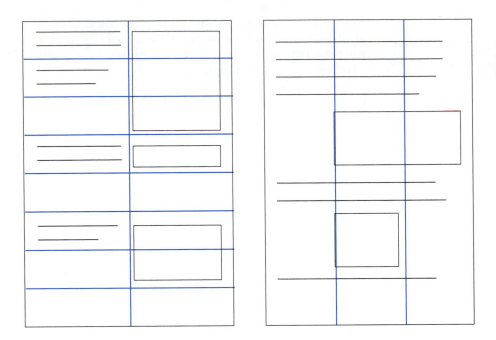

The following notes list all the format decisions that the author made for Figure 17.6:

1. Type, font, size, position of header text
2. Position, width, length of header rule
3. Size, type, font, position, and grammatical form (*-ing* word) of level 1 heads
4. Size, type, font, and position of instructional text; space between head and next line of text (leading)
5. Size, type, font, and position of numeral for instructional step
6. Punctuation following the numeral
7. Position of second line of text
8. Space between individual instructions
9. Punctuation and wording of reference to figure
10. Size, type, font, position of notes or warnings
11. Space between text and visual aid
12. Width (in points) and position of frame for visual aid
13. Size, type, font, and position of figure number and brief explanatory text
14. Position, width, length of footer rule
15. Type, font, size, position of footer text
16. Type, font, size, position of page number

Figure 17.6

Page Template
with Decision
Points

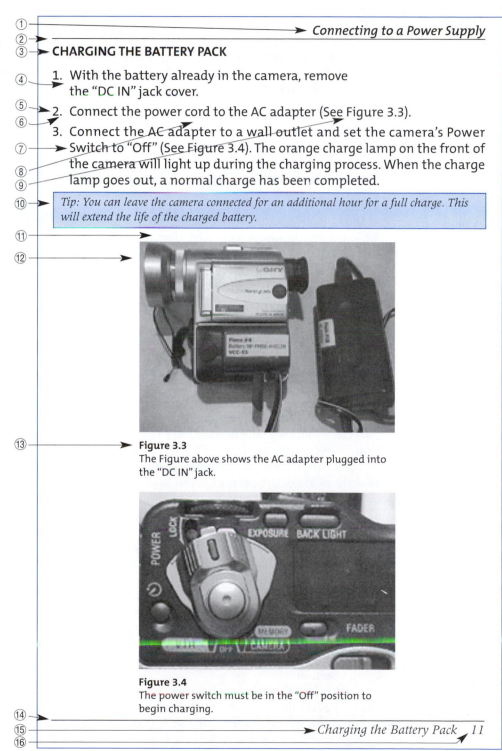

① ➤ *Connecting to a Power Supply*

②

③ ➤ **CHARGING THE BATTERY PACK**

④ ➤ 1. With the battery already in the camera, remove the "DC IN" jack cover.

⑤ ➤ 2. Connect the power cord to the AC adapter (See Figure 3.3).
⑥

3. Connect the AC adapter to a wall outlet and set the camera's Power
⑦ ➤ Switch to "Off" (See Figure 3.4). The orange charge lamp on the front of
⑧ the camera will light up during the charging process. When the charge
⑨ lamp goes out, a normal charge has been completed.

⑩ ➤ *Tip: You can leave the camera connected for an additional hour for a full charge. This will extend the life of the charged battery.*

⑪

⑫

⑬ ➤ **Figure 3.3**
The Figure above shows the AC adapter plugged into the "DC IN" jack.

Figure 3.4
The power switch must be in the "Off" position to begin charging.

⑭
⑮ ➤ *Charging the Battery Pack*   *11*
⑯

Writing the Manual

. . . .

The student sample shown in Figure 17.7 is taken from a user's manual for a video camera. This manual uses a one-column page; instructions are aligned on the left margin and visuals are centered. All the visual aids are clear digital photographs.

Introduction

In the introduction explain the manual's purpose and whatever else the reader needs to become familiar with the product: how to use the manual, the appropriate background, and the level of training needed to use the mechanism.

The introduction tells the purpose of the machine, states the purpose and divisions of the manual, and explains why a user needs the machine. In Figure 17.7, the introduction is entitled "Preface."

Arrange the Sections

A manual has two major sections:

▶ Description of the parts (see Figure 17.7, "Identifying Parts of the Camera")
▶ Instructions for all the sequences (see Figure 17.7, "Recording Video")

The Parts Section

The parts and functions section provides a drawing of the machine with each part clearly labeled. The description explains the function of each item. This section answers the question: What does this part do? or What happens when I do this? An easy, effective way to organize the parts description is to key the text to a visual aid of the product. Most appliance manuals have such a section.

The Sequences Section

The sequences section enables users to master the product. Arrange this section by operations, not parts. Present a section for each task. Usually the best order is chronological, the order in which readers will encounter the procedures. Tell first how to assemble, then how to check out, then how to start, to perform various operations, and to maintain. However, be aware that readers seldom read manuals from beginning to end, so you must enable them to find the information they need. Make the information easy to locate and use by cross-referencing to earlier sections. Never assume that readers will have read an earlier section.

In addition, as mentioned earlier, repeat key instructions or visuals. Do not make readers flip back and forth between pages. Rather, place the appropriate information where readers will use it (Rubens).

Figure 17.7

Excerpts from
an Operator's
Manual for a
Digital Mini
Video Camera

Table of Contents

TABLE OF CONTENTS

iii

(continued)

Figure 17.7

(continued)

IDENTIFYING PARTS OF THE CAMERA

LCD Screen   Projects camera image for recording or camcorder image for review. Swivels to any position.

LCD Open Button   Opens LCD screen when pushed

Lens Cap   Covers lens; protects lens from scratches

Microphone   Records sound during recording

Viewfinder   Shows image as it will be recorded

Open/Eject Button   Opens tape cover and ejects tape after being pushed

Tip: While using the LCD screen is easier to use than the viewfinder, it does consume more power from the battery pack.

2

Preparing the Camera

RECORDING VIDEO
PREPARING THE CAMERA

This camera uses DV Mini Video Cassette Tape. They can be purchased in two-packs from either the campus bookstore or local department stores.

1. Remove the lens cap.
2. Connect the power source.
3. Insert a video cassette.
4. Press the small green button on the Power Switch and slide it to the "CAMERA" position (see Figure 5.1).
5. Press "OPEN" on the LCD panel and open the LCD screen to 90 degrees (see Figures 5.2 and 5.3).

Figure 5.1
The power switch is in the "CAMERA" position. The start/stop button is also circled.

Figure 5.2
The "OPEN" button for the LCD screen is located just below the Power Switch.

16
(continued)

Figure 17.7

(continued)

Recording Video

SHOOTING FOOTAGE

1. Press the round Start/Stop button with the red dot in the center of the Power Switch to start recording.

> *Tip: As soon as you press the Start/Stop button, the red recording lamp on the front of the camera will light up to indicate that the camera is recording (see Figure 5.4).*

2. Press the red Start/Stop button again to stop recording (see Figure 5.1).

Figure 5.3
The image above shows the LCD screen fully opened.

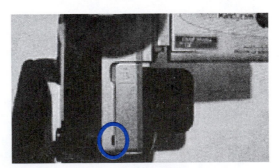

Figure 5.4
The recording lamp is located next to the A/V jack cover. The lamp turns red while recording.

17

Assume Responsibility

All manual writers have an ethical responsibility to be aware of the dangers associated with running a machine. Keep in mind that if you leave out a step, the operator will probably not catch the error, and the result may be serious. Also, you must alert readers to potentially dangerous operations by inserting the word WARNING in capital letters and by providing a short explanation of the danger. These warnings should always appear before the actual instruction. Sometimes generic warnings, which apply to any use of the mechanism, are placed in a special section at the front of the manual.

Other Sections

Manuals traditionally have several other sections, although not all of them appear in all manuals or in the exact arrangement shown here. These sections are the front matter, the body, and the concluding section. The front matter could include such elements as

- Title page
- Table of contents
- Safety warnings
- A general description of the mechanism
- General information, based on estimated knowledge level of the audience
- Installation instructions

The body could include this element:

- A theory of operation section

The concluding section could include such elements as

- Maintenance procedures
- Troubleshooting suggestions
- A parts list
- The machine's specifications

Test the Manual

Usability testing helps writers find the aspects of the manual that make it easier or harder to use, especially in terms of the speed and accuracy with which users perform tasks (Craig). You need to plan, conduct, and evaluate a usability test (Brooks).

Planning a Usability Test

Planning the test is selecting what aspects of the manual you want to evaluate, what method you will use, and who will be the test subjects.

Select the aspects of the manual that you want to study. The most important question is, Does this manual allow the readers to use the object easily and confidently? Consider using some or all of these questions (Bethke et al.; Queipo):

> ▶ *Time*
>   How long did it take to find information? to perform individual tasks? to perform groups of tasks?
> ▶ *Errors*
>   How many and what types of errors did the subject make?
> ▶ *Assistance*
>   How often did the subject need help?
>   At what points did the subject need help?
>   What type of help did the subject need?
> ▶ *Information*
>   Was the information easy to find? easy to understand? sufficient to perform the task?
> ▶ *Format*
>   Is the format consistent?
>   Are the top-down areas (headings, introductions, highlighters) helpful?
>   Is the arrangement on the page helpful?
> ▶ *Audience Engagement*
>   Is the vocabulary understandable?
>   Is the text concrete enough?
>   Is the sequence "natural"? Does it seem to the learner that this is the "route to follow" to do this activity?

Select the method you will use to find the answers to your questions. Some questions need different methods. One method often is not sufficient to derive all the information you want to obtain. Test methods (Sullivan) include

> ▶ *Informal observation*—watching a person use the manual and recording all the places where a problem (with any of the topic areas you selected to watch for) arose.
> ▶ *User protocols*—the thoughts that the user speaks as he or she works with the manual and that an observer writes down, tapes, or video records.
> ▶ *Computer text analysis*—subjecting the text to evaluation features that a software program can perform, including word count, spelling, grammar, and readability scores (i.e., at what "grade level" is this material?)
> ▶ *Editorial review*—knowledgeable commentary from a person who is not one of the writers of the text.
> ▶ *Surveys and interviews*—a series of questions that you ask the user after he or she has worked with the manual.

Your goal is to match the test method with the kind of information you want. For instance, an editorial review would produce valuable information on consistency. User protocols or survey/interviews would help you determine if the vocabulary was at the appropriate level or if the page arrangement was helpful. Observation would tell you if the information given was sufficient to perform the task.

Select the test subjects. The test subjects are most often individuals who are probable members of the manual's target audience but who have not worked on developing the manual.

Conducting a Usability Test

Conducting the test is administering it. The key is to have a way to record all the data—as much feedback as possible as quickly as possible. You can use several methods:

▶ If you do an informal observation, you can use a tally sheet (Rubin) that has three columns—Observation, Expected Behavior or User Comment, and Design Implications. Fill out the observation column as you watch and the other two columns later.

▶ For user protocols, you can design a form like the one you use for informal observations, or you can audio or video record, although the difficulties with taping methods—setup procedure, use of the material after the session—require a clear decision on your part of whether you will really use the data you record.

▶ Computer and editorial analysis will tell you about the features of the text but will not tell the audience's reaction; these tests are relatively simple to set up, although telling an editor what to look for and setting a grammar checker to search for only certain kinds of problems are essential.

▶ For surveys and interviews, you can design a form (as outlined in Chapter 5) and administer it after the subject has finished the session. Sample questions include:

Were you able to find information on X quickly?
Did the comments in the left margin help you find information?
Did you read the introductions to the sequences?
Did the introduction to each sequence make it easier for you to grasp the point of sequence?

A typical way to record the answers is either yes/no/comments or some kind of recording scale (1 = highly agree, 5 = highly disagree).

Evaluating a Usability Test

Evaluating the test is determining how to use the results of the test (Sullivan). Your results could indicate a problem with

▶ The text (spelling, grammar, sufficiency of information)
▶ The text's design (consistency, usefulness of column arrangement or high-lighting techniques)
▶ The "learning style" of the audience (sequence of the text, basic way in which they approach the material)

If you have determined beforehand what is an acceptable answer (e.g., this procedure should take X minutes; this word is the only one that can be used to refer to that object), you will be able to make the necessary changes. For more help on this topic, consult John Craig (see Works Cited).

Worksheet for Preparing a Manual

☐ **Consider the audience.**
How much do readers know about the general terms and concepts?
Where and when will they use the manual?
What should they be able to do after reading the manual?
List all the terms a user must comprehend. Define each term.

☐ **Determine a schedule.**
On what date is the last version of the manual due? By what date must each stage be completed?
Who will review each stage?
How long will each review cycle take?

☐ **List where you can obtain the knowledge you need to write the manual.**
a person? reading? working with the mechanism?

☐ **Analyze the procedures a user must follow to operate the product.**
What must be done to install it, to turn it on, to turn it off, and to do its various tasks?
List the sequence for presenting the processes.
Choose an organizational pattern for the sequence—chronological or most important to least important.
Create a flow chart for each procedure the machine follows.
Create a decision tree for each procedure the user follows.
Name each part and its function.
For a complicated product, you will discover far too many parts to discuss. Group them in manageable sections. Decide which ones your audience needs to know about.

☐ **Choose a visual aids strategy.**
Will you use drawings or photographs?
Will you use a visual aid for each instruction?
Will you use a visual aid on each page?
Will you use callouts?

☐ **Create a storyboard for your manual.**

☐ **Design pages by preparing a style sheet of up to four levels of heads, captions for visual aids, margins, page numbers, and fonts (typefaces).**
Select rules, headers, and footers as needed to help make information easy to find on pages.

☐ **Write step-by-step instructions.**
Clearly label any step that could endanger the person (WARNING!) or the machine (CAUTION!).

☐ **Field-test the manual.**
Select the features of the manual you want to field-test.
Select a method of testing those features. Be sure to create a clear method for recording answers. Determine what you think are acceptable results for each feature (e.g., How long should it take to perform the process?).
Select subjects to use the manual.

Examples

The excerpts shown in Example 17.1 are several sections of an operator's manual for a telephone. The entire manual is a one-page foldout, with seven sections including one in Spanish. Presented here are the table of contents, the parts and functions section (1-B, "Location of Controls"; 3, "Speed Dialer"; and 4, "One-Touch Dialer,"), two sections of operating instructions (3B and 4B), and one page of troubleshooting. These pages represent sections you will find in almost all manuals written for consumers. Example 17.2 is a usability report, written to explain the results of a usability test on a website.

Example 17.1

Excerpts From an Operator's Manual

Source: Reprinted by permission of Panasonic.

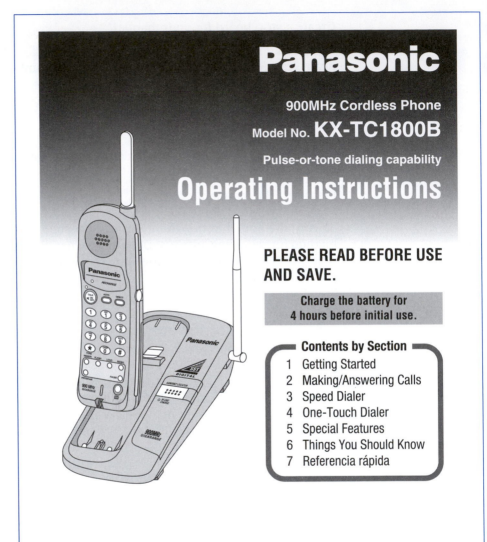

Panasonic

900MHz Cordless Phone

Model No. **KX-TC1800B**

Pulse-or-tone dialing capability

Operating Instructions

PLEASE READ BEFORE USE AND SAVE.

Charge the battery for 4 hours before initial use.

Contents by Section

1 Getting Started
2 Making/Answering Calls
3 Speed Dialer
4 One-Touch Dialer
5 Special Features
6 Things You Should Know
7 Referencia rápida

1 | Getting Started ➡

1-B Location of Controls

Handset

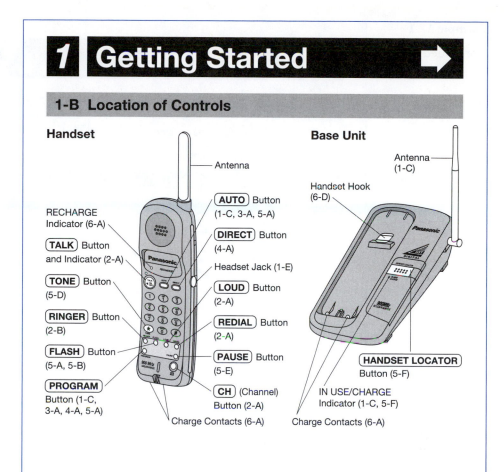

Antenna

RECHARGE
Indicator (6-A)

(TALK) Button
and Indicator (2-A)

(TONE) Button
(5-D)

(RINGER) Button
(2-B)

(FLASH) Button
(5-A, 5-B)

(PROGRAM)
Button (1-C,
3-A, 4-A, 5-A)

(AUTO) Button
(1-C, 3-A, 5-A)

(DIRECT) Button
(4-A)

Headset Jack (1-E)

(LOUD) Button
(2-A)

(REDIAL) Button
(2-A)

(PAUSE) Button
(5-E)

(CH) (Channel)
Button (2-A)

Charge Contacts (6-A)

Base Unit

Antenna
(1-C)

Handset Hook
(6-D)

(HANDSET LOCATOR)
Button (5-F)

IN USE/CHARGE
Indicator (1-C, 5-F)

Charge Contacts (6-A)

(continued)

Example 17.1

(continued)

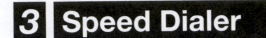

3 | Speed Dialer   Section 3

3-A Storing Phone Numbers in Memory

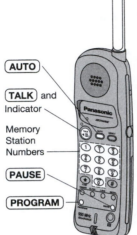

AUTO

TALK and
Indicator

Memory
Station
Numbers

PAUSE

PROGRAM

You can store up to 10 phone numbers in the handset. The dialing buttons (⓪ to ⑨) function as memory stations. **The TALK indicator light must be off before programming.**

1 Press (PROGRAM).
  • The TALK indicator flashes.

2 Enter a phone number up to 22 digits.

3 Press (AUTO).

4 Press a memory station number (⓪ to ⑨).
  • A beep sounds.
  • To store other numbers, repeat steps 1 through 4.

• If a pause is required for dialing, press (PAUSE) where needed. Pressing (PAUSE) counts as one digit (5-E).

If you misdial

Press (PROGRAM) to end storing. ➡ Start again from step 1.

To erase a stored number

Press (PROGRAM) ➡ (AUTO) ➡ the memory station number (⓪ to ⑨) for the phone number to be erased.
  • A beep sounds.

3-B Dialing a Stored Number

Press (TALK) ➡ (AUTO) ➡ The memory station number (⓪ to ⑨).
• If your line has rotary or pulse service, any access numbers stored after pressing (TONE) will not be dialed.

4 | One-Touch Dialer

4-A Storing a Phone Number in the DIRECT Button

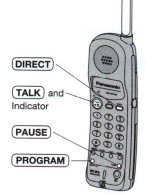

(DIRECT)

(TALK) and
Indicator

(PAUSE)

(PROGRAM)

A phone number stored in the (DIRECT) button can be dialed with a one-touch operation. **The TALK indicator light must be off before programming.**

1 Press (PROGRAM).
  • The TALK indicator flashes.

2 Enter a phone number up to 22 digits.
  • If you misdial, press (PROGRAM), and start again from step 1.

3 Press (DIRECT).
  • A beep sounds.

  • If a pause is required for dialing, press (PAUSE) where needed. Pressing (PAUSE) counts as one digit (5-E).

4-B Dialing the Stored Number in the DIRECT Button

Press (TALK) ➡ (DIRECT).
• If your line has rotary or pulse service, any access numbers stored after pressing (TONE) will not be dialed.

To erase a stored number: press (PROGRAM) ➡ (DIRECT).

(continued)

Example 17.1

(continued)

Things You Should Know   Section 6

6-E Before Requesting Help

Problem	Remedy
The unit does not work.	• Check the settings (1-C). • Charge the battery fully (6-A). • Clean the charge contacts and charge again (6-A). • Install the battery properly (1-C, 6-B). • Place the handset on the base unit and unplug the AC adaptor to reset. Plug in and try again. • Re-insert the battery and place the handset on the base unit. Try again.
An alarm tone sounds.	• You are too far from the base unit. Move closer and try again. • Place the handset on the base unit and try again. • Plug in the AC adaptor. • Raise the base unit antenna.
Static, sound cuts in/out, fades. Interference from other electrical units.	• Locate the handset and the base unit away from other electrical applicances (6-C). • Move closer to the base unit. • Raise the base unit antenna. • Press (CH) to select a clearer channel.
The unit does not ring.	• To ringer volume is set to OFF. Press (RINGER) while the TALK indicator light is off (2-B).
While storing a number, the unit starts to ring.	• To answer the call, press (TALK). The program will be cancelled. Store the number again.
You cannot store a phone number in memory.	• You cannot store a number while the unit is in the talk mode. • Do not pause for over 60 seconds while storing. • Move closer to the base unit.
Previously programmed information is erased.	• If a power failure occurs, programmed information may be erased. Reprogram if necessary.
You cannot redial by pressing (REDIAL).	• If the last number dialed was more than 32 digits long, the number will not be redialed.
(HANDSET LOCATOR) does not function.	• The handset is too far from the base unit or is engaged in an outside call.
The RECHARGE indicator flashes or the unit beeps intermittently.	• Charge the battery fully (6-A).
You charged the battery fully, but the RECHARGE indicator flashes.	• Clean the charge contacts and charge again (6-A). • Install a new battery (6-B).
The IN USE/CHARGE indicator light does not go out while charging.	• This is normal.
If you cannot solve your problem	• Call our customer call center at 1-800-211-PANA (7262).

Example 17.2

Report on NALC
Branch 728
Website Usability

1. EXECUTIVE SUMMARY

This report contains the results of a usability test for NALC Branch 728 website created by Hilary Peterson. I performed the test with three participants, all of whom indicated that they have at least a moderate familiarity with the Internet and with website use in general. All the test subjects are members of the target audience for the site, but had not seen the website before the test. I gave the test subjects a time limit of 15 minutes to complete three specific tasks; during that time, I observed the subjects and made careful notes of their actions and comments for later analysis. After the test, each subject completed a survey to express his overall response to the experience.

Analysis of the data I collected shows that the website is very effective overall. It is easy to navigate, and the page layout allows for easy scanning of the pages for important information. There is a need, however, to rethink the name of one of the lines, Get Smart!, and how it relates to the Letter Carrier Perfect Page to which it connects.

I make recommendations for the problem area.

2. INTRODUCTION

The goal of this report is to document the findings of a usability test of NALC Branch 728's website, created by Hilary Peterson. As the author of the report, I also conducted the usability test. The website is available at the following URL: http://fp1.centurytel.net/nalcbranch728. To serve and inform the members of Branch 728 and of other NALC branches in the area, there is a need for a useful, well-written website that includes an on-line version of the Branch's newsletter as well as a wealth of information for its users.

This report will interest the Web master, since it will indicate any improvements that should be made to the website. It will also interest my client, the editor of *The Sawdust Cities Satchel* and vice president of Union Branch 728. He will use this report to determine whether or not to request funding for the website.

These pages include a brief description of my methodology along with summaries of my test results, observations, and results of the post-test survey. I express my conclusions based on these data as well as my recommendations for improvement where I found problem areas during the test.

3. METHODOLOGY

3.1 Participant Selection

The primary audience for NALC Branch 728's website is members of Branch 728 who have at least a moderate familiarity with the Internet and with

(continued)

Example 17.2

(continued)

website use in general. The website was designed with this type of user in mind, so the pages are meant to be clean, simple, and easy to navigate.

My client chose three members of Branch 728 to participate in the usability test. The subjects possess varying levels of computer skills, but all of the subjects indicated that they were at least fairly familiar with the Internet. None of the subjects had ever seen or used the Branch 728 website before.

3.2 Task Selection

I gave each of the test subjects three test scenarios and three tasks to complete; these appear in Appendix A.

With question 1 of Task One, I aimed to test the readability of the homepage, where the answer to question 1 is found.

With question 2 of Task One, my goal was to test the usefulness of the bookmarking feature on the Constitution and By-Laws page.

Task Two tested the "scannability" of a very long page of text, the Get Smart! page. I also wanted to find out whether or not I could reasonably assume that most of my audience members were aware that "Get Smart!" is the new name for the Letter Carrier Perfect handbook.

Task Three tests the function and visibility of the Outside Links page.

3.3 Test Procedure

I scheduled 15 minutes for each participant to complete the assigned tasks. The subjects performed the test individually. Before the test, I gave an introduction to the test, emphasized the subject's role, and told each participant what to expect during the test. This introduction appears in Appendix A.

Next, I gave each subject scenarios and instructions for completing the tasks, encouraging them to talk aloud while they were performing the test. Hearing the participants say why they chose to click on a particular link helped to show the effectiveness of the links.

As the test's administrator, I did not interact with the test subjects or give hints when they were lost. Instead, I took careful notes on the subjects' behavior, demeanor, comments, and activities on screen.

3.4 Post-Test Survey

To obtain each participant's overall feelings about his experience, I administered a short post-test survey, shown in Appendix B.

4. RESULTS

4.1 Observations

All three of the test subjects completed the assigned tasks within the time limit of 15 minutes. The subjects completed the tasks in four minutes, six minutes, and nine minutes. The fastest test participant completed the tasks in the fewest possible number of steps. The slowest participant completed the tasks within the 15 minute time limit, but he did not use the most efficient path through the site.

All three of the subjects stayed on the homepage at the beginning of the test. The answer to the first question in Task One was there, and two of the three subjects found it within one minute. One subject did not find the answer right away, and he browsed unsuccessfully through a different page before returning to the homepage and finding the answer.

Question 2 of Task One was answered successfully by all three subjects within an appropriate amount of time. The subjects started by immediately clicking on the link that would take them to the correct page. However, two of the three subjects skipped over the bookmarking feature at the top of the Constitution and By-Laws page, which would have been the quickest way to complete the task. Of the two that did not immediately use the bookmarks, one scanned the long page of text and found the answer to the question. The other participant started to scan the page and then went back to the top and successfully used the bookmark to more quickly find the desired information.

4.2 Survey Results

The results of the post-test survey indicate that, overall, the website was easy to navigate, well designed, and easy to scan for important information. All of the test subjects indicated that they felt comfortable when using the site. All three test participants commented on the easy-to-use navigation bars. Two of the participants mentioned that "Get Smart!" and "Letter Carrier Perfect" are not yet synonymous for all of the people who will use this site, and that the links to the Get Smart! page should also indicate a connection to Letter Carrier Perfect. The post-test survey appears in Appendix B.

5. CONCLUSIONS

5.1 Positive Aspects

Two of the three test subjects found the answer to question 1 of Task One within a minute. They did this by scanning the homepage for the necessary information. I suspect that the subject who did not find the answer immediately felt pressured during the test and would have been more successful in a different situation. Question 2 of Task One went well for all of the test subjects. On a scale of 1–5 (1 = difficult; 5 = easy), two participants rated Task One a 5.

Task Three also went well for all of the participants. On a scale of 1–5 (1 = difficult; 5 = easy), completing Task Three was a 5 for all participants.

(continued)

Example 17.2

(continued)

On the post-test survey, organization of the site, moving from page to page, ease of reading and scanning text, and page layout were rated at very easy or very effective by all subjects. All three participants indicated that they felt comfortable using the site and that it is a site that they would use often.

5.2 PROBLEM AREAS

The main problem the participants had during the test was completing Task Two. The task instructions ask the participant to find the on-line version of "Letter Carrier Perfect," the name of which has recently been changed to "Get Smart!" I discovered during the test that this name change is not yet known to all of the website's audience members, and two of the subjects were not at all confident that clicking on the Get Smart! link would bring them to an on-line version of Letter Carrier Perfect.

6. RECOMMENDATIONS

Determine whether to continue using a link named "Get Smart!" to connect to a page that is an on-line version of "Letter Carrier Perfect."

Keeping the "Get Smart!" link is probably a good idea, because it will help to reinforce the name change of this handbook from "Letter Carrier Perfect" to "Get Smart!"

7. APPENDIX A

7.1 NALC Branch 728 Website Usability Test Introduction and Instructions

Thank you for participating in this project. Your input will help us to create a more useful website for Branch 728's members. You will be able to use this site in the near future to read *Satchel* articles and to take note of upcoming Branch meetings and activities, among other things.

The purpose of this test is to evaluate the usefulness and effectiveness of Branch 728's website.

To accomplish this, we will ask you to perform specific tasks. During the test, we will not make suggestions or be able to help in any way. Just use the site as you normally would to complete the tasks, and we will take notes for later analysis.

Please talk through your motions during the test. For example, let us know why you are choosing to click on a particular link. Then we'll have you fill out a short survey about your experience.

Please remember that the purpose of this test is to evaluate the usability of our website. We are not testing your personal abilities.

Task Scenarios and Instructions

Task One   As a member of NALC Branch 728, you are preparing for the next union meeting. You have decided to use the Branch's new website to gather some information. These are the questions you need to answer:

1. When is the next union meeting?
2. According to the Branch By-Laws, how many members constitute a quorum?

Task Two   As a Letter Carrier, you would like to find out more about your rights and responsibilities concerning safety, service, and mail security. The on-line version of Letter Carrier Perfect is arranged into three distinct sections. What are the three sections?

Task Three   You have taken some time to view the new Branch 728 website, but you need some more information from the Wisconsin State AFL-CIO. Use the Branch 728 website to connect to the Wisconsin State AFL-CIO website.

8. APPENDIX B

8.1 Post-Test Survey (highlighted numbers reflect the most common participant response)

1. What would have helped you complete these tasks more successfully?

2. What parts of the website did you find most helpful?

Please circle the number that best corresponds to your feelings about each question.

1. Completing Task One (next union meeting/quorum questions) was
   difficult   1   2   3   4   5   **easy**
2. Completing Task Two (Get Smart!/Letter Carrier Perfect question) was
   difficult   1   2   3   4   5   **easy**
3. Completing Task Three (connecting to WI State AFL-CIO site) was
   difficult   1   2   3   4   5   **easy**

(continued)

Example 17.2

(continued)

4. Organization of the site was
   not effective   1   2   3   4   5   **very effective**
5. Moving from page to page was
   confusing   1   2   3   4   5   **easy**
6. The text on the website was generally
   difficult to read   1   2   3   4   5   **easy to read**
7. Scanning the pages for important information was
   confusing   1   2   3   4   5   **easy**
8. When I was looking for the pages I needed, I
   felt lost   1   2   3   4   5   **felt I knew where I was**
9. While working on the tasks, I
   felt confused   1   2   3   4   5   **felt comfortable**
10. I thought the layout of most of the pages was
    difficult to understand   1   2   3   4   5   **easy to understand**
11. NALC Branch 728's webpage seems like a site that I would
    never use   1   2   3   4   5   **use often**

Exercises

▶ You Analyze

1. Collect one or two professional (VCR, stereo, CD, automobile, appliance, or computer) manuals and bring to class. Analyze them for page design, visual logic, text-visual interaction, sequence of parts, and assumptions made about the audience. Discuss these topics in groups of two to four, and then report to the class the strategies that you find most helpful and are most likely to use in Writing Assignment 1.

▶ Group

2. In groups of two to four, analyze the page layout of one of the manuals that appears in this chapter. Write a brief description of and reaction to this layout and share your reactions with the group. Your instructor will ask some groups to report their results.

▶ You Create

3. Using any machine or software program you know well, write a parts description.
4. Using any machine or software program you know well, create a flow chart for the sequences you want a reader to learn. Convert that flow chart into step-by-step instructions.

5. Review the types of decisions included in creating a template (pp. 447–449). Then create your own design for what you wrote in Exercise 3 or 4.

6. For the manual you are creating for Writing Assignment 1 or for the section you wrote for Exercise 4, create a storyboard.

7. Write the introduction to the parts description and sequences you created in Exercises 3 and 4.

8. For the manual you are creating for Writing Assignment 1, complete the following exercises. Your instructor will schedule these steps at the appropriate time in your project.

   a. Consider how consistently it handles all details of format.

   b. Consider how precisely it explains how to perform an action. Read closely to see whether everything you need to know is really present.

   c. Conduct a field test by asking a person who knows almost nothing about the product to follow your manual. Accompany the tester, but do not answer questions unless the action is dangerous or the tester is hopelessly lost (say, in a software program). Note all the problem areas, and then make those changes. Discuss changes that would help the user.

Writing Assignments

1. Write an operator's manual. Choose any product that you know well or one you would like to learn about. The possibilities are numerous—a bicycle, a sewing machine, part of a software program such as FrontPage® or Dreamweaver®, a computer system, any laboratory device, a welding machine. If you need to use high-quality photographs or drawings, you may need help from another student who has the necessary skills. Your manual must include at least an introduction, a table of contents, a description of the parts, and the instructions for procedures. You might also include a troubleshooting section. Give warnings when appropriate. Complete this chapter's worksheet or the appropriate exercises.

2. Write a learning report for the writing assignment you just completed. See Chapter 5, Writing Assignment 7, pages 134–135, for details of the assignment.

Web Exercise

Create a mini-manual to publish on a website. Use a simple machine, say, a flashlight. Include one section describing the parts and one section presenting the appropriate sequences for operating. Include several visual aids.

Works Cited

Bethke, F. J., W. M. Dean, P. H. Kaiser, E. Ort, and F. H. Pessin. "Improving the Usability of Programming Publications." *IBM Systems Journal* 20.3 (1981): 306–320.

Brooks, Ted. "Career Development: Filling the Usability Gap." *Technical Communication* 38.2 (April 1991): 180–184.

Cohen, Gerald, and Donald H. Cunningham. *Creating Technical Manuals: A Step-by-Step Approach to Writing User-Friendly Manuals.* New York: McGraw-Hill, 1984.

Cooper, Marilyn. "The Postmodern Space of Operator's Manuals." *Technical Communication Quarterly* 5.4 (1996): 385–410.

Craig, John S. "Approaches to Usability Testing and Design Strategies: An Annotated Bibliography." *Technical Communication* 38.2 (April 1991): 190–194.

Queipo, Larry. "Taking the Mysticism Out of Usability Test Objectives." *Technical Communication* 38.2 (April 1991): 185–189, 190–194.

Redish, Virginia. "Writing for People Who Are 'Reading to Learn to Do.'" *Creating Usable Manuals and Forms: A Document Design Symposium.* Technical Report 42. Pittsburgh, PA: Carnegie-Mellon Communications Design Center, 1988.

Riney, Larry A. *Technical Writing for Industry: An Operations Manual for the Technical Writer.* Englewood Cliffs, NJ: Prentice-Hall, 1989.

Rubens, Phillip M. "A Reader's View of Text and Graphics: Implications for Transactional Text." *Journal of Technical Writing and Communication* 16.1/2 (1986): 73–86.

Rubin, Jeff. "Conceptual Design: Cornerstone of Usability." *Technical Communication* 43.2 (May 1996): 130–138.

Sullivan, Patricia. "Beyond a Narrow Conception of Usability Testing." *IEEE Transactions of Professional Communication* 32.4 (December 1989): 256–264.

Section

3

Professional Communication

Chapter 18   Oral Presentations

Chapter 19   Letters

Chapter 20   Job Application Materials

Chapter 18 Oral Presentations

Chapter 18
In a Nutshell

Oral reports range from brief answers to questions at meetings, to hour-long speeches to large audiences. Follow these guidelines:

▶ Plan your presentation. Determine your audience. Determine whether your slides will carry information to explain or will function as a helpful outline. Use a simple template that does not draw attention to itself. Choose an organizational pattern, such as narrative or problem-solution, as appropriate.

▶ Speak in a normal voice. Help yourself speak normally by not memorizing—practice enough so you can speak from notes.

▶ Arrange your speech in a narrative fashion. Use topic sentences to begin sections so that you are constantly telling the audience where they are in the sequence.

▶ Practice with any technology (laptops, computer slide presentations) before you give the speech.

▶ Be presentable. Dress appropriately; if you don't know what a professional should wear in this situation, ask someone who does. Avoid irritating mannerisms (smacking lips, shaking keys in pockets, saying "um" repeatedly).

Throughout your career, you will give oral reports to explain the results of investigations, propose solutions to problems, report on the progress of projects, make changes to policy, create business plans, justify requests for such items as more employees and equipment, or persuade clients to purchase your services and merchandise. Sometimes these presentations are impromptu, but more often they are scheduled and therefore require careful planning. Since the introduction of PowerPoint presentational software, oral reports are almost always accompanied by a visual presentation. Edward Tufte says, "In corporate and government bureaucracies, the standard method for making a presentation is to talk about a list of points organized onto slides projected up on the wall" (3). Tufte calculates that trillions of slides are produced yearly (3).

This chapter explains how to plan and deliver an oral presentation.

Planning the Presentation

Planning includes decisions about audience, situation, organizational pattern, and presentation.

Plan for Your Audience

Because of the popularity of the oral report, many listeners have been subjected to the same kind of presentation many times. Indeed, the use of PowerPoint is so common that most oral reports are simply called "PowerPoints" or "Power-Point reports." Therefore, an audience may often anticipate that any presentation is likely to be another dreary time-waster (Miller; Tufte 23).

The key to preventing audience apathy or even a downright hostility is to focus on your audience, not on the technology of your presentation. The central goal for your presentation is, simply, that you be relevant. One speech expert says, "People will pay close attention to something they perceive as having relevance to their own lives and concerns"(Bacall). In order to remain aligned to your audience, consider differences between listeners and readers, and ask the same audience analysis questions that you ask in report writing.

Presenting to a Listening Audience

Speakers Use Personal Contact. A presentation allows you to have personal contact with listeners. You can make use of personality, voice, and gestures, as well as first-person pronouns, visuals, and feedback from listeners. If you are a person speaking to people, your audience will react positively

Listeners Are Present for the Entire Presentation. That listeners are present for the entire report may seem advantageous, but it also may make communication more difficult. Many listeners want to hear only selected parts of a report—the parts that apply directly to them. If, for instance, your listeners are the plant manager and her staff, the plant manager would probably prefer a capsule version of the report, which a short abstract would provide, and would rather leave the details for staff members to examine. The oral report gives her no choice but to listen to all of your detailed information, a situation that might put her in a negative frame of mind.

Ask Audience Analysis Questions

In speeches the audience analysis questions are the same as for reports or other documents. Ask these questions (based on Laskowski):

▶ Who are they?
▶ How many will be present?
▶ What is their knowledge level?
▶ What task will your presentation help them complete?
▶ What do they need?
▶ Why are they there?

Consider how you would give a speech on your new data analysis software to these two audiences: three experienced managers seated around a conference table, waiting to decide whether to place an order with you; and 50 sales reps seated in a lecture hall, eager to familiarize themselves with the product prior to making sales calls. Obviously, these situations would require two very different presentations. The answers to the audience analysis questions, then, will help you develop a presentation that keeps your audience attentive to your points, so that they walk away feeling that their time with you has been well spent and productive.

Plan for the Situation

Presentations are made in many venues, from small rooms in which people sit around a conference table to large auditoriums packed with conference attendees. Spend time investigating the physical layout of the room. Follow these guidelines (based in part on Jacobs):

▶ Determine the size of the room and where you will stand in relation to the audience, the screen, and the computer controls.
▶ Determine the location of the electrical outlets and the electrical cords on the floor.
▶ Learn how sound carries in the room. Will you have to use a microphone? If so, do you know how to adjust it so that your voice carries well without ringing or buzzing?

▶ Determine whether you will have to bring a disk to use in a computer already present in the room, or whether you need to bring your laptop.

▶ If you have to bring your own laptop, determine how to hook it up to the overhead projector system located in the room.

Plan Your Organizational Pattern

The organizational pattern that you choose depends on the needs of your audience and the actual content of the material. Common organizational patterns are problem-solution, goal-methods-results-discussion, narrative (see Shaw, Brown, and Bromiley), as well as the traditional approach of introduction, body and conclusion (Tracy).

The narrative approach, advocated by Shaw, Brown, and Bromiley, has three stages: set the stage, introduce the dramatic conflict, resolve the conflict. To set the stage, the speaker defines the current situation by analyzing factors that affect the situation—market forces, forces that affect change, company objectives and strengths. To introduce dramatic conflict, the speaker explains the challenges that face the company in the current situation. What are the obstacles (the bad guys) in this situation? poorly functioning technology? new competitor? market share losses? To resolve the conflict, the speaker must show how the audience can overcome the obstacles to win—This change will cause the technology to function correctly. This strategy will offset the new competitor's appeal. This strategy will induce the consumers to purchase again.

Shaw, Brown, and Bromiley (see also Tufte) feel that the story method is very effective, especially in contrast to "bullet-point" presentations. Bullet-point presentations have two problems: first, listeners tend to remember only the first and last items in a list; and second, listeners are unable to clarify the relationships of the items in a list. For instance, items in a list can have only three relationships: sequence (this point follows that point), priority (from most to least important or vice versa), and membership (all these items are the same kind of thing). Missing is the ability of the listeners to easily understand critical relationships such as cause or contrast.

The story structure "defines relationships, a sequence of events, cause and effect, and a priority among items—and those items are likely to be remembered as a complex whole" (4). As a result, the act of writing the speech will cause the speaker to think clearly about the complexity of the ideas, and so give the listeners access to the speaker's thought process. The upshot will be listeners who grasp the significance of the main point because they have been engaged by the story that expounds it.

Plan Your Presentation

To plan the presentation, determine your relationship to the slides, create a storyboard, and finally the series of slides.

Determine Your Relationship to the Slides

A presentation consists of two things—your voice and personal presence, and the information on the slides—and you need to determine their relationship. Because the slide can contain only a small amount of information, it is not the main source of information in the speech—you are. However, if you do not control the relationship, the slides can easily take over, basically drawing all the audience attention away from you and the points you have to make. To do so, use the advice of many experienced speakers—deliver quality content to the audience (Bacall; Stratten; Tufte 22). In order to deliver quality content, decide which combination of information from you and information from the slide will best convey the information to the audience.

What is the purpose of the information on the slide? Understand that it is not the slide that provides the complexity—You do. According to Tufte, people learn because information is placed into context. People need not just bits of information but also a narrative that interrelates them, explaining causal assumptions or analytical structure (6). The slide can provide the bits but you supply the context.

It is helpful to see the slide in either "the foreground" or "the background." If it is in the foreground, it provides information, either visual aid or text, that you explain. Project a visual in order to explain it. For instance, if you have a table of sales made during a quarter, you might show the table with its columns of numbers, and then point out items in the table and explain the relationships that are important to the audience. If you practice with the program, you will even be able to highlight the items that you wish to interrelate. Project text in order to emphasize. If, for instance, you want the audience to remember a certain point, you can project just that point on the screen and then discuss it.

With foreground material, then, you speak to the slide. The idea is that you will explain in some detail the implication of what the audience sees on screen, and the combination of the visual and your explanation is the point to get. For instance, Figure 18.1 illustrates the credentials of two researchers. The speaker projected the slide, then spoke for several minutes explaining in turn the various achievements of each researcher. The point of the slide is to establish the credibility of the researchers, and the speaker filled in details and context about the credentials listed.

When a slide is in the background, you do not speak to it. Instead it is present in order to summarize key points, either what is about to be covered or what has just been covered (Glenn Miller). At its best, this method helps the audience stay on track and follow and remember the points that have been made. If you have four subpoints to make about a topic, projecting them on a slide, then discussing each in turn will help the listeners follow along.

Figure 18.2 shows a slide that functioned as an outline. The speaker projected a point, addressed it, then moved on to the next point. The slide functions to keep the audience aware of their place in the presentation.

Understanding this relation to your slides will help you create a presentation that delivers content to audiences effectively.

Figure 18.1

Slide of
Credentials

Reprinted by
permission of
Dr. Scott Zimmerman.

Figure 18.2

Slide Showing an
Outline

Reprinted by
permission of
Dr. Scott Zimmerman.

Create a Storyboard

A storyboard is a text and graphics outline of your presentation. A storyboard
can be as simple as a two-column table that lists topics on the left and visuals on
the right (Figure 18.3, p. 480). A storyboard can also be a three-column sequence of lines, boxes, and comments. Place the slide's text in the line; sketch
the look of the slide in the box, and add explanatory material in the comments
space (Figure 18.4, p. 481). Variations of storyboards also exist in presentation
programs. PowerPoint, for instance, has an outline function and a slide sorter
function that allow you to see the entire sequence of your presentation (Figure 18.5, p. 481).

Figure 18.3

Sample Text
Storyboard

Topic	Visual Aid
Introduction	
Source of assignment	
Recommendation	List of recommendations
Preview	Outline of main topics
	List of main methods for each type
	(use of a two-column slide)
Section 1	
Method of researching	Cross-sectional view
Section 2	
Three types of laminates	
Advantages of each	List of advantages
Section 3	
Cost	Table of costs
Conclusion	
Summary	

Use storyboards as an integrated outline that shows both the sequence of the topics and the content of the slide or slides devoted to that topic (Korolenko 151–153; Lindstrom 110; "Web").

Create the Slides or Other Visual Aids

After you have the storyboard outline of your presentation, create the slides. A well-designed presentation will help convey your point pleasingly to an attentive audience. But, as Edward Tufte points out, "If your words or images are not on point, making them dance in color won't make them relevant" (22). The following guidelines (based in part on Scoville; Tessler; Welsh) will help you create pleasing slides and sequences that help to convey information rather than to distract attention from the main points. As you create slides, understand the parts of a slide, pay attention to helping your reader, and make effective use of fonts, colors, animations, sound effects, and slide transitions.

Understand the Parts of a Slide. The parts of the slide are the title, the text or graphics, the border, and the background (see Figure 18.6, p. 482).

▸ The title appears at the top, usually in the largest type size. Use it to explicitly identify the contents of the visual. A rule ($\frac{1}{2}$ or 1 point in size) separates the title from the text or graphic.
▸ The text makes the points you want to highlight. Use phrases that convey specific content rather than generic topics.
▸ The graphic consists of a table, chart, or drawing.
▸ The border is a line that provides a frame around the visual.
▸ The background is the color or design that appears behind the text or graphics.

Figure 18.4

Sample
Three-Column
Storyboard

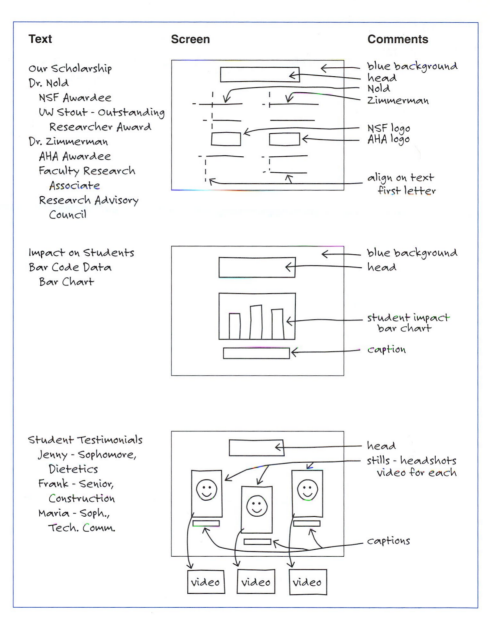

Figure 18.5

PowerPoint
Outline View and
Slide Sorter View

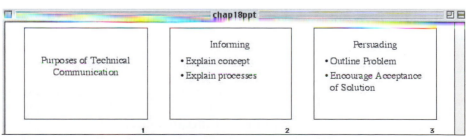

Help the Reader. The most important part of any presentation is the content, but do not try to put all of your ideas on slides. Too much text will cause viewers to read slides, deflecting attention from you. Display only text that the audience should read carefully. Follow these guidelines to design your slides:

▶ Use a landscape (horizontal) layout for your text rather than a portrait (vertical) layout. Landscape makes the longer lines of text easier to read, and columns easier to use.

▶ Title each slide so your audience will have a quick reference to the topic at hand.

▶ Create (or choose) a template or "master" to establish a visual logic. A template establishes rules for consistently presenting parts of the slide—the title, the text, the visual aids, the background. For instance, make all titles 24-point Arial, black, and centered at the top. Make all text 18-point Arial, black, flush left. Center all visuals with a caption in 14-point Arial below the visual. Figure 18.7 illustrates a template in action. Notice that slices a and c, which function as section introductions, use one sentence, a question, with a yellow font. However, slides b and d, the content of the section, have a title and a bulleted list, in a white font. Notice that the template used for Figure 18.8, a version of slide d, undercuts the seriousness of the content with its use of unbusiness-like fonts.

For visual aids (tables, graphs, pictures):

▶ Simplify the graph so that it makes only one point.

▶ Use graphs for dramatic effect. A line graph that plunges sharply at one point calls attention to the drop. Your job is to interpret it.

▶ Use tables for presenting numbers. Be prepared to point out the numbers you want the audience to notice.

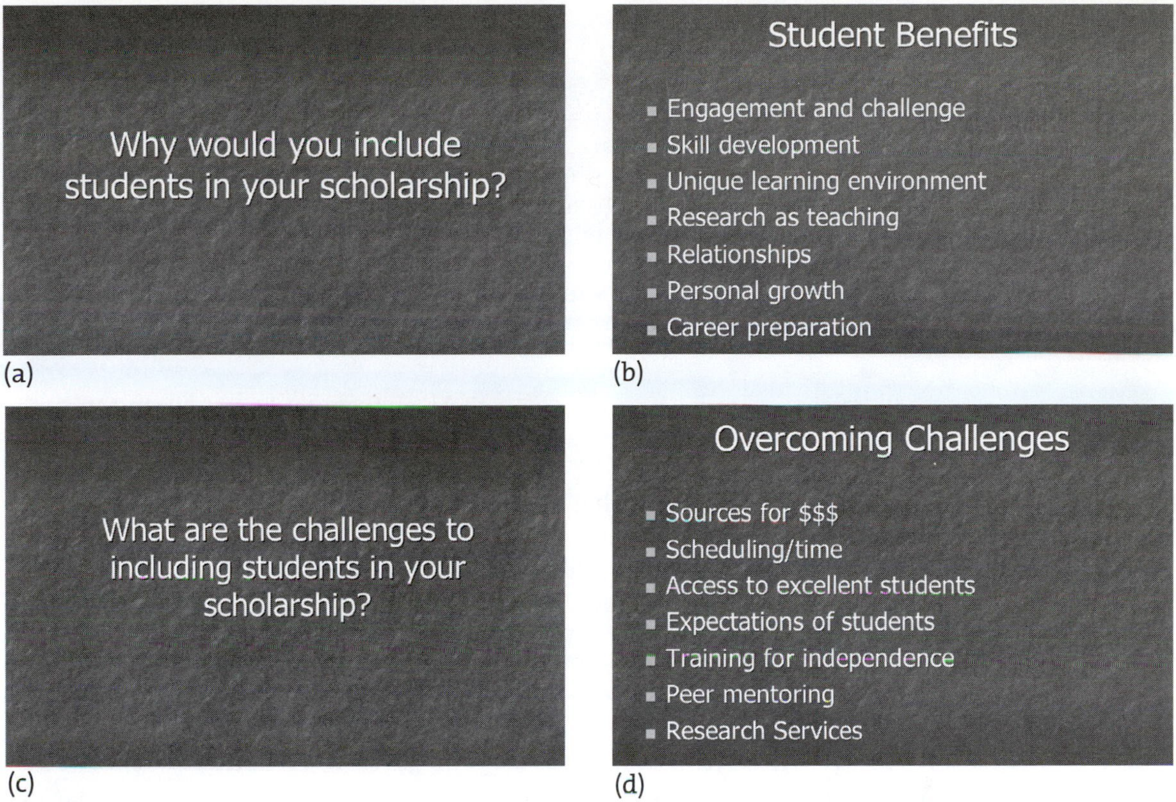

(a)

(b)

(c)

(d)

Figure 18.7

Slide Template

Figure 18.8

Font May Change Meaning

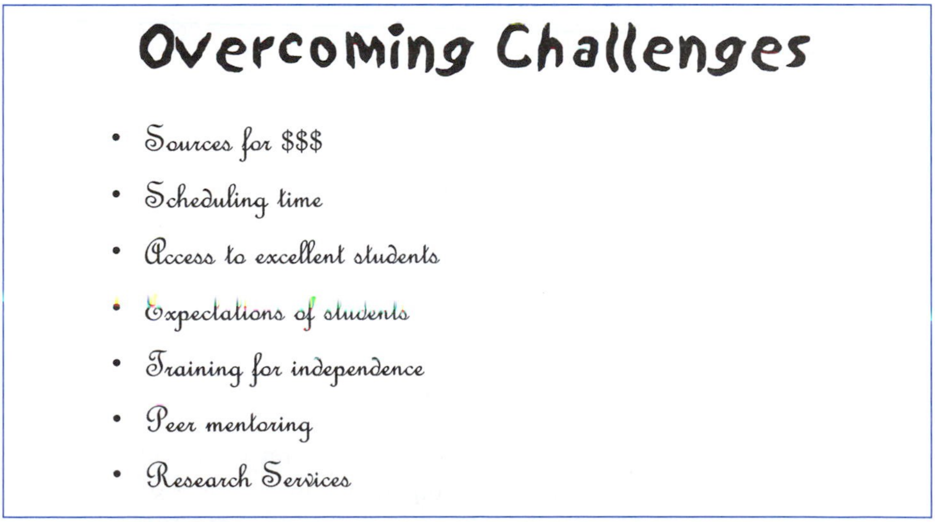

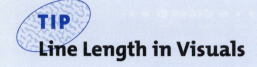

Line Length in Visuals

Many texts (see George Miller) encourage using no more than seven lines and limiting lines to seven words. It is more helpful to think about whether you want the audience to read a long quote or to present a group of short lines, summarizing what you are saying. Shorter text lines keep the words in the background as you speak, focusing attention on you.

▶ Use pictures to illustrate an object that you want to discuss, for instance, the control panel of a new machine.

For text visuals (visuals that use only words):

▶ Use initial capitals followed by lowercase letters.
▶ Use 18-point type for body text, 24-point type for titles.

Keep running text to a minimum. Try to keep text to no more than six to eight lines per slide, fewer if possible. If you project a long quote, stop speaking and let people read it themselves (Stratten).

Fonts.   Keep your text font simple yet elegant, subtle yet striking. Fonts can portray a wide range of emotions, from casual to authoritative, from serious to comic. Selecting and using a font that will elicit the desired response from your audience is important. Follow these three guidelines when selecting fonts:

▶ Use only one font, preferably a sans serif font like Helvetica or Arial.
▶ For impact, make sparing use of different sizes, boldfacing, or italics. For instance, a long quote when italicized, is difficult to read. Too much boldfacing or too many different sizes gives a cluttered, "ransom note" effect that is very distracting.
▶ Use larger font sizes and/or different colors for your titles. When using larger fonts, be sure to use those sizes that are easy to read (18 to 24 points), but not so large that they become distracting.

Colors.   Use color to enhance your presentation. Color combinations should help viewers focus on key points, not on the combination itself. To use color effectively, consider these guidelines:

▶ Use color intelligently to establish visual logic. Give each item (title, test, border) in the template its own color. Use only one color for emphasizing key words. (See "Focus on Color," pp. 166–173.)

▶ Use a background color; blue is commonly used.

▶ Use contrasting colors—white or yellow text on green or blue background.

▶ Use green and red sparingly. Ten percent of the population is color blind, and can't distinguish between red or green. It is best to assume that at least one person in every ten of your audience will be limited in his or her ability to translate color.

▶ Avoid hard-to-read color combinations, such as yellow on white and black on blue. Violet can also be very hard to read.

▶ Select combinations with an awareness of technology. Colors that look well together in a sign or in print, as in a magazine or newspaper, will probably not work the same way projected onto a screen. For example, ambient light, which is what is produced by an LCD projector, will affect contrast greatly; it will turn a dark color, such as burgundy or a deep green, into a pastel.

▶ Use what is known to work well. Yellow backgrounds with black lettering work well in most situations (think about school buses). Other good combinations are deep-blue backgrounds with yellow letters, or gray backgrounds with black letters.

▶ Evaluate templates before using them. Programs such as PowerPoint have a variety of background templates with color schemes that are handsome and exciting at first glance, but when you make your selection, always try to keep your audience in mind. Look at it from their standpoint—will this combination help them understand your point? If you're not certain about the combination, don't use it.

Animation. Animation is making text move. In slides, text can appear and depart from the slide by many animated routes—for instance, text can slide to the left or right, up or down, or disintegrate or blossom out from the center. Animation can emphasize important points visually while you explain them verbally. However, the idea is to enhance the production, not provide gimmicky entertainment. Follow these guidelines:

▶ Use only one text animation. Remember that the audience should be tracking as closely as possible with the presenter. To keep the audience tracking, pick only one primary reading transition. For example, a simple "wipe-right" text animation—the text appears to move off the screen left to right—will keep the reader's eye going in the normal reading direction.

▶ Treat previous lines carefully. Fading or subduing the previous bullets when the new information appears will help to keep the audience focused, but select a subdued tone for the previous bullets. Using an entirely different color will only draw attention away from the new text you are trying to introduce.

▶ Use animated graphics to make complex points. The multimedia, electronic presentation can use graphics effectively to show the progression of complicated points. Process steps, time lines, and flow charts all benefit from animated graphics. Keep in mind that being consistent is important when using animated graphics. Your audience will appreciate consistent use of color contrast and special effects.

Sound Effects. Sound effects can accompany animation. As the word appears on the screen, audiences can hear sounds such as the clacking of an old type-writer or brakes squealing. Remember, the point is to enhance the contents of the slides, not distract the audience from them. Use these guidelines:

- Use no sound at all.
- Use the same sound for each transition. In other words, develop an audio logic so that each instance of the sound indicates a repetition of a step in the sequence (new slide or new line of text).
- Use subdued, undramatic sounds for transition. The first time you use a ricocheting bullet sound, you may get a reaction from your audience, but the second time you use it, the reaction may be greatly diminished, and the third time may have no effect at all.
- If you have a special point that you want to emphasize, or if you want to use a sound effect for some comic relief in a deadly serious situation, go ahead, but use these strategies sparingly.

Slide Transitions. Slide transitions are like animations. Like scene cuts in a movie, they help move one set of visual data out of view and move a new set into view. The goal of using this device is to create a sequence logic—each instance of the event means that another type of data is about to appear. If your audience concentrates on the transition rather than on the message, you've lost your audience. Follow these guidelines:

- Be consistent. Use the same transition for the same kind of event. Make each slide move off to the left or right or whatever you've chosen.
- Use only one or two simple transitions.
- Select transitions in order to aid the viewer. For example, a transition that "wipes up" will help to guide the watcher's eye back to the top of the slide so that the subject of the slide is immediately identified. A simple fade-to-black between sections of a presentation signals that a new topic is being considered.

Visual Aids and Slides. The visual aids that you use on your slides are the same that you would use in a hardcopy report: outlines; slides or drawings; tables and graphs. In addition, consider paper handouts.

Many slides are, in effect, outlines. They are lists of words in some kind of hierarchy (level 1, level 2, etc.). Like all outlines, this type of slide shows listeners the sections and subsections of the presentation. And like all outlines, they project very little depth or indication of relationship of ideas. This device, while common, is a background device (see above).

Tables and graphs can present data in a way that enables listeners to grasp relationships right away. An oral explanation of the relationship among the percentages that affect a pay increase is hard to follow, but a table or graph makes it clear. Note, however, that the space on a slide is restricted, so complex tables (as are often produced as part of scientific experiments) can be hard to project. (See Chapter 7 for more on tables and graphs.)

Paper handouts are often useful for a presentation. PowerPoint has a function that allows a speaker to print out and distribute the entire presentation in a series of "thumbnails," small versions of the actual slide. This device, although commonly used, is essentially just an outline, with none of the attendant contextualization that the speaker provides. You can supply some of that context by using PowerPoint's Notes function and handing out copies of the slides with your commentary in the note window. One speech expert suggests this: "If what you say when you expand the bullet points is useful for the audience to take away, put it in the handout" (Stratten).

In addition, paper handouts can more effectively show complex text, numbers, and data graphics (Tufte 22). A handout can replace or supplement projected visual aids. Pass out copies of a key image, perhaps a table. Listeners can make notes on it as you speak.

Making an Effective Presentation

To make an effective presentation, develop your introduction, navigate the body, develop your conclusion, rehearse your presentation, and deliver your presentation.

Develop the Introduction

The introduction establishes both the tone and the topic of the speech. Your tone is your attitude toward the listeners and the subject matter. Use the introduction to establish the relevance of the presentation to the audience. Be serious, but not dull. Avoid being so intense that no one can laugh, or so flip that the topic seems insignificant. Be explicit about your purpose and the sections of the presentation. In other words, follow the old advice of "Tell 'em what you are going to tell 'em" (Bacall; Tracy).

Follow these guidelines:

▶ You do not need to begin your report with a joke, a quotation by an authority, or an anecdote, but a well-chosen light story often helps relax both you and the audience.
▶ Explain why your report is important to your audience.
▶ Present your conclusions or recommendations right away. Then the audience will have a viewpoint from which to interpret the data you present.
▶ Explain how you assembled your report.
▶ Indicate your special knowledge of or concern with the subject.

> ❱ Identify the situation that required you to prepare the report (or the person who requested it).
> ❱ Preview the main points so your listeners can understand the order in which you will present your ideas.

Navigate the Body

Many studies have shown that listeners simply do not hear everything the speaker says. Therefore, you should give several minutes to each main idea—long enough to get each main point across, but not long enough to belabor it.

Use Transitions Liberally

Clear transitions are very helpful to an audience of listeners. Your transitions remind them of the report's structure, which you established in the preview. Indicate how the next main idea fits into the overall report and why it is important to know about it. For instance, a proposal may seem very costly until the shortness of the payback period is emphasized.

Emphasize Important Details

Presumably, if you have created a storyboard, you know the details that you want to emphasize for the audience, and you have placed them on slides. Choose details that are especially meaningful to the audience. Explain any anticipated changes in equipment, staff, or policy, and show how these changes will be beneficial.

Impose a Time Limit

Find out how long the audience expects the presentation to last and fit your speech into that time frame. If they expect 15 minutes and you talk for 15 minutes, they will feel very good. Generally, speak for less time than is required. It is much better to present one or two main ideas carefully than to attempt to communicate more information than your listeners can comfortably grasp.

Develop a Conclusion

The conclusion section restates the main ideas presented in the body of the report. Follow these guidelines:

> ❱ As you conclude your report, you should actually say, "In conclusion . . ." to capture (or recapture) your listeners' interest.
> ❱ For a proposal, stress the main advantages of your ideas, and urge your listeners to take specific action.
> ❱ For a recommendation report, emphasize the most significant data presented for each criterion, and clearly present your recommendations.

▶ Use a visual to summarize the important data.

▶ End the report by asking whether your listeners have any questions.

Rehearse Your Presentation

During rehearsals, go straight through the speech, using note cards. If it is a formal presentation, when you practice, wear the same clothes you will wear in the actual presentation.

Practice Developing a Conversational Quality

When you make your speech, sound like a person speaking to people, and use both voice and gestures to emphasize important points. Even the best information will fall on deaf ears if it is delivered like a robotic time-and-weather announcement. Rehearse until you feel secure with your report, but always stop short of memorizing it. If you memorize, you will tend to grope for memorized words rather than concentrating on the listeners and letting the words flow.

Practice Handling Your Technology and Visual Aids

Understand how to open and navigate your presentation. If necessary, have the presentation on several media. Often, speakers have the same file both on a disk and on their laptops. If you are unfamiliar with the technology, practice opening the files from both a disk and a laptop. Because technological arrangements in new places can be difficult to navigate, have a backup plan in case your technology does not work. Practice giving the presentation so that you know how to open the software program and advance and reverse the slides. Practice talking to the audience and looking at the screen only for those slides that you will use as foreground slides.

If you have paper visual aids or overhead transparencies, arrange these in the order they will be needed, and decide where you will place them when you are finished with them. If a listener asks you to return to a visual, you want to be able to find it easily. If you are using handouts, decide whether to distribute them before or during the presentation. Distributing them before the presentation eliminates the need to interrupt your flow of thought later, but because the listeners will flip through the handouts, they may be distracted as you start. Distributing them during the presentation causes an interruption, but listeners will focus immediately on the visual.

Rehearse

Practice your presentation at least once under conditions similar to those in which you will make the presentation, particularly for reports to large groups. Use a room of approximately the same size, with the same type of equipment for projecting your voice and your visuals. If you have never used a microphone, now is the time to practice with one.

Deliver Your Presentation

You will increase your effectiveness if you use notes and adopt a comfortable extemporaneous style.

Use Notes

Experienced speakers have found that outlines prepared on a few large note cards (5 by 8 inches, one side only) are easier to handle than outlines on many small note cards. Some speakers even prefer outlines on one or two sheets of standard paper, mounted on light cardboard for easier handling.

The outline should contain clear main headings and subheadings. Make sure your outline has plenty of white space so you can keep your place.

Adopt a Comfortable Style

The extemporaneous method results in natural, conversational delivery and helps you concentrate on the audience. Using this method, you can direct your attention to the listeners, referring to the outline only to jog your memory and to ensure that ideas are presented in the proper order. Smile. Take time to look at individual audience members and to collect your thoughts. Instead of rushing to your next main point, check whether members of the audience understood your last point. Your word choice may occasionally suffer when you speak extemporaneously, but reports delivered in this way still communicate what you want to say better than those memorized or read.

The following suggestions will help as you face your listeners and deliver the presentation:

1. Look directly at each listener at least once during the report. With experience, you will be able to tell from your listeners' faces whether you are communicating well. If they seem puzzled or inattentive, repeat the main idea, give additional examples for clarity, or solicit questions. Don't proceed in lockstep through your notes. Adapt.

2. Make sure you can be heard, but also try to speak conversationally. You should feel a sense of your voice as a round, full tone, projecting with conviction. You should also feel that your voice fills the space of the room, with the sound of your voice bouncing back slightly to your own ears. The listeners should get the impression that you are talking to them rather than just presenting a report. Inexperienced speakers often talk too rapidly.

3. Try to become aware of—and to eliminate—your distracting mannerisms. No one wants to see speakers brush their hair, scratch their arms, rock back and forth on the balls of their feet, smack their lips. If the mannerism is pronounced enough, it may be all the audience remembers of your presentation. Stand firmly on both feet without slumping or swaying.

Globalization and Oral Presentations

Giving a presentation to a foreign or non-English-speaking audience is easier if you give some thought to relating to an audience whose culture is not your own. A key idea for your planning is that although English is commonly studied as a second language, "English proficiency within a given audience can vary widely, so the best approach is to simplify and clarify content at every turn" (Zielinski). In order to simplify and clarify content, follow these tips: Use simple sentences, make clear transitions, avoid digressions, reduce use of potentially confusing pronouns, restate key points, pause periodically, use subject-verb-object word order, repeat phrases using the exact wording. If you call it a "plan" the first time, continue to use that word; don't switch to "proposal" or "map" or "vision" (Zielinski).

Also, be aware that the international audience's reaction to you may differ greatly from what you are used to. For example, in Japan, it is not unusual for audience members to close their eyes in order to convey concentration and attentiveness, while in the United States closed eyes are a sign that you are lulling the audience to sleep. Applause is a generally universal sign of approval, but whistling in Europe is a negative reaction to your presentation. Finally, know that other cultures have a different sense of acceptable personal space than Americans have. Middle Easterners and Latin Americans tend to stand much closer than Americans find comfortable, while many Asian cultures stand quite far away from each other. Keep this in mind if you have others onstage with you or if you will be going into the audience for your presentation (McKinney, "International").

Be aware of body language conventions. Hand gestures that are accepted in the United States, such as the A-OK symbol (the circle formed with your index finger and thumb), or the thumbs-up gesture, are considered obscene in some countries. Pointing with a finger can be impolite; use a fully extended hand. In some countries, emphatic gestures are poorly received. Body language that is unwittingly offensive can cause an audience to focus on what is inappropriate and lose the content of your presentation (Zielinski).

Plan for differences in technology. Bring pictures of the equipment that you will need during your presentation. Bring a voltage converter. Remember that many countries have differently sized standard paper and may use a two-hole instead of a three-hole punch. Most importantly, have a backup plan and keep a sense of humor (McKinney, "Public," "Professional").

For further reference, check out these websites:

Executive Planet at <www.executiveplanet.com/> is a guide to all aspects of conducting international business: etiquette, customs, and culture. If you'll be presenting in a foreign country, you can find out the details of greetings, business attire, and meeting formalities as well as general information for many different countries.

(continued)

(continued)

For insight into the cultural dynamics of other countries and regions and
how this may impact your business dealings abroad, refer to "International
Business Etiquette and Manners" at <www.cyberlink.com/>. Your presentations will be more effective if you know of the appropriate way you
should act.

Although addressed to potential conference speakers, this article offers an
invaluable overview of preparing to give a presentation to an international
audience: Gagnon, Michael, and Raymond Wallace. "Making a Presentation
in English at a European Conference." Federation of European Chemical
Societies. Division of Chemical Education. July 2001. 7 Feb. 2004 <http://
216.239.37.104/search?q=cache:df8EOxSS4-8J:www.chemsoc.org/pdf/
enc/fecs/fecsedgagan.pdf+%22visual+aids%22+tips+%22international+
audience%22&hl=en&ie=UTF-8>.

Works Cited

McKinney, C. "Public Speaking: Bilingual Help." 2003. Advanced Public
Speaking Institute. 31 Jan. 2004 <www.public-speaking.org/public-speaking-
bilingual-article.htm>.

———. "Public Speaking: International Perspective on Humor." 2003. Advanced Public Speaking Institute. 31 Jan. 2004 <www.public-speaking
.org/public-speaking-international-article.htm>.

———. "Public Speaking: Professional Photographs." 2003. Advanced Public
Speaking Institute. 31 Jan. 2004 <www.public-speaking.org/public-speaking-
equipmentphotos-article.htm>.

Zielinski, Dave. "Going Global, Part I." 3M United States. *Presentation* (n.d.)
6 Feb. 2004 <www.3m.com/meetingnetwork/presentations/pmag_
going_global_1.html>.

4. To point out some aspect of a visual projected by an overhead projector, lay a
pencil or an arrow made of paper on the appropriate spot of the transparency.

5. When answering questions, make sure everyone hears and understands
each question before you begin to answer it. If you cannot answer a question
during the question-and-answer session, say so, and assure the questioner
that you will find the answer and provide it at a later time.

Worksheet for Preparing an Oral Presentation

- ☐ Identify your audience.
  What is your listeners' level of knowledge about the topic?
  What is their level of interest in the entire speech?
  Why are they attending?
  What do they need?

- ☐ Create an outline showing the main point and subpoints.
  Which strategy will best help the audience? Problem-solution?
  Narrative? IMRD?

- ☐ Assign a time limit to each point.

- ☐ Create a storyboard.
  What visual aid will illustrate each point most effectively?

- ☐ Decide whether you need any kind of projection or display equipment.
  laptop? LCD projector? flip chart?

- ☐ Review the speaking location. Do you know how to make your technology (laptop, disk, projector) interact with the technology resident at the site?

- ☐ Determine your relationship to the slides. Will they be foreground or background for you?

- ☐ Prepare clearly written note cards—with just a few points on each.

- ☐ Rehearse the speech several times, including how you will actually handle the technology.

Worksheet for Evaluating an Oral Presentation

- ☐ Answer these questions:
  1. Clarity
     Did the speaker tell you the point early in the speech? Could you tell when the speaker moved to a new subpoint?
  2. Tone
     Did the speaker sound conversational? Did the speaker go too fast? go too slow? speak in a monotone?
  3. Use of technology
     Did the speaker interact effectively with the slides?
     Did the slides help you understand the content, or were they distracting?

Exercises

▶ You Create

1. Create a PowerPoint presentation of two or three slides that illustrates a problem in one of your current projects. Give a brief speech (2 to 3 minutes) explaining the problem. Alternate 1: Prepare two or three PowerPoint slides that illustrate the solution or its effects, and present the entire problem and solution to the class in a 4- to 5-minute speech. Alternate 2: Prepare the PowerPoint presentation in groups of two to four. Select a speaker for the group. Give the speech.

2. Report on a situation with which you are involved. Your work on an assignment for this class is probably most pertinent, but your instructor will provide his or her own requirements. Depending on the available time, draw a visual on the board, make a transparency, create a handout, or prepare one or two PowerPoint slides. In 2 minutes, explain the point of the visual aid. Class members will complete and/or discuss the evaluation questions above.

3. For Exercises 1 and 2, each member of the audience should prepare a question to ask the speaker. Conduct a question-and-answer session. When the session is finished, discuss the value and relevance of the questions that were asked. What constitutes a good question? Also evaluate the answers. What constitutes a good answer?

4. Make a storyboard for the speech you will give for the following Speaking Assignment. Divide a page into these three columns and fill them in, following this example:

Point	Visual Aid	Time
Method of extrusion	Cross section of laminate	2 minutes

5. Use the generic graph on page 495 to make a 2-minute speech. Title the graph and explain its source, its topic, and the significance of the pattern. Choose any topic that would change over time.

6. Give a brief speech in which you freely use technical terms. The class will ask questions that will elicit the definitions. If there is time, redeliver the speech at a less technical level.

Speaking Assignment

Your instructor may require an oral presentation of a project you have written during the term. The speech should be extemporaneous and should conform to an

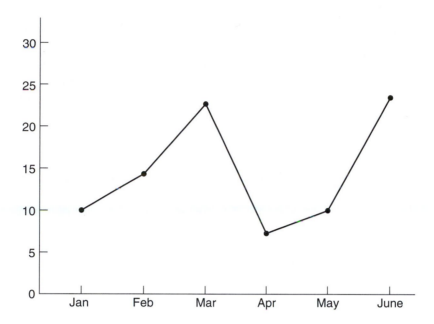

agreed-upon length. Outline the speech, construct a storyboard, make your visuals, and rehearse. Follow your presentation with a question-and-answer session.

Writing Assignment

Write a learning report for the speaking assignment you just completed. See Chapter 5, Writing Assignment 7, pages 134–135, for details of the assignment.

Web Exercises

1. Create a PowerPoint version of a document you have previously written for this class. Upload the PowerPoint to the Web. Give a 4- to 5-minute speech using the on-line slides. Alternate: Upload the PowerPoint presentation created above. Have classmates, as assigned by your teacher, read and evaluate the report. Use the evaluation sheet for the appropriate type of report in the appropriate chapter.

2. Using screens that you download from the Web, create a PowerPoint presentation to your classmates in which you do either of the following:

   a. Explain the effective elements of a well-designed screen.
   b. Explain how to use the screen to perform an activity (order plane tickets, contact a sales representative, perform an advanced search).

Works Cited

Bacall, Robert. "How Attention Works for Audiences." PowerPointAnswers.com. 13 Oct. 2003 <www.powerpointanswers.com/article1035.php>.

Jacobs, Kathryn. "Overcoming Stage Fright." 2002. PowerPointAnswers.com. 14 Nov. 2003 <www.powerpointanswers.com/article1002.html>.

Korolenko, Michael. *Writing for Multimedia: A Guide and Sourcebook for the Digital Writer.* Belmont, CA: Wadsworth, 1997.

Laskowski, Lenny. "A.U.D.I.E.N.C.E. Analysis: It's Your Key to Success." 2001. PowerPointers.com 13 Oct. 2003 <www.powerpointers.com/showarticle .asp?articleid=248>.

Lindstrom, Robert L. *The* Business Week *Guide to Multimedia Presentations.* New York: McGraw-Hill, 1994.

Meng, Brita. "Get to the Point." *Macworld* 5.4 (1988): 136–143.

Miller, George. "The Magical Number Seven, Plus or Minus Two: Some Limits on Our Capacity for Processing Information." *Psychological Review* 63 (1956): 81–97.

Miller, Glenn. "Presentation Disasters: Conference Style." 08 Mar. 2004. PowerPointAnswers.com. 1 May 2004 <www.powerpointanswers.com/ article1036.php>.

Scoville, Richard. "Slide Rules." *Publish!* 4.3 (1989): 51–53.

Shaw, Gordon, Robert Brown, and Philip Bromiley. "Strategic Stories": How 3M Is Rewriting Business Planning." *Harvard Business Review* 76 (May–June 1998): 42–44. Reprint 98310.

"Storyboard" WWWNetSchool. 13 Oct. 2003 <www.thirteen.org/edonline/ software/earthinflux/orgc.html>.

Stratten, Scott. "Business Tip: Giving Effective PowerPoint Presentations." 13 Oct. 2003 <http://virtual.yosemite.ce.ea.s/itolhurst/ESGIS/Presentational/ Business_Tip.htm>.

Tessler, Franklin. "Step-by-Step Slides." *Macworld* 5.12 (1988): 148–153.

Tracy, Larry. "Preparing a Presentation." 2000. PowerPointers.com. 11 Oct. 2003 <www.powerpointers.com/showarticle.sap?articleid=216>.

Tufte, Edward. *The Cognitive Style of PowerPoint.* Cheshire, CT: Graphics Press, 2003.

"Web Resources in Multimedia Storyboards." 1999–2003. Teacher Resource Center. Georgia Department of Education. 2 May 2004 <www.glc.k12.ga.us/ trc/cluster.asp?mode=browse&intPathID=7801>.

Welsh, Theresa. "Presentation Visuals: The Ten Most Common Mistakes." *intercom* 43.6 (1996): 22–43.

Chapter

19 Letters

Chapter 19
In a Nutshell

Letters are presented in an agreed-upon set of ways. Study Figure 19.1 on page 499 to find out what these conventions are and how they fit on the page. That figure presents the parts in block format, which arranges all the items at the left margin. Using block format is an easy way to present yourself as a credible professional.

Letters represent you or your company in professional, often legal and emotional, situations. The key to all types of letters is to treat the reader appropriately, using the "you" attitude and speaking to readers in clear, understandable, nonconfrontational words. Write several short paragraphs rather than fewer long ones. Treat readers as you would want yourself or people close to you treated.

Business letters are an important—even a critical—part of any professional's job and are written for many reasons to many audiences. They may request information from an expert, transmit a report to a client, or discuss the specifications of a project with a supplier. Letters represent the firm, and their quality reflects the quality of the firm. This chapter introduces you to effective, professional letter writing by explaining the common formats, the standard elements, the planning required, and several common types of business letters.

Three Basic Letter Formats

The three basic formats are the block format, the modified block format, and the simplified format (*Merriam; Webster's*).

Block Format

In the *block format,* place all the letter's elements flush against the left margin. Do not indent the first word of each paragraph. The full block format, shown in Figure 19.1 or Example 20.3 (p. 536), is widely used because letters in this format can be typed quickly.

Modified Block Format

The *modified block format* (an example appears in Figure 19.3, p. 508) is the same as the full block format with two exceptions: The date line and closing signature are placed on the right side of the page. The best position for both is five spaces to the right of the center line, but flush right is acceptable. A variation of this format is the *modified semiblock.* It is the same as the modified block, except that the first line of each paragraph is indented five spaces.

Simplified Format

The *simplified format* (see Figure 19.4, p. 509, for an example) contains no salutation and no complimentary close, but it almost always has a subject line. It is extremely useful for impersonal situations and for situations where the identity of the recipient is not known. In personal situations, writers start the first paragraph with the recipient's name.

Elements of a Letter

Internal Elements

This section describes the elements of a letter from the top to the bottom of a page.

Figure 19.1

Block Format

Heading

> 4217 East Eleventh Avenue
> Post Office Box 2701
> Austin, TX 78701
>
> *(skip 2 lines)*

Date

> February 24, 2005
>
> *(skip 2 lines)*

Inside address

> Ms. Susan Wardell
> Director of Planning
> Acme Bolt and Fastener Co.
> 23201 Johnson Avenue
> Arlington, AZ 85322
> *(double-space)*

Salutation, mixed punctuation

> Dear Ms. Wardell:
> *(double-space)*

Subject line

> SUBJECT: ABC CONTRACT (optional)
> *(double-space)*

Body paragraphs flush left

> ———————————————————————
> ———————————————————————
> ———————————————————————
>
> *(double-space between paragraphs)*
>
> ———————————————————————
> ———————————————————————
> ———————————————————————
>
> *(double-space)*

Closing, mixed punctuation

> Sincerely yours,

Signature

> *[signature: John K. Palmer]*   *(skip 3 lines)*

Typed name
Position in company

> John K. Palmer
> Treasurer
> *(double-space)*

Typist's initials
Enclosure line

> abv
> enc. (2)
> *(skip 1 or 2 lines; depends on letter length)*

Copy line

> c: Ms. Louise Black

TIP
Punctuation

Letter items are punctuated by either *open* or *mixed* patterns. You may choose either. Be consistent. In open format, put no punctuation after the salutation and complimentary close. In mixed format, put a colon after the salutation and a comma after the complimentary close.

Heading

The heading is your address.

> 4217 East Eleventh Avenue
> Post Office Box 2701
> Austin, TX 78701

- Spell out words such as *Avenue, Street, East, North,* and *Apartment* (but use *Apt.* if the line would otherwise be too long).
- Put an apartment number to the right of the street address. If, however, the street address is too long, put the apartment number on the next line.
- Spell out numbered street names up to *Twelfth.*
- To avoid confusion, put a hyphen between the house and street number (1021-14th Street).
- Either spell out the full name of the state or use the U.S. Postal Service Zip Code abbreviation. If you use the Zip Code abbreviation, note that the state abbreviation has two capital letters and no periods and that the Zip Code number follows one space after the state (NY 10036).
- Note on letterhead: place the date two lines below the last line of the letterhead, in the position required by the format (e.g., flush left for block).

Date

Dates can have one of two forms: February 24, 2006, or 24 February 2006.

- Spell out the month.
- Do not use ordinal indicators, such as 1st or 24th.

Inside Address

The inside address is the same as the address that appears on the envelope.

> Ms. Susan Wardell
> Director of Planning
> Acme Bolt and Fastener Co.
> 23201 Johnson Avenue
> Arlington, AZ 85322

▶ Use the correct personal title (Mr., Ms., Dr., Professor) and business title (Director, Manager, Treasurer).
▶ Write the firm's name exactly, adhering to its practice of abbreviating or spelling out such words as *Company* and *Corporation*.
▶ Place the reader's business title after his or her name or on a line by itself, whichever best balances the inside address.
▶ Use the title *Ms.* for a woman unless you know that she prefers to be addressed in another way.

Attention Line

Attention lines are generally used only when you cannot name the reader ("Attention Human Resources Manager"; "Attention Payroll Department").

▶ Place the line two spaces below the inside address.
▶ Place the word *Attention* against the left margin. Do *not* follow it by a colon.

Salutation

The salutation always agrees with the first line of the inside address.

▶ If the first line names an individual (Ms. Susan Wardell), say, "Dear Ms. Wardell:" If the name is "gender neutral" (Robin Jones), say "Dear Robin Jones:"
▶ If the first line names a company (Acme Bolt and Fastener Co.), use the simplified format (see Figure 19.4, p. 509) with a subject line or repeat the name of the company ("Dear Acme Bolt and Fastener Co.:").
▶ If the first line names an office (Director of Planning), address the office, use an attention line, or use a subject line.

Dear Director of Planning: (*or*)
Attention Director of Planning (*or*)
SUBJECT: ABC CONTRACT

▶ If you know only the first initial of the recipient, write "Dear S. Wardell," or use an attention line.
▶ If you know only a Post Office box (say, from a job ad), use a subject line.

Box 4721 ML
The Daily Planet
Gillette, WY 02716
Subject: APPLICATION FOR OIL RIG MANAGER

Subject Line

Use a subject line to replace awkward salutations, as explained above, or to focus the reader's attention.

SUBJECT: **Request to Extend Deadline**

- Follow the word *Subject* with a colon.
- For emphasis, capitalize or boldface the phrase.
- Use of the word *Subject* is optional, especially in simplified format. If you do not use *Subject,* capitalize the entire line.

Body

Single-space the body, and try to balance it on the page. It should cover the page's imaginary middle line (located 5½ inches from the top and bottom of the page). Use several short paragraphs rather than one long one. Use 1-inch margins at the right and left.

Complimentary Closing and Signature

Close business letters with "Sincerely" or "Sincerely yours." Add the company name if policy requires it.

Sincerely yours,
ACME BOLT AND FASTENER CO.

John K. Palmer

John K. Palmer
Treasurer

- Capitalize only the first word of the closing.
- Place the company's name immediately below the complimentary closing (if necessary).
- Allow three lines for the handwritten signature.
- Place the writer's title or department, or both, below his or her typed name.

Optional Lines

Place optional lines below the typed signature.

- Place the typist's initials in lowercase letters, flush left.
- Add an enclosure line if the envelope contains additional material. Use "Enclosure:" or "enc:". Place the name of the enclosure (résumé, bid contract) after the colon, or put the number of enclosures in parentheses.

enc: (2)
Enclosure: résumé

- If copies are sent to other people, place "c:" (for copy) at the left margin and place the names to the right.

c: Joanne Koehler

Succeeding Pages

For succeeding pages of a letter, place the name of the addressee, the page number, and the date in a heading.

Susan Wardell   -2-   February 24, 2006

Envelopes

The standard business envelope is 9½ by 4³⁄₁₆ inches. Place the stamp in the upper right corner. Place your address (the same one that you used in the heading) in the upper left corner.

Place the address anywhere in the "read area" of the U.S. Postal Service's optical character recognition (OCR) machines (Figure 19.2). The U.S. Postal Service (USPS) recommends the following descending order:

▶ Attention line
▶ Company name
▶ Street address; on the street address line, also add directions (N, NE, S, etc.), designator (St., Ave., Rd.), and sublocation (Apt., STE [suite], RM [room]). *Note:* The postal service will deliver the mail to the address directly above the city and state.
▶ City and state
▶ Zip Code to the right of the state; use all 9 digits if you know them

Figure 19.2

OCR Area of Envelope

Planning Business Letters

In planning your letter, you must consider your audience, your tone, and your format.

Globalization and Letters

Basically, writing letters to a foreign audience requires that you be clear and direct. Jargon, idioms, and humor can all be misinterpreted or misunderstood, so keep your writing free of them. In most countries, any business communication is formal, so keep your tone professional. Refrain from sounding casual and what may seem to an international audience as overly familiar.

Letter-writing conventions differ from country to country; make an effort to learn what format your audience expects. For example, outside the United States, the date is usually written as day/month/year. So, March 9, 2004, would be 9/3/04 or 9 March 2004. In France, Spain, and Italy, the months are not capitalized when they are written out.

Also, salutations differ from country to country. In Germany, business is kept on a formal level and you should not use given names unless you are directly invited to do so. You would address a letter to Herr Schmidt and refer to him as Herr Schmidt throughout the letter. In China, people are addressed with their family (last) names preceding their given (first) names. They do use Mr. or Ms. at all times, so make sure that you are using the correct last name. And in Japan, it is customary to address a letter using a person's last name and their job title, or else with a –san at the end for a more general salutation ("International").

Let's say that you need to write a letter to a colleague in Spain. Using this sample format from "Writing Letters in Spanish," you would address the letter to either Señor (Sr.) or Señora (Sra.) and could add Don or Doña as an additional title of respect:

On separate line

Sociedad Anonima is corporation; Calle is street; San Bernado is the street name; 15 is the building number, 3rd floor, suite C; 28015 Madrid is the postal code and city

Sra. Doña
Maribel Muñoz Franco
Editores Internacionales S.A.
Calle San Bernado, 15-3°-C
28015 MADRID

If you know the addressee, the greeting of "Estimada Señora Muñoz:" is appropriate. If not, you may use "Muy Señores Mios:" as a more formal way to greet your colleagues. Many endings exist to close your letter. The following are all adequate for most situations:

A la espera de sus prontaa noticias, le saluda atentamente,
Sin otro particular, le saluda atentamente,
Le saluda atentamente,
Atentamente,

For further reference, check out this website:
The Business Start Page, at www.bspage.com, contains much general business information. Under the Addresses, you will find the elements of international addresses and gives tips on the correct salutations. The site also has a section for international e-mail etiquette.

Works Cited

"International Addresses and Salutations." The Business Start Page. 6 Feb. 2004
<www.bspage.com/address.html>.

"Writing Letters in Spanish." 2004. AskOxford.com. 6 Feb. 2004 <www.askoxford
.com/languages/es/spanish_letters/?view=uk>.

Consider Your Audience

Before you begin to write, consider the audience's knowledge level and its specific need for the information. The audience could be a customer who can identify the defective part but who may not know much about how the part fits into the larger system in the machine. The audience could be an expert engineer or a manager who understands the general theory of this kind of project (say, constructing a commercial building) but knows little about this particular project.

Also be aware that the audience has various reasons for needing your letter. Assessing that need allows you to write a more effective letter. Your letter may be shown to someone other than just the recipient, usually a person more distant from the situation. And your letter could be used as a basis for decisions, even legal action.

Consider Your Tone

Use Plain English

You want to sound natural; you are, after all, one human being addressing another. Plain conversational English makes your point better than "businessese." Even if your letter has legal implications, you should use a relaxed, clear tone. Consider this brief passage:

> Pursuant to our discussion of February 3 in reference to the L-19 transistor, please be advised that we are not presently in receipt of the above-mentioned item but expect to have it in stock within one week. Enclosed herewith please find a brochure regarding said transistor as per your request.

Here is the paragraph rewritten in a more direct, conversational style that makes the contents much easier to grasp.

> I've enclosed a brochure on the L-19 transistor we talked about on February 3. Our shipment of L-19s should arrive within a week.

Use the "You" Approach

The "you" approach is based on the writer's recognition that the recipient is a person who appreciates being approached in a personal way. Applying this approach requires only that you use "I," "we," and "you." Consider the following example, and note how the writer addresses the customer's dissatisfaction both by showing empathy and by proposing solutions to the problem.

Dear Mr. Hillary:

After my January 27th visit to your complex to investigate the poorly performing laser printers, I talked to our technical support and to Megacorp's customer representative. We have several suggestions for solving the problem. If they do not work, we will investigate the relatively more difficult task of replacing all ten printers.

Technical support suggests that you discontinue using the printer driver in your desktop publishing program. Hold the shift key down while you select print and you will default to the word processing print driver. In tests we performed here we found that the word processor driver prints about 8–10 times faster. This simple procedure change should solve most of the problems.

The difficulty you have in printing eps images in your desktop publishing program is more complicated. Our simple recommendation is that you convert them to GIF images. We tried this and the documents printed about 3 times faster. Our more involved recommendation is that we upgrade the RAM in the printers to 4 MB. Because of the difficulties you have experienced and because of the length of time you have been our customer, we will install them free and charge only our cost.

If none of these suggestions works, we will have to begin to negotiate to return the printers to Megacorp or trade them in. My brief contact with the customer rep at Mega indicates that this option will be more difficult.

I will contact you next week on Tuesday to see if our suggestions have had any effect.

Sincerely,

Marian Goodrich

Marian Goodrich
District Sales Manager

Consider Format

Format can affect the way your audience accepts your message. Use one of the basic formats or one that your company requires. Readers expect you to be a knowledgeable professional; using the correct form helps reinforce that impression. Make careful design decisions. Choose several short paragraphs rather than one long paragraph. Use bullets and indentations to help readers grasp key points easily. Review Chapter 6 to see how format can help presentation.

Ethics and Letters

Letters are written to communicate something in a more formal way than is an e-mail or a memo. Understanding the letter's format and audience is essential if the writer is to establish his or her *ethos* (the character that the reader perceives). Business letters are written for a specific purpose and therefore are to the point. Sometimes this terseness can be misconstrued, so the letter's structure directly affects how the information it contains is interpreted. Letters containing good news are, not surprisingly, easier to write than letters with bad news and structuring it is simple: Start with the good news. However, relaying bad news makes considering the audience reaction much more important. Communicating bad news in a way that develops an effective *pathos* (sense of the emotional impact of the letter) also creates a positive *ethos* (perception of the character of the writer) for the sender. The letter writer must take ethical responsibility for his or her epistolary communication.

Types of Business Letters

The rest of this chapter examines several types of business letters and suggests how to structure their contents.

Transmittal Letters

A *transmittal letter* (Figure 19.3, p. 508) conveys a report from one firm to another. (Transmittal correspondence is explained in Chapter 14.) To write a transmittal letter, follow these guidelines:

▶ Identify the report enclosed.
▶ Briefly explain the report's purpose and scope.
▶ Explain any problems encountered.
▶ Acknowledge the people who helped.

General Information Letters

General information letters can deal with anything. They serve to keep the writer in touch with the reader (a common public relations device), to send information, or to reply to requests. Figure 19.4 (p. 509) shows an example. To write such a letter, follow these guidelines:

▶ Use a context-setting introduction.
▶ If there is an acceptance or rejection, state it clearly.
▶ Use formatting to highlight the main point.
▶ Add extra information as needed, but keep it brief.

VINZ CONSULTING
EVERYTHING ABOUT THE SHOP
1021 Portland Drive
East Pines, MD 20840-1461
(307) 432-8866

October 27, 2006

Mr. Charles Lindsay
Mountain Milling
3266 Crestview Drive
Charleston, WV 25301

Dear Mr. Lindsay:

Attached is my final report on the type of milling machine you should purchase for your plant. I recommend that you purchase Ironton's #02119-BTUA.

As we discussed on my site visit last month, I have researched the appropriate literature on this subject, talked to several sales reps, and observed three different demonstrations of the 02119 and its two competitors. You were particularly concerned about size and power— the 02119 will do the job for you.

I have enjoyed our work together and look forward to working with you in the future. I have found your staff particularly helpful in filling my several requests about your plant's capacity and materials flow.

Sincerely,

Steve Vinz

Steve Vinz
Project Manager

Closing and signature
set near middle of
page

Figure 19.4

General
Information
Letter in
Simplified Format

Maxwell and Goldman
3227 Girard Avenue South
Minneapolis, Minnesota 55408
608-385-1944 / fax 608-385-1945
www.maxgold.com

July 14, 2006

Mr. Duwan James
James Corporation
4810 River Heights Drive
St. Paul, Minnesota 55106

No salutation

Duwan, here is the background information on the Adjustable Speed
Drive. The cost of this system, as we discussed earlier, is $1000.00
installed.

1. The Adjustable Speed Drive can operate as a clutch to inch and jog
   your conveyor to the exact assembly position. The operator can
   control the speed instantly from zero to maximum speed, and any
   speed in between.

2. The conveyor speed can be varied simply and easily while the motor
   remains at a constant operating speed. The operator controls the
   speed by hand with the control lever.

3. This speed drive system offers a speed range of 0 to 160 feet per
   minute (fpm) for the conveyor. This is approximately 2 miles per
   hour (mph). You indicated that your average speed is 60 fpm. This
   speed can be locked in for normal runs or sped up for resupplying
   the line, and again, slowed down for positioning.

4. The drive system is a compact package weighing 45 pounds and
   having overall dimensions of 10" x 12" x 32". It won't overload or
   clutter the conveyor frame. The whole system operates from one
   power cord and requires no special maintenance. The unit is sealed
   and prelubricated. The only maintenance necessary would be a
   periodic check of mounting hardware.

Duwan, if you need more information or have any more questions, call
or fax to the numbers above, or email me at goldmans@maxgold.com.

No closing or typed
name

Shana Goldman

Worksheet for Writing a Business Letter

Planning

☐ **Analyze the audience.**
Who will receive this letter?
Why do they need it? What will they do as a result of receiving it?

☐ **Name your goal for the reader.**
What do you want to happen after the reader reads the letter?

☐ **Choose a format for the reader.**
Use the simplified format for more impersonal or more routine situations.

Generating

☐ **State your main points succinctly.**

☐ **Compose with a "you" attitude.**

Finishing

☐ **Reread the letter slowly, word for word, to weed out any errors in spelling and grammar or problems with style (wrong tone, "garden path" sentences, see Chapter 4).**

☐ **Reread the letter to make sure your facts are accurate.**

☐ **Review each of the following for standard form:**
Your address. Do not use abbreviations, except for the state.
The recipient's name, title, corporation title, and address.
The salutation. Repeat the recipient's name.
If you do not know the recipient's name, use a subject line.
The complimentary closing.
Your typed signature (three spaces below the closing).
Your signature (between the closing and the typed signature).

Exercises

▶ You Revise

1. Rewrite this passage using plain English and a "you" attitude.

There was a question asked to me in regard to the complete fulfillment of contract 108XB (Manual Effector Arm Robot A). Complete documentation of same has not been fulfilled. The specifications are interpreted by this office to mean that no such documentation was required.

▶ You Analyze

2. Analyze this letter.

> A shredded conveyor belt? What a disaster! And here we were, so happy the last time we talked. Well, while it's always something, there's a silver lining in every cloud, so let's talk about what happened.
>
> Why did this happen? We haven't had a conveyor belt shred at a customer site in 31 years. You're the first. Have you checked your operating procedures? What do you do for training? These things are practically indestructible — who runs your machines? What do you know about them?
>
> Anyhow, if that's the cloud, the silver lining is that you get one free. It's in the mail. COD.
>
> Then I reviewed the problem with our design engineer. She feels that the belt exactly fills the specifications and that the fault probably is with your staff, but there is a slight possibility that there could be a problem with the metal "hooks" that join the two ends of the rubber. As your employees install the new belt, make them check those hooks. They should not "wobble." If they do, call me.
>
> Our sales representative can get to your place on Friday, June 19. If anything else strange comes up, let her know; she can fix anything — she's a great gal.
>
> Hope there's no hard feelings. Your business is important to us.

▶ You Create

3. a. Write two passages. In the first passage, try to be overtechnical, acting as if you expect that anyone would know the terms and concepts you use. In the second passage, rewrite your text so that it assumes that the technical language is foreign to the reader.

   b. In groups of two to four, read your two versions, then discuss your results with the class.

   c. Write a memo that tells what you learned from this assignment.

Writing Assignments

1. Write a general information letter to your instructor to give her or him the background details of a report you will write. Explain items that the instructor needs to know to read the report as a knowledgeable member of the corporate community.

2. As part of a research project, write a letter of inquiry to a professional. Ask him or her for information about your topic. Your questions should be as specific as you can make them. Ask questions such as "How does Wheeler Amalgamated extrude the plastic used in the cans for Morning Bright orange juice?"

Avoid questions such as "Can you send me all the information you have on the extruding process and any other processes of interest?"

3. As part of an assignment that requires a formal report, write a transmittal letter. Follow Figure 19.3.

4. Write a learning report for the writing assignment you just completed. See Chapter 5, Writing Assignment 7, pages 134–135, for details of the assignment.

Web Exercise

Analyze two or three company homepages for the "you" attitude. What happens (or doesn't happen) on the screen to make readers feel that they are being addressed personally?

Works Cited

Merriam Webster's Secretarial Handbook. Ed. Sheryl Lindsell-Roberts. 3rd ed. Springfield, MA: Merriam, 1993.

United States Postal Service (USPS). *Addressing for Optical Character Recognition.* Notice 165. June 1981.

United States Postal Service (USPS). *Here's How to Address Your Mail for the Best Mail Service.* Notice 36SUC380. Washington, DC, n.d.

Webster's New World Office Professionals' Desk Reference. Ed. Anthony S. Vlamis. New York: Macmillan, 1999.

Job Application Materials

Chapter 20
In a Nutshell

The goal of the *letter of application* and the résumé is to convince someone to offer you a job *interview.*

Basic letter strategies. Relate to the potential employer's needs. Show how you can fill those needs. If, in the job announcement, an employer lists several requirements, your letter should include a paragraph on each. In those paragraphs, present a convincing and memorable detail: "At Iconglow I was in charge of the group that developed the on-line Help screens. Under my direction, we analyzed what topics were needed and which screen design would be most effective."

Write in small chunks, putting the employer's keywords at the beginning of each chunk. Pay close attention to spelling and grammar—mistakes could cost you an interview.

Basic résumé strategies. Design your *résumé* so that key topics jump out. Include sections on

- Your objective (one brief line).
- How to contact you.
- Your education (college only).
- Your work history (most relevant jobs at the top; list job title, employer, relevant duties, and responsibilities).

Most résumés place the major heads at the left margin and indent the appropriate text about one inch.

Basic interview strategies. At the *interview,* you talk to people who have the power to offer you the job. Impress them by knowing about their company and by telling the truth—if you don't know the answer, say so.

This chapter explains the process of producing an effective résumé and letter. You must analyze the situation, plan the contents of the résumé and letter, present each in an appropriate form, and perform effectively at an interview.

Analyzing the Situation

To write an effective résumé and letter of application, you must understand your goals, your audience, the field in which you are applying for work, your own strengths, and the needs of your employers.

Understand Your Goals

Your goals are to get an interview and to provide topics for discussion at that interview. If you present your strengths and experiences convincingly in the letter and résumé, prospective employers will ask to interview you. To be convincing, you must explain what you can do for the reader, showing how your strengths fill the reader's needs.

The letter and résumé also provide topics for discussion at an interview. It is not uncommon for an interviewer to say something like "You say in your résumé that you worked with material requirements planning. Would you explain to us what you did?"

Understand Your Audience

Your audience could be any of several people in an organization—from the human resources manager to a division manager, one person or a committee. Whoever they are, they will have only a limited amount of time to read your letter and résumé and so will want to see immediately your qualifications stated in a professional manner.

The Reader's Time

Employers read letters and résumés quickly. A manager might have 100 résumés and letters to review. On the initial reading, the manager spends only 30 seconds to 3 minutes on each application, quickly sorting them into "yes" and "no" piles.

Skill Expectations

Managers want to know how the applicant will satisfy the company's needs. They look for evidence of special aptitudes, skills, contributions to jobs, and achievements at the workplace (Harcourt and Krizar). Suppose, for instance, that the manager placed an ad specifying that applicants need "experience in materials resource planning." Applications that show evidence of that experience probably will go into the "yes" pile, but those without evidence will go into the "no" pile.

Professional Expectations

Managers read to see if you write clearly, handle details, and act professionally. Clean, neat documents written in clear, correct English and formatted on high-quality paper demonstrate all three of these skills.

Assess Your Field

Find out what workers and professionals actually do in your field, so that you can assess your strengths and decide how you may fill an employer's needs. Answer the following questions:

1. What are the basic activities in this field?

2. What skills do I need to perform them?

3. What are the basic working conditions, salary ranges, and long-term outlooks for the areas in which I am interested?

Talk to professionals, visit your college placement office, and use your library. To meet professionals, set up interviews with them, attend career conferences, or join a student chapter or become a student member of a professional organization. Your college's placement service probably has a great deal of career and employer information available.

In your library two helpful books, among many that describe career areas, are the *Dictionary of Occupational Titles (DOT)* and the *Occupational Outlook Handbook (OOH)*, both issued by the U.S. Department of Labor. The *DOT* presents brief but comprehensive discussions of positions in industry, listing the job skills that are necessary for these positions. You can use this information to judge the relevance of your own experience and course work when considering a specific job. Here, for instance, is the entry for manufacturing engineer:

> **012.167-042 MANUFACTURING ENGINEER (profess. & kin.)** Plans, directs, and coordinates manufacturing processes in industrial plant: Develops, evaluates, and improves manufacturing methods, utilizing knowledge of product design, materials and parts, fabrication processes, tooling and production equipment capabilities, assembly methods, and quality control standards. Analyzes and plans work force utilization, space requirements, and workflow, and designs layout of equipment and workspace for maximum efficiency [INDUSTRIAL ENGINEER (profess. & kin.) 012.167-030]. Confers with planning and design staff concerning product design and tooling to ensure efficient production methods. Confers with vendors to determine product specifications and arrange for purchase of equipment,

materials, or parts, and evaluates products according to specifications
and quality standards. Estimates production times, staffing requirements, and related costs to provide information for management decisions. Confers with management, engineering, and other staff regarding
manufacturing capabilities, production schedules, and other considerations to facilitate production processes. Applies statistical methods to estimate future manufacturing requirements and potential.
GOE: 05.01.06 STRENGTH: L GED: R5 M5 L5 SVP: 8 DLU: 89

The *OOH* presents essays on career areas. Besides summarizing necessary job
skills, these essays contain information on salary ranges, working conditions,
and employment outlook. This type of essay can help you in an interview. For
instance, you may be asked, "What is your salary range?" If you know the appropriate figures, you can confidently name a range that is in line with industry standards.

Assess Your Strengths

To analyze your strengths, review all your work experience (summer, part-time,
internship, full-time), your college courses, and your extracurricular activities
to determine what activities have provided specific background in your field.
   Prepare this analysis carefully. Talk to other people about yourself. List every
skill and strength you can think of; don't exclude any experiences because they
seem trivial. Seek qualifications that distinguish you from your competitors.
Here are some questions (based in part on Harcourt and Krizar) to help you analyze yourself.

1. What work experience have you had that is related to your field? What
   were your job responsibilities? In what projects were you involved? With
   what machinery or evaluation procedures did you work? What have your
   achievements been?

2. What special aptitudes and skills do you have? Do you know advanced testing methods? What are your computer abilities?

3. What special projects have you completed in your major field? List
   processes, machines, and systems with which you have dealt.

4. What honors and awards have you received? Do you have any special college achievements?

5. What is your grade point average?

6. How have you paid for your college expenses?

7. What was your minor? What sequence of useful courses have you completed? A sequence of three or more courses in, for example, management,
   writing, psychology, or communication might have given you knowledge
   or skills that your competitors do not possess.

8. Are you willing to relocate?

9. Are you a member of a professional organization? Are you an officer? What projects have you participated in as a member?

10. Can you communicate in a second language? Many of today's firms are multinational.

11. Do you have military experience? While in the military, did you attend a school that applies to your major field? If so, identify the school.

Assess the Needs of Employers

To promote your strengths, study the needs of your potential employers. At your college's library or placement service, you can find many helpful volumes that describe individual firms. Read annual reports and company brochures, and visit company websites. You can easily discover the names of persons to contact for employment information and details describing the company, as well as its location(s) and the career opportunities, training and development programs, and benefits it offers.

Planning the Résumé

Your résumé is a one-page (sometimes two-page) document that summarizes your skills, experiences, and qualifications for a position in your field. Plan it carefully, selecting the most pertinent information and choosing a readable format.

Information to Include in a Résumé

The information to include in a résumé is that which fills the employer's needs. Most employers expect the following information to appear on applicants' résumés (Harcourt and Krizar; Hutchinson and Brefka):

▶ Personal information: name, address, phone number
▶ Educational information: degree, name of college, major, date of graduation
▶ Work history: titles of jobs held, employing companies, dates of employment, duties, a career objective
▶ Achievements: grade point average, awards and honors, special aptitudes and skills, achievements at work (such as contributions and accomplishments)

Résumé Organization

Traditionally, the information required on a résumé has usually been arranged in chronological order, emphasizing job duties. Because employers are accustomed to this order, they know exactly where to find information they need and can focus easily on your positions and accomplishments (Treweek).

The chronological résumé has the following sections:

▶ Personal data
▶ Career objective
▶ Summary (optional)
▶ Education
▶ Work experience

Personal Data

The personal data consist of name, address, telephone number (always found at the top of the page), place to contact for credentials, willingness to relocate, and honors and activities (usually found at the bottom of the page). If appropriate add an e-mail address and personal website url.

List your current address and phone number. Tell employers how to acquire credentials and letters of reference. If you have letters in a placement file at your college career services office, give the appropriate address and phone numbers. If you do not have a file, indicate that you can provide names on request.

Federal regulations specify that you do not need to mention your birth date, height, weight, health, or marital status. You may give information on hobbies and interests. They reveal something about you as a person, and they provide topics of conversation at a surprising number of interviews.

Career Objective

The career objective states the type of position you are seeking or what you can bring to the company. A well-written objective reads like this: "Management Consulting Position in Information Systems" or "Position in Research and Development in Microchip Electronics" or "To use my programming, testing, and analysis skills in an information systems position."

Summary

The summary, an optional section, emphasizes essential points for your reader (Parker). In effect, it is a mini-résumé. List key items of professional experience, credentials, one or two accomplishments, and one or two skills. If you don't have room for the summary in the résumé, consider putting it into your accompanying letter.

SUMMARY OF QUALIFICATIONS

Strong operations and client relationship management background with proven expertise in leading an operations team for multimillion-dollar retail organization. Well-developed customer relations skills that build lasting client loyalty. Proven new business development due to excellent prospecting and client rapport building skills. Able to develop processes that increase productivity, profitability, and employee longevity.

Education

The education section includes pertinent information about your degree. List your college or university, the years you attended it, and your major, minor, concentration, and grade point average (if good). If you attended more than one school, present them in reverse chronological order, the most recent at the top. You can also list relevant courses (many employers like to see technical writing in the list), honors and awards, extracurricular activities, and descriptions of practicums, co-ops, and internships. You do not need to include your high school.

EDUCATION

Bachelor of Science, University of Wisconsin–Stout, May 2002.
Major: Food Systems and Technology; Emphasis: Food Science
Minor: Chemistry

Associate of Applied Science Degree, Georgia Military College, Brunswick, Georgia 1995.

ACADEMIC ACCOMPLISHMENTS

Phi Theta Kappa—International Honor Society of the Two Year College
Academic National Honor Society

Work Experience

The work experience section includes the positions you have held that are relevant to your field of interest. List your jobs in reverse chronological order—the most recent first. In some cases, you might alter the arrangement to reflect the importance of the experience. For example, if you first held a relevant eight-month internship and then took a job as a dishwasher when you returned to school, list the internship first. List all full-time jobs and relevant part-time jobs—as far back as the summer after your senior year in high school. You do not need to include every part-time job, just the important ones (but be prepared to give complete names and dates).

Each work experience entry should have four items: job title, job description, name of company, and dates of employment. These four items can be arranged in several ways, as the following examples show. However, *the job description is the most important part* of the entry. Describe your duties, the projects you worked on, and the machines and processes you used. Choose the details according to your sense of what the reader needs.

Write the job description in the past tense, using "action" words such as *managed* and *developed*. Try to create pictures in the reader's mind (Parker). Give specifics that he or she can relate to. Arrange the items in the description in order of importance. Put the important skills first. The following example illustrates a common arrangement of the four items in the entry:

PROFESSIONAL EXPERIENCE

Job title
Company details

Sales/Marketing Director, Information Services Group, LLC, Milwaukee, WI 2000–2001

Duties and
accomplishments

Established client database that led to strong relationships with key accounts.

Extensive coordination over all advertising including: writing, proofreading, detail organization, layouts, designs, and productions.

Personally responsible for several major accounts, doubling sales revenues for fiscal year 2000.

Developed new accounts due to excellent prospecting and follow-through abilities.

Extensive telemarketing, cold calls, and sales presentations.

Organized, developed, and implemented new employee handbook; responsible for material and design.

Order of Entries on the Page

In the chronological résumé, the top of any section is the most visible position, so you should put the most important information there. Place your name, address, and career objective at the top of the page. In general, the education section comes next, followed by the work section. However, if you have had a relevant internship or full-time experience, put the work section first. Figure 20.1 shows a chronological résumé.

Writing the Résumé

. . . .

Drafting your résumé includes generating, revising, and finishing it. Experiment with content and format choices. Ask a knowledgeable person to review your drafts for wording and emphasis. Pay close attention to the finishing stage, in which you check consistency of presentation and spelling.

The résumé must be easy to read. Employers are looking for essential information, and they must be able to find it on the first reading. To make that information accessible, use highlight strategies explained in Chapter 6: heads, boldface, bulleting, margins, and white space. Follow these guidelines (and compare them to the sample résumés in this chapter):

▶ Usually limit the résumé to one page.
▶ Indicate the main divisions at the far left margins. Usually, boldface heads announce the major sections of the résumé.

Figure 20.1

Sample Résumé

Michelle L. Stewart

2837 Main Street   (715) 421-8765
Eau Claire, WI 54701   michstew27@yahoo.com

CAREER OBJECTIVE

To obtain a position in the food industry as a Consumer Scientist.

SUMMARY OF QUALIFICATIONS

Strong operations and client relationship management background with proven expertise in leading an operations team for multimillion-dollar retail organization. Well-developed customer relations skills that build lasting client loyalty. Proven new business development due to excellent prospecting and client rapport building skills. Able to develop processes that increase productivity, profitability, and employee longevity.

EDUCATION

Bachelor of Science Degree, University of Wisconsin–Stout, May 2002
Major: Food Systems and Technology; Emphasis: Food Science
Minor: Chemistry
Associate of Applied Science Degree, Georgia Military College, Brunswick, Georgia 1995.

ACADEMIC ACCOMPLISHMENTS

Phi Theta Kappa—International Honor Society of the Two Year College Academic National Honor Society

PROFESSIONAL EXPERIENCE

Sales/Marketing Director, Information Services Group, LLC, Milwaukee, WI 2000–2001

Established client database that led to strong relationships with key accounts.

Extensive coordination over all advertising including: writing, proofreading detail organization, layouts, designs, and productions.

Personally responsible for several major accounts, doubling sales revenues for fiscal year 2000.

Developed new accounts due to excellent prospecting and follow-through abilities.

(continued)

Figure 20.1

(continued)

Extensive telemarketing, cold calls, and sales presentations.

Organized, developed, and implemented new employee handbook; responsible for material and design.

Sales Professional, IKON Technology Services, Milwaukee, WI 1999–2000

Responsible for maintaining client relationships with several key business accounts.

Coordinated delivery of products and services to ensure on-time completion of projects.

Operations/Merchandise Manager, Best Buy Co., Inc, Madison, WI 1997–1999

Managed the daily processes of operations team including: development for three-person staff, flow delegation, implementation of new systems, processing of all transactions, and effective cost management.

Managed major departmental merchandising reorganizations according to company standards.

Created new performance standards and assessment tools that ranked employee performance and provided training and employee coaching to help employees meet new goals.

Primarily responsible for scheduling staff of 105+ and maintaining monthly labor budgets based on sales volume/store performance/job functions resulting in improved productivity.

Trained staff on company procedures to increase their ability to provide excellent customer service resulting in an increase in team morale, customer loyalty, and profitability.

Coordinated human resource activities including: hiring, training, performance evaluations, and team development.

Administered employee benefits, compensation, and payroll; mediated employee legal conflict/disputes; and dealt with employee terminations.

Developed new processes to reduce controllable expenses, resulting in an increase in net profit.

RELATED EXPERIENCE

Front End Operations Manager, Sam's Club, Madison, WI 1996–1997

Pharmacy Department Manager, Wal-Mart Corporation, St. Mary's, GA 1992–1996

- Boldface important words such as job titles or names of majors; use underlining sparingly.
- Use bulleted lists, which emphasize individual lines effectively.
- Single-space entries, and double-space above and below. The resulting white space makes the page easier to read.
- Control the margins and type size. Make the left margin 1 inch wide.
- Use 10- or 12-point type.
- Treat items in each section consistently. All the job titles, for example, should be in the same relative space and in the same typeface and size.
- Print résumés on good-quality paper; use black ink on light paper (white or off-white). Avoid brightly colored paper, which has little positive effect on employers and photocopies poorly.
- Consider using a résumé software program. Actually a database, it provides spaces for you to fill with appropriate data and offers several designs for formatting the page.

Planning a Letter of Application

The goal of sending a letter of application is to be invited to an interview. To write an effective letter of application, understand the employer's needs, which are expressed in an ad or a job description. Planning a specific letter requires you to analyze the ad or description and match the stated requirements with your skills.

Analyze the Employer's Needs

To discover an employer's needs, analyze the ad or analyze typical needs for this kind of position. To analyze an ad, read it for key terms. For instance, a typical ad could read, "Candidates need 1+ years of C++. Communication skills are required. Must have systems analysis skills." The key requirements here are 1+ years of C++, communication skills, and systems analysis.

If you do not have an ad, analyze typical needs for this type of job. A candidate for a manufacturing engineer position could select pertinent items from the list of responsibilities printed in the *Dictionary of Occupational Titles* (see pp. 515–516).

Match Your Capabilities to the Employer's Needs

The whole point of the letter is to show employers that you will satisfy their needs. If they say they need 1+ years of C++, tell them you have it.

As you match needs with capabilities, you will develop a list of items to place in your letter. You need not include them all; discuss the most important or interesting ones.

Ethics and Résumés

In writing a résumé, you want to engender confidence in your abilities, and avoid either underselling or overselling your experience. Recruiters are often looking through scores of résumés in search of measurable accomplishments that sound relevant for the position being filled. Using creative words to enhance the sound of an otherwise mundane job may result in your résumé being passed over, as the manager spends extra seconds trying to decipher the hidden meanings. What you say about your experience should be defensible and logical, and can be creative, but not outlandish. The résumé that honestly and straightforwardly presents the candidate's experience with a positive spin has the best chance of being read and landing you an interview (Truesdell). For example, stretching dates of employment to cover a jobless period is lying, and can cost you the offer or get you fired later on. Saying that you "specialized in retail sales and assisted in a 10 percent increase in sales at the store level," when you worked at the local video store and the store saw a 10 percent increase in sales is embellishing, but is not a lie. Claiming that you were store manager when you were not is lying (Trunk).

Résumé padding, telling lies on résumés, is becoming more and more common. Lying about experience or accomplishments on a résumé is risky. Résumé padding may get you a job for the short term if the employer hiring you does not check your résumé carefully, but résumé padding can come back to haunt you later (Callahan). *Inventing experiences, educational degrees, and accomplishments shouldn't be done due to the damage it can cause the writer down the road, even if the writer does not have a moral problem with lying.* False information on your résumé sits like a land mine waiting to explode (Callahan). In the recent past, lies on résumés ruined prestigious and lengthy careers. Companies have fired successful CEOs upon discovering they had misrepresented university credentials (Callahan). When share prices plummet as a result, investors may even sue the CEO for fraud. An otherwise stellar career that may never have needed a made-up degree can be brought to an unceremonious and humiliating end. Being a candidate without a master's degree or with a gap in employment is not out of the ordinary. Being a candidate who got caught stretching dates of employment to cover gaps or inventing a degree that never existed is inexcusable, ruinous, and unethical (Callahan; Truesdell).

Works Cited

Callahan, David. "Résumé Padding." *The Cheating Culture.* 26 Feb. 2004 <www. cheatingculture.com/resumepadding.htm>.

Truesdell, Jason. "Honesty." *Tech.Job.Search.* 26 Feb. 2004 <www.jagaimo.com/ jobguide/resume/g-honest.htm>.

Trunk, Penelope. "Resume Writing: Lies v. Honesty." 2 June 2003. *The Brazen Careerist.* 26 Feb. 2004 <www.bankrate.com/smm/news/career/ 20030602a1.asp?prodtype=advice>.

Writing a Letter of Application

A letter of application has three parts: the introductory application, the explanatory body, and the request conclusion. You may organize the letter in one of two ways—by skills or by categories. This section first reviews the parts of a letter of application and then presents the same letter organized by skill and by category.(See Figure 20.2, pp. 528–529.)

Apply in the Introduction

The introductory application should be short. Inform the reader that you are applying for a specific position. If it was advertised, mention where you saw the ad. If someone recommended that you write to the company, mention that person's name (if it is someone the reader knows personally or by name). You may present a brief preview that summarizes your qualifications.

Apply

Tell source

Qualification preview

> I am interested in applying for the patient services manager position recently advertised in your Web homepage. I will complete a bachelor's degree in Dietetics in May 2003 from the University of Wisconsin–Stout. The skills I have developed from my academic background support my strong interest in working with your leading food and facility management services. I feel that my career goals and strong beliefs in assisting others to achieve a higher quality of life make me an excellent candidate for this position.

Convince in the Body

The *explanatory body* is the heart of the letter. Explain, in terms that relate to the reader, why you are qualified for the job. This section should be one to three paragraphs long. Its goal is to show convincingly that your strengths and skills will meet the reader's needs. Write one paragraph or section for each main requirement.

Base the content of the body on your analysis of the employer's needs and on your ability to satisfy those needs. Usually the requirements are listed in the ad. Show how your skills meet those requirements. If the ad mentioned "experience in software development," list details that illustrate your experience. If you are not responding to an ad, choose details that show that you have the qualifications normally expected of an entry-level candidate.

The key to choosing details is "memorable impact." The details should immediately convince readers that your skill matches their need. Use this guideline: In what terms will they talk about me? Your details, for instance, should show that you are the "development person." If you affect your reader that way, you will be in a positive position.

Globalization and Job Applications

Applying for a job overseas may open the way for an exciting professional and personal adventure. Understanding how to reformat your résumé to fit the needs of your potential employers will make the process faster and easier.

The term *curriculum vitae* (CV) is often used in other countries. A CV is generally the same as a résumé—a document detailing your education and work experience. However, different countries, employers, and cultures call for different areas of information and different levels of detail. Most countries have specific formats that they find acceptable. If you are unsure about which format you should use, ask; or use the standard, reverse-chronological format—put the most recent job highest in the list and work backward. Many countries expect information on résumés and CVs considered unprofessional or illegal in the United States: date of birth, marital status, and nationality ("Résumés").

When applying to a Japanese company, you may submit either a two-page, standard format résumé in Japanese, called a *rirekisho*, or a cover letter and résumé in English. Center your name and contact information at the top of the page and begin your résumé with a summary of your major qualifications. Your next section will be Employment Experience, followed by Education. Finally, you will end your résumé with personal information—date of birth, marital status, and nationality (Thompson).

If you are interested in working and living in the United Kingdom, you will format your résumé a bit differently. Include a cover letter addressed to a specific person. Your résumé will start with your personal information: name, contact information, date of birth, marital status, and nationality. Three major sections follow. The Profile section describes your professional designation, your immediate ambitions, and, in bulleted list form, your relevant skills and work-related achievements. Your Employment History section begins with your current position and then provides the name, location, and focus of the companies you have worked for. In the Education section, list your schools in reverse chronological order and include degrees awarded, additional courses and training, and any special skills (Thompson).

If you do write your résumé in the language of the country to which you are applying, find a native speaker of the language to review it. If it is written in a language other than English, one of your goals is to show that you are familiar with culturally appropriate language. Furthermore, if you e-mail your résumé, keep in mind that the American standard paper size is 8½ by 11 inches, whereas the European A4 standard is 210-297 mm. Be sure to reformat your document in your word processing program for those parameters to ensure that the receiver doesn't lose any of your information.

For further reference, check out these websites:

Goinglobal at <www.goinglobal.com> contains lots of information on applying, interviewing, and working abroad. You can find advice on how you can expect to fit into the culture whether you want to work in Australia, Belgium, or China.

Jobweb at <www.jobweb.com> is the on-line complement to the *Job Choices* magazine series. It has good, general information on all aspects of your job search, including an international search. It includes sample résumés and CVs and has articles written by professionals.

Monster Workabroad at <http://workabroad.monster.com> includes practical advice on getting a job overseas. You'll find specific job information as well as general information on your country of choice. You can also receive newsletters and chat with other international job hunters.

Works Cited

"Résumés/CVs." Goinglobal.com. 7 Feb. 2004 <www.goinglobal.com/topic/resumes.asp>.

Thompson, Mary Anne. "Writing Your International Résumé." Jobweb. 31 Jan. 2004 <www.jobweb.com/Resources/Library/International/Writing_Your_185_01.htm>.

Skills Section

Source of skill

As a student, I have learned many cost control methods. The courses Institutional Food Purchasing and Food Beverage Cost Controls have allowed me to become familiar with proper operating techniques, accounting for the monitoring of costs in relation to patient food service. I understand the qualities that a good manager possesses and believe that it is essential to strive to achieve and maintain a high level of patient satisfaction.

Convince

I have also had experience in food safety. Through the educational aspects of Microbiology and Food Science, I've developed a strong background in microorganisms and deciphering what control methods to use that will ensure the quality of the organization and its services. I have acquired a solid background in relation to Hazard Analysis Critical Control Point (HACCP) practices and methods.

Use of skill

Skill activities

I am knowledgeable about menu delivery systems and the proper procedures that must be followed to assure compliance with regulations. After completing my JACHO certification, I am aware that this organization sets the standards by which health care quality is measured. I will thus work to continuously improve the safety and quality of care provided to the public through the health care accreditation standards of JACHO.

Request an Interview

In the final section, ask for an interview and explain how you can be reached. The best method is to ask, "Could I meet with you to discuss this position?" Also explain when you are available. If you need two days' notice, say so. If you can't possibly get free on a Monday, mention that. Most employers will try to

Figure 20.2

Response to Ad
for Patient
Services Manager

Clear application

Use of key words at
the beginning of the
paragraph.

Details of work
experience and
classes used to create
interest.

Use of key words at
the beginning of the
paragraph.

1427 Crestview Street
Menomonie, WI 54751

November 5, 2006

Johnson-United Hospital
2715 Jamestown Avenue
Gaithersburg, Maryland 20878

Subject: Patient Services Manager Position

I am interested in applying for the patient services manager position recently advertised in your Web homepage. I will complete a bachelor's
degree in Dietetics in May 2007 from the University of Wisconsin–Stout.
The skills I have developed from my academic background support my
strong interest in working with your leading food and facility management services. I feel that my career goals and strong beliefs in assisting
others to achieve a higher quality of life make me an excellent candidate for this position.

As a student, I have learned many cost control methods. The courses Institutional Food Purchasing and Food Beverage Cost Controls have allowed me to become familiar with proper operating techniques,
accounting for the monitoring of costs in relation to patient food service. I understand the qualities that a good manager possesses and believe that it is essential to strive to achieve and maintain a high level of
patient satisfaction.

I have also had experience in food safety. Through the educational aspects of Microbiology and Food Science, I've developed a strong background in microorganisms and deciphering what control methods to
use that will ensure the quality of the organization and its services. I
have acquired a solid background in relation to Hazard Analysis Critical Control Point (HACCP) practices and methods.

I am knowledgeable about menu delivery systems and the proper procedures that must be followed to assure compliance with regulations.
After completing my JACHO certification, I am aware that this organization sets the standards by which health care quality is measured. I will
thus work to continuously improve the safety and quality of care provided to the public through the health care accreditation standards of
JACHO.

I would welcome the opportunity to meet with you and discuss my qualifications for the position. I have enclosed a copy of my résumé. If you have any questions or would like to talk with me, I can be reached by phone at (715) 555-1224 or e-mail at michshan@uw.edu. Thank you for considering me for this position. I look forward to hearing from you soon.

Sincerely,

Shannon M. Michaelis

Shannon M. Michaelis, R.D.

Enclosure: résumé

Asks for interview.

work around such restrictions. If no one is at your house or dorm in the morning to answer the phone, tell the reader to call in the afternoon. A busy employer would rather know that than waste time listening to a phone ring. Thank your reader for his or her time and consideration. Readers appreciate the gesture; it is courteous and it indicates that you understand that the reader has to make an effort to fulfill your request.

Request

How to contact writer
Thank you

I would welcome the opportunity to meet with you and discuss my qualifications for the position. I have enclosed a copy of my résumé. If you have any questions or would like to talk with me, I can be reached by phone at (715) 555-1224 or e-mail at michshan@uw.edu. Thank you for considering me for this position. I look forward to hearing from you soon.

Select a Format

To make a professional impression, follow these guidelines:

- Type the letter on 8½-by-11-inch paper.
- Use white, 20-pound, 100 percent cotton-rag paper.
- Use black ink.
- Use block or modified block format explained in Chapter 19.
- Sign your name in black or blue ink.
- Proofread the letter carefully. Grammar and spelling mistakes are irritating at best; at worst, they are cause for instant rejection.
- Mail the letter, folded twice, in a business envelope.

Examples 20.1–20.3 (pp. 534–537) show three application letters organized by skills. Examples 20.4 and 20.5 on pages 537–538 show two résumé styles.

Interviewing

The employment interview is the method employers use to decide whether to offer a candidate a position. Usually the candidate talks to one or more people (either singly or in groups) who have the authority to offer a position. To interview successfully, you need to prepare well, use social tact, perform well, ask questions, and understand the job offer (Stewart and Cash).

Prepare Well

To prepare well, investigate the company and analyze how you can contribute to it (Spinks and Wells). To investigate the company, read company literature, annual reports, descriptions in *Moody's*, items from *Facts on File, F&S Index, Wall Street Journal Index,* or *Corporate Report Fact Bank,* and the company's website. After you have analyzed the company, assess what you have to offer. Answer these questions:

- What contributions can you make to the company?
- How do your specific skills and strengths fit into its activities or philosophy?
- How can you further your career goals with this company?

Use Social Tact

To use social tact means to behave professionally and in an appropriate manner. Acting too lightly or too intensely are both incorrect. First impressions are extremely important; many interviewers make up their minds early in the interview. Follow a few common sense guidelines:

- Shake hands firmly.
- Dress professionally, as you would on the job.
- Arrive on time.
- Use proper grammar and enunciation.
- Watch your body language. For instance, sit appropriately; don't lounge or slouch in your chair.
- Find out and use the interviewers' names.

Perform Well

Performing well in the interview means to answer the questions directly and clearly. Interviewers want to know about your skills. Be willing to talk about yourself and your achievements; if you respond honestly to questions, your answers will not seem like bragging. For a successful interview, follow these guidelines:

- Be yourself. Getting a job based on a false impression usually ends badly.
- Answer the question asked.

⟩ Be honest. If you don't know the answer, say so.

⟩ If you don't understand a question, ask the interviewer to repeat or clarify it.

⟩ In your answers, include facts about your experience to show how you will fit into the company.

Ask Questions

You have the right to ask questions at an interview. Make sure you have addressed all pertinent issues (Spinks and Wells). If no one has explained the following items to you, ask about them:

⟩ Methods of on-the-job training

⟩ Your job responsibilities

⟩ Types of support available—from secretarial to facilities to pursuit of more education

⟩ Possibility and probability of promotion

⟩ Policies about relocating, including whether you get a promotion when you relocate and whether refusing to relocate will hurt your chances for promotion

⟩ Salary and fringe benefits—at least a salary range, whether you receive medical benefits, and who pays for them

Understand the Offer

Usually a company will offer the position—with a salary and starting date—either at the end of the interview or within a few days. You have the right to request a reasonable amount of time to consider the offer. If you get another offer from a second company at a higher salary, you have the right to inform the first company and to ask whether they can meet that salary. Usually you accept the offer verbally and sign a contract within a few days. This is a pleasant moment.

Writing Follow-Up Letters

After an interview with a particularly appealing firm, you can take one more step to distinguish yourself from the competition. Write a follow-up letter. It takes only a few minutes to thank the interviewer and express your continued interest in the job.

> Thank you for the interview yesterday. Our discussion of Ernst and Young's growing MIS Division was very informative, and I am eager to contribute to your team.
>
> I look forward to hearing from you.

Worksheet for Preparing a Résumé

☐ Write out your career objective; use a job title.

☐ List all the postsecondary schools you have attended.

☐ List your major and any minors or submajors.

☐ List your GPA if it is strong.

☐ Complete this form. Select only relevant courses or experiences.

College Courses	Skills Learned	Projects Completed

☐ List extracurricular activities, including offices held and duties.

☐ Complete this form for all co-ops, internships, and relevant employment.

Job Title	Company	Dates	Duties	Achievements

☐ List your name, phone number, current address, and permanent address if it is different. If appropriate, add e-mail address and personal website url.

☐ Review standard résumé format. See pages 521–522.

☐ Choose a layout design.

Worksheet for Writing a Letter of Application

☐ State the job for which you are applying.

☐ State where you found out about the job.

☐ Complete this form:

Employer Need (such as "program in C++")	Proof That You Fill the Need (show yourself in action: "developed two C++ programs to test widget quality")

☐ Select a format; the block format is suitable.

☐ **Write compelling paragraphs:**
An introduction to announce that you are an applicant
A body paragraph for each need, matching your capabilities to the need
Select details that cause "measurable impact"—ones that cause readers to remember you because you can fill their needs
A conclusion that asks for an interview

☐ **Purchase good-quality paper and envelopes, and get a new cartridge for your printer if you produce the letter yourself.**

Worksheet for Evaluating a Letter of Application

Answer these questions about your letter or a peer's:
a. Are the inside address and date handled correctly? Are all words spelled out?
b. Do the salutation and inside address name the same person?
c. Is there a colon after the salutation?
d. Does the writer clearly apply for a position in paragraph 1?
e. Does each paragraph deal with an employer need and contain an "impact detail"?
f. Does the closing paragraph ask for an interview? in an appropriate tone?
g. Would you ask this person for an interview? Why or why not?

Examples

The following examples illustrate ways in which applicants can show how their skills meet the employers' needs.

Example 20.1

Letter Organized by Skills

Slightly modified block format uses indented paragraphs to save space.

Specific application

Each body paragraph has in first line a key word or phrase from the ad.

Brief narrative demonstrates skills.

Two examples illustrate skills.

Narrative illustrates skills.

Request for contact

1503 West Second Street
Menomonie, WI 54751

ABC Global Services
1014 Michigan Avenue
Chicago, IL 60605

November 3, 2007

SUBJECT: I/T Specialist–Programmer Position

I would like to apply for the I/T Specialist–Programmer Position for ABC Global Services in Chicago. I learned of this position through the University of Wisconsin–Stout Placement and Co-op Office. Because of my past co-op experiences and educational background, I feel I am an ideal candidate for this position.

I am very familiar with all the steps of the application life cycle and processes involved in each of these steps. In my Software Engineering course, I and four other students developed a Math Bowl program that is used at the annual Applied Math Conference. We analyzed the problem through an analysis document, created a design document, coded the Math Bowl program from the design document, and maintained the software. In my past co-op with IBM, I also was involved in requirements planning, design reviews, coding, testing, and maintenance for my team's projects.

I have experience with low-level languages such as C++ and Assembler Language through my work experience and course study. In my Computer Organization class, I developed a CPU simulator in C++, which manipulated the Assembler Language. During my Unisys co-op in the Compiler Products department, I developed test programs in C, which manipulated the low-level compiler code.

I have strong customer relations and communication skills. Every day I deal with students and faculty at my job as a Lab Assistant at the Campus Computer Lab. It is my job to help them learn the available software and troubleshoot user's problems.

Enclosed is my résumé for your consideration. I am interested in talking with you in person about this position and a possible interview. Please call me at (715) 233-3341 at your convenience. Thank you for your consideration.

Sincerely,

Heather Miller

Heather Miller

enc: résumé

Example 20.2

Letter Organized by Skills

Standard block format

Dana Runge
461-19th Avenue West Apt. 1
Menomonie, WI 54751

October 29, 2006

Rachel Rizzuto
Campus Relations Specialist
Target Headquarters
1000 Nicollet Mall
Minneapolis, MN 55403

Dear Ms. Rizzuto:

Specific application

I would like to apply for the Business Analyst position registered at the University of Wisconsin–Stout Placement and Co-op Office. I believe that my past retail experience and education make me an ideal candidate for the position.

Each body paragraph has in first line a key word or phrase from the ad.

I have excellent analytical skills that were developed further from my education and work experience. In my Managerial Accounting course, my group researched the Pepsi Corporation's performance through their 2005 Annual Report. We applied that research by converting numbers and statistics into financial ratios and analysis. We then compared that information to industry norms and made conclusions. Our results were reported to the class as a PowerPoint presentation.

Narrative demonstrates skills.

I have great experience with planning and organizing. As the Events Coordinator for the Stout Retail Association, I researched, planned, and implemented a three-day trip to Chicago. I organized everything, from the hotel and transportation to the tours and activities. I had to make several contacts and reservations, as well as create a detailed time and activity agenda. I was responsible for the schedule of sixty members.

Narrative demonstrates skills.

I demonstrated clear and effective communication for a group country report in my Environmental Science class. I led the group in outlining what each member was responsible for doing, as well as the time line for completion dates. We set up meetings and exchanged communication channels, such as phone numbers and e-mail addresses. Our group had a clear understanding of what each needed to do as an individual, as well as in a group. We then compiled our information into a class presentation.

Request for contact

A copy of my résumé is enclosed for your review. I am available for an interview at your convenience. If you have any questions, please feel free to call

(continued)

Example 20.2

(continued)

(715-237-1421) or e-mail me (rungdan@uws.edu). Thank you for your time and consideration. I look forward to your reply.

Sincerely,

Dana Runge

Enclosure: résumé

Example 20.3

Letter Organized by Skills

Standard block format

421 Main Street North
Apartment 12
Menomonie, WI 54751

November 5, 2006

Kelly Services
370 Wabasha Street North
St. Paul, MN 55102-1306

ATTENTION: Human Resources Department

SUBJECT: KSW/260A/SDA

Specific application

I would like to apply for the Human Resource Assistant Position for Ecolab in St. Paul. I learned of this position from the Monster.com website on the Internet. I will graduate in August 2007 from the University of Wisconsin–Stout with a Bachelor of Science degree in Service Management.

First line of each body paragraph contains key word or phrase from the ad.

I have obtained strong organizational and process management skills as an office assistant for the University of Wisconsin–Stout Dining Services. As an aid to the Director, Assistant Director, and Accounts Payable Specialist, I have gained the ability to multitask and maintain excellent communication skills.

Examples demonstrate skills.

My day-to-day experience with Microsoft Office Suite involves creating and maintaining reports in Excel, planning and preparing presentations in PowerPoint, and designing various documents in Word.

I have extensive experience in preparing and maintaining presentation and other collateral recruiting materials for the annual Resident Advisor Resource

Examples
demonstrate skills.

Fair and the Employee Orientations for Dining Services. I have created informational documents, designed motivational brochures, and prepared electronic presentations in order to enhance employees' and students' views of Dining's available services. By explaining the services we offer and recruiting students for employment at these events, I have developed excellent interpersonal and communication skills.

A copy of my résumé is enclosed for your consideration. Through my work experience and educational background, I am confident that I can be an asset to your company. Could I meet with you at your convenience to discuss my qualifications for this position? You may contact me during the evening at (715) 237-1142. Thank you for your time and consideration. I look forward to hearing from you.

Sincerely,

Cara Robida

Cara Robida

Enclosure: résumé

Request for interview

Example 20.4

Résumé for an Internship

Rodney C. Dukes
4807 Oknol Street
Colfax, WI 54703
(715) 477-0012
dukrod@uw.edu

OBJECTIVE To obtain an Internship as Training Document Writer

EDUCATION University of Wisconsin–Stout (Recipient—Baldrige Award 2001), Menomonie, Wisconsin, Technical Communications Major, Junior, Applied Field—Biomedical Engineering, GPA 3.3, Financed 75% of Education

Chicago Police Academy, Chicago Illinois, Officer Training Program
May 1991 State-Certified Peace Officer

Brief, specific objective

Educational history, most recent listed at top

Education appears first because it contains most relevant skills.

(continued)

Example 20.4

(continued)

Courses chosen to
demonstrate skills
needed for position
of training

RELATED COURSES	Course Names and Titles • TRHRD-360—Training Systems in Business and Industry • GCM-141—Graphic Communications and Electronic Publishing • ENGL-415—Technical Writing • ENGL-247—Critical Writing • ENGL-207—Writing for the Media • PSYC-379—Public Relations • SPCOM-236—Listening

Experience history,
most recent job listed
first

EXPERIENCE	**Technical Writer** (June 2003–August 2003), Luther Hospital, Eau Claire, Wisconsin • Researched current operational manuals • Developed specifications for manual • Communicated with superiors on a regular basis • Tested design samples and submitted prototypes for review • Evaluated the effectiveness of the tutorial • Created ACLS Manikin tutorial for the CPR Instructor training

Consistent
presentation
of each job

Security Officer, Luther Hospital, Eau Claire, WI (Jan. 2001–Present)
• Employee, patient, and patron safety
• Monitor and transport Behavioral Health patients
• Secure Helipad for MAYO 2
• Patrol of hospital and property

Bulleted list begins
with action verbs.

Drug Enforcement Agent, Ho Chunk Nation, Black River Falls, WI
• (Oct. 1999–Sept. 2000)
• Enforced the Drug and Alcohol Policies
• Conducted and supervised urinalysis testing and training of employees
• Submitted reports to the Compliance Director and Justice Department

STRENGTHS	**Computer Skills,** Microsoft Word, Excel, FrontPage, PowerPoint
REFERENCES	Available upon request.

Example 20.5

Résumé
Emphasizing Job
Experience of
Working
Professional

Source: Reprinted
by permission of
Amy Reid.

Objective indicating
skills she can bring to
position

Experience history
listed first because it
is most relevant to
objective.

Most recent job at top
of section

Longer paragraphs
typical of
experienced
professionals who
have many skills

Amy Reid
N687 W28471 Harness Avenue
Germantown, WI 53022
262-555-3801
reids@graphdes.com

OBJECTIVE   To acquire a position that utilizes my copywriting, editing, and proofing skills in a print- or Web-based graphic design, advertising, or publishing environment.

EXPERIENCE   **GS Design, Inc.—Milwaukee, Wisconsin**

Project Manager, November 1996 to present
Coordinate production of *Hog Tales*® magazine—a 48-page bimonthly publication for members of the Harley Owners Group. Responsibilities include meeting with the editor, who works for Harley-Davidson Motor Company: prepping the photographs and provided copy; meeting with in-house copy-writers to have them write new content or edit the provided; edit the copywriters' copy, and prep it for the designers and layout staff; do simple layouts; review and proof all layouts; coordinate the layouts and original materials to be turned over to the editor for review; coordinate multiple rounds of revisions and proof them; coordinate final production of all pages; work closely with pre-press and the printer; review and final proof the color proofs from pre-press; attend the press checks at Perry-Judd's, Inc. in Baraboo; and archive all the proofs and materials upon completion of the edition.

Proofreader, December 1988 to present
Review all in-house projects and proposals for accurate grammar, punctuation, spelling, and content prior to being printed or posted on the Web. Edit copy as needed, or critique and meet with copywriters when more in-depth editing is required.

Production Manager, November 1996 to April 1999
Scheduled and assigned projects with in-house copywriters, designers, and production staff, and monitored progress to meet deadlines throughout process to completion. Met with clients and vendors to coordinate production schedules and kept tasks on track through project delivery. Coordinated bilingual and trilingual translations for projects as required.

(continued)

Example 20.5

(continued)

Graphic Designer, December 1988 to November 1996
Managed projects from concept through final print production. Responsibilities included designing, critiquing, typesetting, illustrating, and coordinating photography.

Econoprint—Milwaukee, Wisconsin
Desktop Publisher, May 1988 to September 1988
Met with customers, typeset projects per specs, and pasted-up galley copy and graphics. Occasional tasks included bindery work and customer service.

Education placed lower in résumé because job skills are more important for an experienced professional.

EDUCATION AND SKILLS

University of Wisconsin–Stout, Menomonie, Wisconsin
Bachelor of Science, May 1988
Major: Art
Concentration: Graphic Design
Minor: Technical Writing

American Management Association, Keye Productivity Center
"How To Be A Better Proofreader," September 1997

Proficient in Macintosh-based Quark XPress and Microsoft Word; limited working knowledge of Adobe Photoshop, Adobe Illustrator, Microsoft Excel, and Clients & Profits.

Professional honors

HONORS

ADDY Award, Milwaukee Ad Club

Collateral Material—Invitation (design and production of personal wedding invitation and collateral), March 2001

Featured in article in *Publishing & Production Executive,* **April 2000**

Professional coworkers or managers

REFERENCES

Jeff Prochnow
GS Design, Inc.
414-555-9821
proch@graphdes.com

Marc Tebon
GS Design, Inc.
414-555-8436
Tebon@graphdes.com

Additional references available upon request.

Exercises

▶ Group

1. Your instructor will arrange you in groups of two to four by major. Each person should photocopy relevant material from one source in the library. Include at least the *Dictionary of Occupational Titles* and the *Occupational Outlook Handbook*. In class, make a composite list of basic requirements in your type of career. Use that list as a basis for completing the Writing Assignments that follow.

▶ You Analyze

2. Analyze one of the letters in the Sample Documents section of the Instructor's Resource Manual (Examples 46–51). Comment on format and effectiveness of tone, detail, and organization.

3. Analyze this letter. It responded to an ad for a consumer scientist. The general requirements included abilities to evaluate and analyze new products; to write reports; and to assist with training, test kitchen organization, cooking demonstrations, and developing recipes and guides.

2837 Main Street
Eau Claire, WI 54701

April 9, 2006

Wolf Appliance Company, LLC
Attention: Human Resources
P.O. Box 44848
Fitchburg, Wisconsin 53719

Subject: Consumer Scientist Position

I am extremely interested in becoming a part of Wolf Appliance Company as a Consumer Scientist. In May 2006, I will be graduating from the University of Wisconsin—Stout with a Bachelor of Science degree in Food Systems and Technology, with a concentration in Food Science. I am also knowledgeable in the development and design of marketing tools for new products.

My background includes extensive experience with analyzing data, evaluating results, and developing reports. Through course work, I have had to collect, analyze, and develop scientific reports in various formats utilizing MS Office, adapting the report to the specific situation. As an operations manager, I also had the responsibility of gathering data, reporting the information, and explaining variances to the Corporate Office for profit and loss statements.

I developed excellent organizational, training, and evaluating techniques as an operations manager for a $40-million-a-year business. With an average staff of 105 employees, I was responsible for ensuring the training and evaluations were done promptly and accurately.

As a merchandise manager, I was responsible for managing several re-merchandising projects simultaneously. The projects I was responsible for were successful because of my organization and leadership abilities. The enclosed résumé outlines my credentials and accomplishments in further detail.

If you are seeking a Consumer Scientist who is highly self-motivated and is a definite team player wanting to be a part of a successful team, then please consider what I have to offer. I would be happy to have a preliminary discussion with you or member of your committee. You can reach me at (715) 421-8765. I look forward to talking with you and exploring this opportunity further.

Sincerely,

Michelle Stewart

4. Analyze an ad and yourself by filling out the third point in the Worksheet for Writing a Letter of Application (pp. 532–533).

5. Analyze yourself by using the fifth and seventh points in the Worksheet for Preparing a Résumé (p. 532).

6. After completing Writing Assignment 2, read another person's letter. Ask these questions: What do you like about this letter? What do you dislike about this letter? How would you change what you dislike?

7. After completing Writing Assignment 1, read the ad and résumé of a classmate. Read the ad closely to determine the employer's needs. Read the résumé swiftly—in a minute or less. Tell the author whether he or she has the required qualifications. Then switch résumés and repeat. This exercise should either convince you that your résumé is good or highlight areas that you need to revise.

▶ You Revise

8. Revise this rough draft.

1221 Lake Avenue
Menomonie WI 54751

March 21, 2006

Human Resources
Polaris Industries Inc.
301 5th Avenue SW
Roseau MN 54826

Subject: Inquiring about a summer co-op in the mechanical design area.

My name is Josh Buhr and I am a junior at UW—Stout majoring in Mechanical Design. I would like this co-op to learn and grow in my field.

Working with your company as a design engineer, I would be able to fulfill many of the requirements that you have listed in your ad. I am able to design and lay out drawings on AutoCAD in 3-D so I can easily verify that there is freedom of movement between the parts. With AutoCAD 14 I can even animate the object to see even clearer the movement between parts.

If I find a problem with one of the designs I will be able to talk to the people necessary and bring up the topic in a productive manor. The course at Stout "discussion," has taught me how to work in groups and even how to seat people to get the most out of everyone's mind, in a positive atmosphere.

I also need to mention that the AutoCAD experience at Stout has given me many needed skills for design. In the second semester of AutoCAD, I learned and used geometric tolerancing on a complete assembly and layout of a two cycle engine. This engine was fully dimensioned to clearance fits and I even made the piston and a few other parts animated to see that everything was in a freedom of movement.

If you like the qualifications that I have listed above please feel free to call me at (715) 235-4037 to set up an interview. I would be more that happy to meet with you. I have wanted to work with your company for several years after I bought my SJ 500, and was never passed since.

Sincerely,

Josh Buhr

9.   Revise this rough draft.

871 17th Avenue East
Menomonie, WI 54751
February 12, 2005

Parker Hannifin Corporation
Personnel Department
2445 South 25th Avenue
Broadway, IL 60513

Dear Personnel Department:

I wrote to apply for the entry level Plant Engineering position that you advertised in the Minneapolis Star Tribune January 27, 2005.

I will graduate from the University of Wisconsin—Stout in December with a Bachelor of Science degree in Industrial Technology with a concentration in Plant Engineering.

I have been in charge of projects dealing with capital equipment justification and plant layout while on my co-op with Kolbe and Kolbe Millwork, Company. Some specific work that I completed include: installation of a portable blower system and light design of jigs, conveyor beds and machines. I worked in the machine shop my first month there and have a working knowledge of equipment and procedures in that environment.

I am a self-motivated individual who prides in excelling in everything I take on. This quality is reflected in my résumé through my increased job responsibilities topping with my co-op "experience" and my 3.5 cumulative grade point average.

I would appreciate an opportunity to interview with you. Please contact me at the above address or call me at (715) 471-1627 after 2:00 p.m. on weekdays, or I can return your call. Thank you for your time and consideration. I look forward to hearing from you.

Sincerely,

Keith Munson

▶ You Create

10. Create a work experience résumé entry (pp. 539–540) for one job you have held. Select an arrangement for the four elements. Select details based on the kind of position you want to apply for. Alternate: Create a second version of the entry but focus one version on applying for a technical position and one version on applying for a managerial position.

11. Write a paragraph that explains a career skill you possess.

Writing Assignments

1. Using the worksheet on page 532 as a guide, write your résumé following one of the two formats described in this chapter.

2. Find an ad for a position in your field of interest. Use newspaper Help Wanted ads or a listing from your school's placement service. On the basis of the ad, decide which of your skills and experiences you should discuss to convince the firm that you are the person for the job. Then, using the worksheet on pages 532–533 as a guide, write a letter to apply for the job.

3. Write a learning report for the writing assignment you just completed. See Chapter 5, Assignment 7, pages 134–135, for details of the assignment.

Web Exercise

Visit at least two websites at which career positions are advertised relative to your expertise. Do one or more of the following, depending on your instructor's directions:

a. Apply for a position. Print a copy of your application before you send it. Write a brief report explaining the ease of using the site. Include comments about any response that you receive.

b. Analyze the types of positions offered. Is it worth your time and energy to use a site like this? Present your conclusions in a memo or an oral report, as your instructor designates. Print copies of relevant screens to use as visual aids to support your conclusions.

c. As part of a group of three or four, combine your research in part b into a large report in which you explain to your class or to a professional meeting the wide range of opportunities available to the job seeker.

Works Cited

Dictionary of Occupational Titles. 4th ed. Rev. Washington, DC: U.S. Dept. of Labor, 1991. On-line version available: <www.oalj.dol.gov/libdot.htm>.

Harcourt, Jules, and A. C. "Buddy" Krizar. "A Comparison of Résumé Content Preferences of Fortune 500 Personnel Administrators and Business Communication Instructors." *Journal of Business Communications* 26.2 (1989): 177–190.

Hutchinson, Kevin L., and Diane S. Brefka. "Personnel Administrators' Preferences for Résumé Content: Ten Years After." *Business Communication Quarterly* 60.2 (1997): 67–75.

Parker, Yana. *The Résumé Catalog: 200 Damn Good Examples.* Berkeley, CA: Ten Speed Press, 1996.

Spinks, Nelda, and Barron Wells. "Employment Interviews: Trends in the Fortune 500 Companies—1980–1988." *The Bulletin of the Association for Business Communications* 51.4 (1988): 15–21.

Stewart, Charles J., and William B. Cash, Jr. *Interviewing Principles and Practices* 8th ed. Dubuque, IA: Brown, 1997.

Treweek, David John. "Designing the Technical Communication Résumé." *Technical Communications* 38.2 (1991): 257–260.

Focus on
Electronic Résumés

Electronic résumés are changing the job search. The candidate still submits a résumé, and the employer still reads it, but the "electronic way" has a key difference—technology intervenes to do much of the initial sorting. As a result, "keyword strategies" are very important. This section explains briefly how the sorting works, keyword strategies, and on-line, scannable, and ASCII résumés.

How the Sorting Works

Using one of the methods explained below, the candidate submits a résumé that, because it is electronic, is put into a searchable database. When an employer wants to find candidates to interview, he or she searches the database with a software program that seeks those keywords that the employer says are important. For example, the employer might want someone who can design websites and who knows Dreamweaver and HTML programming. Every time the search program finds a résumé with those words in it, it pulls the résumé into an electronic "yes" pile, which the human can then read.

Gonyea and Gonyea explain the process this way: "If the computer finds the same word or words [that describe the candidate the company is attempting to find] anywhere in your résumé, it considers your résumé to be a match, and will then present your résumé, along with others that are also considered to be a match, to the person doing the searching" (62).

Thus, your use of effective keywords is the key to filling out such a form. As Gonyea and Gonyea say, "To ensure that your résumé will be found, it is imperative that you include as many of the appropriate search words as are likely to be used by employers and recruiters who are looking for someone with your qualifications" (62).

Keyword Strategies

Keywords require a radical change in presenting your résumé. Your odds of being one of the "hits"

in the search are increased by including a lot of keywords in your résumé. In addition to using keywords as you describe yourself in the education and work history sections of your résumé, you should also include a keyword section right in your résumé. Some of the major on-line résumé services, such as Monster.com, require you to add one.

Put the keyword section either first or last in your résumé, or in the box supplied by the résumé service. Use words that explain skills or list aspects of a job. The list should include mostly nouns of the terms that an employer would use to determine if you could fill his or her need—job titles, specific job duties, specific machines or software programs, degrees, major, and subjective skills, such as communication abilities. You can include synonyms; for instance, in the list below, Web design and DreamWeaver are fairly close in meaning, because you use one to do the other, but including both increases your chances of the scanner's choosing your résumé.

A short list might look like this:

C++, software engineering, HTML, programmer, needs analysis, client interview, team, Web design, Photoshop, AuthorIT, design requirements.

Remember, the more "hits" the reading software makes in this list, the more likely that your résumé will be sent on to the appropriate department.

On-Line Résumés/Job Searches

The Web has dramatically changed the methods of advertising jobs and responding to advertisements. Job-posting sites allow employers to post employment opportunities; résumé-posting sites allow candidates to post their information. The exact way in which sites work varies, but all of them work in one of two ways. On job-posting sites, like America's Job Bank, employers post ads, listing job duties and candidate qualifications, and candidates respond to those ads. On résumé-posting sites, like

(continued)

(continued)

Monster.com, candidates post résumés in Web-based databases, and employers search them for viable applicants.

You have two options. You can begin to read the "Web want ads," and you can post your résumé.

"Web Want Ads" are posted at the job-posting site. For instance, America's Job Bank <www.ajb .dni.us> lists job notices posted by state employment offices, and Internet Career Connection: Help Wanted-USA <http://iccweb.com/HelpWantedUSA/ hwusa.asp> posts ads for companies around the world in all lines of work. You simply access the site and begin to read. In addition, many of the ré-sumé-posting services have a want ad site. For instance, Monster.com <www.monster.com> has an extensive listing of jobs in all categories. In all of these sites, job seekers can search by city, by job type, by level of authority (entry level, manager, executive). Candidates can search free of charge, but companies pay the sites to post the ads.

Post Your Résumé

Many sites provide this service; usually, it is free. Each site has you create your résumé at the site. You open an account, then fill in the form that the site presents to you. For instance, you will be asked for personal information (e.g., name and address) and also such typical items as job objective, work experience, desired job, desired salary, and special skills. Usually, filling out the form takes about 30 minutes. The site creates a standard-looking format (like the ones discussed in Chapter 19), which is sent to prospective employers when they ask for it.

Scannable Résumés

Many companies use optical character recognition (OCR) software (McNair; Quible) to scan résumés. First, paper résumés are scanned, turning them into ASCII files, which are entered into a database. Second, when an opening arises, the human resources department searches the database for keywords.

Those résumés that contain the most keywords are forwarded to the people who will decide whom to interview. This development means that job seekers must now be able to write résumés that are scannable and that effectively use keywords.

Scannable résumés are less sophisticated looking than traditional ones, because scanners simply cannot render traditional résumés correctly. These documents contain all the same sections as traditional résumés but present them differently:

Use one column. Many scanners scramble two-column text. Start all heads and text at the same left-hand margin.

Use 10- to 14-point fonts. For "fine" fonts like Times and Palatino, use 11 to 12 points; for "thick" fonts like New Century Schoolbook, use 10 or 11 points. For heads, use 12 to 14 points.

Use the same font throughout the document.

Place your name and address at the top of the page, centered. If you include two addresses (campus and home), place them under each other.

Avoid italics, underlining, and vertical lines.

Do not fold your résumé. Mail it in an envelope that will hold the 8½-by-11-inch page.

ASCII Résumés

Often companies ask you to send your résumé by e-mail. The best way to do so is to send the résumé as an ASCII file, one that contains only letters, numbers, and a few punctuation marks but does not contain formatting such devices as boldfacing and italics (Skarzenski).

Like scannable résumés, ASCII résumés maintain all the traditional sections; you just present them so that they will interact smoothly with whatever software program is receiving them.

The key items to be aware of are:

Keep the line length to fewer than 65 characters. Some software systems have difficulty with longer lines.

Use spaces, not tabs. Some software programs misinterpret tabs.

Send the file to the receiver in two ways. You can send a file as part of an e-mail message or as an attachment to an e-mail message. Some programs can read the message both ways, some only one way. If the recipient's program does not have the capabilities, it will not be able to read your message.

To practice sending an ASCII file with your e-mail program, send yourself and a friend your résumé. You and the friend should be able to print out the résumé easily.

Works Consulted

Besson, Taunee. *Résumés*. 3rd ed. New York: Wiley, 1999.

Gonyea, James C., and Wayne M. Gonyea, *Electronic Résumés: A Complete Guide to Putting Your Résumé On-Line*. New York: McGraw-Hill, 1996.

McNair, Catherine. "New Technologies and Your Résumé." *intercom* 44.5 (1997): 65–75.

Quible, Zane K. "Electronic Résumés: Their Time Is Coming." *Business Communication Quarterly* 58.3 (1995): 5–9.

Skarzenski, Emily. "Tips for Creating ASCII and HTML Résumés." *intercom* 43.6 (1996): 17–18.

Yate, Martin. *Résumés That Knock 'Em Dead*. Holbrook, MA: Adams Media, 1998.

Appendix A

Brief Handbook for Technical Writers

This appendix presents the basic rules of grammar and punctuation. It contains sections on problems with sentence construction, agreement of subjects and verbs, agreement of pronouns with their antecedents, punctuation, abbreviations, capitalization, and numbers.

Problems with Sentence Construction

The following section introduces many common problems in writing sentences. Each subsection gives examples of a problem and explains how to convert the problem into a clearer sentence. No writer shows all of these errors in his or her writing, but almost everyone makes several of them. Many writers have definite habits: They often write in fragments, or they use poor pronoun reference, or they repeat a word or phrase excessively. Learn to identify your problem habits and correct them.

Identify and Eliminate Comma Splices

A *comma splice* occurs when two independent clauses are connected, or spliced, with only a comma. You can correct comma splices in four ways:

1. Replace the comma with a period to separate the two sentences.

Splice
> The difference is that the NC machine relies on a computer to control its movements, a manual machine depends on an operator to control its movements.

551

Correction

> The difference is that the NC machine relies on a computer to control its movements. A manual machine depends on an operator to control its movements.

2. Replace the comma with a semicolon only if the sentences are very closely related. In the following example, note that the word *furthermore* is a conjunctive adverb. When you use a conjunctive adverb to connect two sentences, always precede it with a semicolon and follow it with a comma. Other conjunctive adverbs are *however, also, besides, consequently, nevertheless,* and *therefore.*

Splice

> The Micro 2001 has a two-year warranty, furthermore the magnetron is covered for seven years.

Correction

> The Micro 2001 has a two-year warranty; furthermore, the magnetron is covered for seven years.

3. Insert a coordinating conjunction (*and, but, or, nor, for, yet,* or *so*) after the comma, making a compound sentence.

Splice

> The engines of both cranes meet OSHA standards, the new M80A has an additional safety feature.

Correction

> The engines of both cranes meet OSHA standards, but the new M80A has an additional safety feature.

4. Subordinate one of the independent clauses by beginning it with a subordinating conjunction or a relative pronoun. Frequently used subordinating conjunctions are *where, when, while, because, since, as, until, unless, although, if,* and *after.* The relative pronouns are *which, that, who,* and *what.*

Splice

> Worker efficiency will increase because of lower work heights, lower work heights maximize employee comfort.

Correction

> Worker efficiency will increase because of lower work heights that maximize employee comfort.

Exercises

Correct the following comma splices:

1. Different models of computers, software programs, and text formats are incompatible, data processing and information retrieval is slow and inefficient.

2. Web content development is becoming more necessary as businesses rely on the Internet to provide product information, advertising, and purchasing options for consumers, good writers can convey all this information accurately, stylishly, and in a concise manner.

3. For example, a millimeter is one-thousandth of a meter, therefore, a nanometer is one million times smaller.

4. Positive displacement pumps produce a pulsating flow, their design provides a positive internal seal against leakage.

5. In the printing business there are two main ways of printing, the first is by using offset and the second is by using flexography.

6. Two methods currently exist to enhance fiber performance, one method is to orient the fibers.

Identify and Eliminate Run-On Sentences

Run-on, or fused, sentences are similar to comma splices but lack the comma. The two independent clauses are run together with no punctuation between them. To eliminate run-on sentences, use one of the four methods explained in the preceding section and summarized here.

▶ Place a period between the two clauses.
▶ Place a semicolon between them.
▶ Place a comma and a coordinating conjunction between them.
▶ Place a relative pronoun or subordinating conjunction between them.

Exercises

Correct the following run-on sentences:

1. Biology is not the only field of science that nanotechnology will permeate this revolution can quite possibly influence all sciences and most likely create more.

2. Nonpositive displacement pumps produce a continuous flow because of this design, there is no positive internal seal against leakage.

3. Offset is also known as offset lithography or litho printing the offset process uses a flat metal plate with a smooth printing surface that is not raised or engraved.

4. The countries that import OCC at the highest rates often produce corrugated board of 100% recycled fibers due to the lessened performance of bogus board, a dilemma may be created.

5. Images are engraved around the cylinder the plate is not stretched and distorted like traditional plates.

Identify and Eliminate Sentence Fragments

Sentence fragments are incomplete thoughts that the writer has mistakenly punctuated as complete sentences. Subordinate clauses, prepositional phrases, and verbal phrases often appear as fragments. As the following examples show, fragments must be connected to the preceding or the following sentence.

1. Connect subordinate clauses to independent clauses.

    a. The fragment below is a subordinate clause beginning with the subordinating conjunction *because*. Other subordinating conjunctions are *where, when, while, since, as, until, unless, if,* and *after*.

Fragment
: We should accept the proposal. Because the payback period is significantly less than our company standard.

Correction
: We should accept the proposal because the payback period is significantly less than our company standard.

    b. The following fragment is a subordinate clause beginning with the relative pronoun *which*. Other relative pronouns are *who, that,* and *what*.

Fragment
: The total cost is $425,000. Which will have to come from the contingency fund.

Correction
: The total cost of $425,000 will have to come from the contingency fund.

2. Connect prepositional phrases to independent clauses. The fragment below is a prepositional phrase. The fragment can be converted to a subordinate clause, as in the first example below, or made into an *appositive*—a word or phrase that means the same thing as what precedes it.

Fragment
: The manager found the problem. At the conveyor belt.

Correction 1
: The manager discovered that the problem was the conveyor belt.

Correction 2
: The manager found the problem—the conveyor belt.

3. Connect verbal phrases to independent clauses.

    a. Verbal phrases often begin with *-ing* words. Such phrases must be linked to independent clauses.

Fragment
: The crew will work all day tomorrow. Installing the new gyroscope.

Correction
: Tomorrow the crew will work all day installing the new gyroscope.

    b. Infinitive phrases begin with *to* plus a verb. They must be linked to independent clauses.

Fragment
: I contacted three vendors. To determine a probable price.

Correction
: I contacted three vendors to determine a probable price.

Exercises

Correct the following sentence fragments:

1. The National Nanotechnology Initiative 247 million dollars in federal funding in 1999.

2. The fourth aspect of the quality control function making adjustments to the process, in order to bring specifications into line.

3. Thickness availability of Celotex from $\frac{1}{2}''$ to $2\frac{1}{4}''$.

4. Through orientation and the alkali process secondary fiber performance enhanced to levels above current performance.

5. While cost savings initiatives and quality have become critical for businesses to remain competitive today.

6. The virgin fiber sought by foreign corrugated producers to supply their customers with a higher quality product.

Place Modifiers in the Correct Position

Sentences become confusing when modifiers do not point directly to the words they modify. Misplaced modifiers often produce absurd sentences; worse yet, they occasionally result in sentences that make sense but cause the reader to misinterpret your meaning. Modifiers must be placed in a position that clarifies their relationship to the rest of the sentence.

1. In the sentence below, *that is made of a thin, oxide-coated plastic* appears to refer to *the information*.

Misplaced modifier

The magnetic disk is the part that contains the information that is made of a thin, oxide-coated plastic.

Correction

The magnetic disk, which is made of a thin, oxide-coated plastic, is the part that contains the information.

2. In the sentence below, the modifier says that the horizontal position must be tested, but the meaning clearly is something different.

Misplaced modifier

Lower the memory module to the horizontal position that requires testing.

Correction

Lower the memory module that requires testing to the horizontal position.

Exercises

Correct the misplaced modifiers in the following sentences:

1. ADA noted that VCLDs reduced energy endurance without carbohydrate supplementation.

2. Three topic areas related to printing were formulated to determine Web feasibility.

3. Although EPA legislation contains strong support, it cannot make up for the lack of information in the two other areas for oil field service engineers.

4. Technology must continue to improve in order to fully benefit economically in the recovery and preparation processes.

Use Words Ending in *-ing* Properly

A word ending in *-ing* is either a present participle or a gerund. Both types, which are often introductory material in a sentence, express some kind of action. They are correct when the subject can perform the action that the *-ing* word expresses. For instance, in the sentence below, the *XYZ computer table* cannot *compare* cost and durability.

Unclear

Comparing cost and durability, the XYZ computer table is the better choice.

Clear

By comparing cost and durability, you can see that the XYZ computer table is the better choice.

Exercises

In the following sentences, the participle (the -ing word) is used incorrectly; revise them.

1. While walking across the parking lot, the red convertible had its hood raised.

2. When filling out an on-line document, personal data always appear first.

3. When using a laptop computer, the mouse can be difficult to manage.

4. After comparing the weight and the features of the two laptops, both laptops seem to be of equal value.

5. Reviewing the internship, visual displays floor plans, and merchandising were my main duties.

6. By eliminating unnecessary wording, this would decrease cost by $10.00 per book.

Make the Subject and Verb Agree

The subject and the verb of a sentence must both be singular or both be plural. Almost all problems with agreement are caused by failure to identify the subject correctly.

1. When the subject and verb are separated by a prepositional phrase, be sure you do not inadvertently make the verb agree with the object of the preposition rather than with the subject. In the following sentence, the subject *bar* is singular; *feet* is the object of the preposition *of.* The verb *picks* must be singular to agree with the subject.

Faulty   A bar containing a row of suction feet pick up the paper.

Correction   A bar containing a row of suction feet picks up the paper.

2. When a *collective noun* refers to a group or a unit, the verb must be singular. Collective nouns include such words as *committee, management, audience, union,* and *team.*

Faulty   The committee are writing the policy.

Correction   The committee is writing the policy.

3. Indefinite pronouns, such as *each, everyone, either, neither, anyone,* and *everybody,* take a singular verb.

Faulty   Each of the costs are below the limit.

Correction   Each of the costs is below the limit.

4. When compound subjects are connected by *or* or *nor,* the verb must agree with the nearer noun.

Faulty   The manager or the assistants evaluates the proposal.

Correction   The manager or the assistants evaluate the proposal.

Exercises

Correct the subject-verb errors in the following sentences:

1. Everybody in both classes were late in arriving to the classroom.
2. All the books in the bookstore is available for purchase.
3. When writing a technical manual, analysis of various audiences are very important.

4. The reasons the flat screen monitor should be used is well-documented in research.

5. Hypertext (HTML) is a method in which punctuation markings, spaces, and coding is used to create pictures on a webpage.

Use Pronouns Correctly

A pronoun must refer directly to the noun it stands for, its *antecedent*.

As in subject-verb agreement, a pronoun and its antecedent must both be singular or both be plural. Collective nouns generally take the singular pronoun *it* rather than the plural *they*. Problems result when pronouns such as *they, this,* and *it* are used carelessly, forcing the reader to figure out their antecedents. Overuse of the indefinite *it* (as in "*It* is obvious that") leads to confusion.

Problems with Number

1. In the following sentence, the pronoun *It* is wrong because it does not agree in number with its antecedent, *inspections*. To correct the mistake, use *they*.

Vague

The inspections occur before the converter is ready to produce the part. It is completed by four engineers.

Clear

The inspections occur before the converter is ready to produce the part. They are completed by four engineers.

2. In current practice, it is now acceptable to deliberately misuse collective pronouns in an effort to avoid sexist writing.

Technically correct

Everyone must bring his or her card.

Correct for informal situations

Everyone must bring their card.

Problems with Antecedents

If a sentence has several nouns, the antecedent may not be clear.

1. In the following case, *It* could stand for either *pointer* or *collector*. The two sentences can be combined to eliminate the pronoun.

Vague

The base and dust *collector* is the first and largest part of the lead *pointer. It* is usually round and a couple of inches in diameter.

Clear

The base and dust collector, which is the largest part of the lead pointer, is usually round and several inches in diameter.

2. In the following case, *It* could refer to *compiler* or *software*.

Vague   The new *compiler* requires new *software. It* must be compatible with
                         our hardware.

Clear   The new compiler, which requires new software, must be compatible
                         with our hardware.

Problems with *This*

Many inexact writers start sentences with *This* followed immediately by a verb
("*This* is," "*This* causes"), even though the antecedent of *this* is unclear. Often
the writer intends to refer to a whole concept or even to a verb, but because *this*
is a pronoun or an adjective, it must refer to a noun. The writer can usually fix
the problem by inserting a noun after *this*—and so turn it into an adjective—
or by combining the two sentences into one. In the following sentence, *this*
probably refers either to the whole first sentence or to *virtually impossible,* which
is not a noun.

Vague   Ring networks must be connected at both ends—a matter that could
                         make wiring virtually impossible in some cases. This would not be the
                         case in the Jones building.

Clear   Ring networks must be connected at both ends—a matter that could
                         make wiring virtually impossible in some cases. We can easily fill this
                         requirement in the Jones building.

Exercises

Revise the following sentences, making the pronoun references clear:

1. Probably the computer will begin to have problems in three or four months.
   This is only an estimate but it would not be surprising, due to my experience
   with other computers. This would increase the amount of money spent on the
   computer.

2. The computers are in the warehouse and the boxes enter on a conveyor belt
   and then they load them into rail cars.

3. As more and more nodes are installed, it affects the time it takes for information
   to travel around the ring.

4. The pick-and-place robot places the part in the carton. It is sealed by an electronic heat seal device. [*Carton* is the intended antecedent.]

5. The computer malfunctions are generally minor problems that take 1 or 2 hours
   to fix. This costs the company about $1000.

6. As you can see, the IT department has a substantially higher absentee rate than
   any of the other departments. This shows a definite problem in this depart-

ment. This is the only department that requires such extreme on-call require-ments with little or no extra compensation.

Punctuation

Writers must know the generally accepted standards for using the marks of punctuation. The following guidelines are based on *The Chicago Manual of Style* and the U.S. Government's *A Manual of Style*.

Apostrophes

Use the apostrophe to indicate possession, contractions, and some plurals.

Possession

The following are basic rules for showing possession:

1. Add an *'s* to show possession by singular nouns.

   a machine's parts   a package's contents

2. Add an *'s* to show possession by plural nouns that do not end in *s*.

   the women's caucus   the sheep's brains

3. Add only an apostrophe to plural nouns ending in *s*.

   three machines' parts   the companies' managers

4. For proper names that end in *s,* use the same rules. For singular add *'s;* for plural add only an apostrophe.

   Ted Jones's job   the Joneses' security holdings

   This point is quite controversial. For a good discussion, see *The Chicago Manual of Style,* 15th ed. (Chicago: University of Chicago Press, 2003): 283.

5. Do not add an apostrophe to personal pronouns.

   Theirs   ours   its

Contractions

Use the apostrophe to indicate that two or more words have been condensed into one. As a general rule, do not use contractions in formal reports and busi-ness letters.

   I'll = I will   should've = should have   it's = it is   they're = they are

Plurals

When you indicate the plurals of letters, abbreviations, and numbers, use apostrophes only to avoid confusion. *Chicago* (p. 283) and U.S. (p. 118) disagree on this point.

1. Do not use apostrophes to form the plurals of letters.

    Xs   Ys   Zs

2. Do not use apostrophes to form the plurals of abbreviations and numbers.

    BOMs   1990s

3. Use apostrophes to form the possessive of abbreviations.

    OSHA's decision

Brackets

Brackets indicate that the writer has changed or added words or letters inside a quoted passage.

    According to the report, "The detection distance [5 cm] fulfills the criterion."

Colons

Use colons:

1. To separate an independent clause from a list of supporting statements or examples.

    The jointer has three important parts: the infeed table, the cutterhead, and the outfeed table.

2. To separate two independent clauses when the second clause explains or amplifies the first.

    The original problem was the efficiency policy: We were producing as many parts as possible, but we could not use all of them.

Commas

Use commas:

1. To separate two main clauses connected by a coordinating conjunction (*and, but, or, nor, for, yet,* or *so*). Omit the comma if the clauses are very short.

Two main clauses

The Atlas carousel has a higher base price, but this price includes installation and tooling costs.

2. To separate introductory subordinate clauses or phrases from the main clause.

Clause

If the background is too dark, change the setting.

Phrases

As shown in the table, the new system will save us over a million dollars.

3. To separate words or clauses in a series.

Words

Peripheral components include scanners, external hard drives, and external fax/modems.

Phrases

With this program you can send the fax at 5 P.M., at 11 P.M., or at a time you choose.

Clauses

Select equipment that has durability, that requires little maintenance, and that the company can afford.

4. To set off nonrestrictive appositives, phrases, and clauses.

Appositive

AltaVista, a Web search engine, has excellent advanced search features.

Phrase

The bottleneck, first found in a routine inspection, will take a week to fix.

Clause

The air flow system, which was installed in 1979, does not produce enough flow at its southern end.

Dashes and parentheses also serve this function. Dashes emphasize the abruptness of the interjected words; parentheses deemphasize the words.

5. To separate coordinate but not cumulative adjectives.

Coordinate

He rejected the distorted, useless recordings.

Coordinate adjectives modify the noun independently. They could be reversed with no change in meaning: *useless, distorted recordings.*

Cumulative

An acceptable frequency-response curve was achieved.

Cumulative adjectives cannot be reversed without distorting the meaning: *frequency-response acceptable curve.*

6. To set off conjunctive adverbs and transitional phrases.

Conjunctive adverbs

The vice-president, however, reversed the recommendation.

The crane was very expensive; however, it paid for itself in 18 months.

Therefore, a larger system will solve the problem.

Transitional phrases

On the other hand, the new receiving station is twice as large.

Performance on Mondays and Fridays, for example, is far below average.

Dashes

You can use dashes before and after interrupting material and asides. Dashes give a less formal, more dramatic tone to the material they set off than commas or parentheses do. The dash has four common uses:

1. To set off material that interrupts a sentence with a different idea

The fourth step—the most crucial one from management's point of view—is to ring up the folio and collect the money.

2. To emphasize a word or phrase at the end of a sentence

The Carver CNC has a range of 175–200 parts per hour—not within the standard.

3. To set off a definition

The total time commitment—contract duty time plus travel time—cannot exceed 40 hours per month.

4. To introduce a series less formally than with a colon

This sophisticated application allows several types of instruction sets—stacks, queues, and trees.

Parentheses

You can use parentheses before and after material that interrupts or is some kind of aside in a sentence or paragraph. Compared to dashes, parentheses have one of two effects: They deemphasize the material they set off, or they give a more formal, less dramatic tone to special asides. Parentheses are used in three ways:

1. To add information about an item.

Acronym for a lengthy phrase
A definition

This Computer Numerically Controlled (CNC) lathe costs $20,000.

The result was long manufacturing lead times (the total time from receipt of a customer order until the product is produced).

Precise technical data

This hard drive (20GB, 5400 rpm Ultra ATA/66) can handle all of our current and future storage needs.

2. To add an aside to a sentence.

> The Pulstrider has wheels, which would make it easy to move the unit from its storage site (the spare bedroom) to its use site (the living room, in front of the TV).

3. To add an aside to a paragraph.

> The current program provides the user with the food's fat content percentage range by posting colored dots next to the menu item on a sign in the serving area. To determine the percentage of fat in foods, one must match the colored dot to dots on a poster hanging in the serving area. The yellow dot represents a range of 30–60%. (The green dot is 0–29% and the red dot is 61–100%.) This yellow range is too large.

A Note on Parentheses, Dashes, and Commas

All three of these punctuation marks may be used to separate interrupting material from the rest of the sentence. Choose dashes or parentheses to avoid making an appositive seem like the second item in a series.

Commas are confusing	The computer has an input device, a keyboard and an output device, a monitor.
Parentheses are clearer	The computer has an input device (a keyboard) and an output device (a monitor).
Commas are confusing	The categories that have the highest dollar sales increase, sweaters, outerware, and slacks, also have the highest dollar per unit cost.
Dashes are clearer	The categories that have the highest dollar sales increase—sweaters, outerware, and slacks—also have the highest dollar per unit cost.

Ellipsis Points

Ellipsis points are three periods used to indicate that words have been deleted from a quoted passage.

> According to Jones (1999), "The average customer is a tourist who . . . tends to purchase collectibles and small antiques" (p. 7).

Hyphens

Use hyphens to make the following connections:

1. The parts of a compound word when it is an adjective placed before the noun.

> high-frequency system   plunger-type device   trouble-free process

Do not hyphenate the same adjectives when they are placed after the word:

The system is trouble free.

2. Words in a prepositional phrase used as an adjective.

state-of-the-art printer

3. Words that could cause confusion by being misread.

energy-producing cell   eight-hour shifts   foreign-car buyers
cement-like texture

4. Compound modifiers formed from a quantity and a unit of measurement.

a 3-inch beam   an 8-mile journey

Unless the unit is expressed as a plural:

a beam 3 inches wide   a journey of 8 miles

Also use a hyphen with a number plus *-odd*.

twenty-odd

5. A single capital letter and a noun or participle.

A-frame   I-beam

6. Compound numbers from 21 through 99 when they are spelled out and fractions when they are spelled out.

Twenty-seven jobs required a pickup truck.   three-fourths

7. Complex fractions if the fraction cannot be typed in small numbers.

1-3/16 miles

Do not hyphenate if the fraction can be typed in small numbers.

1½ hp

8. Adjective plus past participle (*-ed, -en*).

red-colored table

9. Compounds made from *half-, all-,* or *cross-.*

half-finished   all-encompassing   cross-country

10. Use suspended hyphens for a series of adjectives that you would ordinarily hyphenate.

10-, 20-, and 30-foot beams

11. Do not hyphenate:

    a. *-ly* adverb-adjective combinations:

       recently altered system

    b. *-ly* adverb plus participle (*-ing, -ed*):

       highly rewarding positions   poorly motivated managers

    c. chemical terms

       hydrogen peroxide

    d. colors

       red orange logo

12. Spell as one word compounds formed by the following prefixes:

anti-	co-	infra-
non-	over-	post-
pre-	pro-	pseudo-
re-	semi-	sub-
super-	supra-	ultra-
un-	under-	

    Exceptions: Use a hyphen

    a. When the second element is capitalized (*pre-Victorian*).

    b. When the second element is a figure (*pre-1900*).

    c. To prevent possible misreadings (*re-cover, un-ionized*).

Quotation Marks

Quotation marks are used at the beginning and at the end of a passage that contains the exact words of someone else.

According to Jones (1999), "The average customer is a tourist who travels in the summer and tends to purchase collectibles and small antiques" (p. 7).

Semicolons

Use semicolons in the following ways:

1. To separate independent clauses not connected by coordinating conjunctions (*and, but, or, nor, for, yet, so*).

   Our printing presses are running 24 hours a day; we cannot stop the presses even for routine maintenance.

2. To separate independent clauses when the second one begins with a conjunctive adverb (*therefore, however, also, besides, consequently, nevertheless, furthermore*).

> Set-up time will decrease 10% and materials handling will decrease 15%; consequently, production will increase 20%.

3. To separate items in a series if the items have internal punctuation.

> Plans have been proposed for Kansas City, Missouri; Seattle, Washington; and Orlando, Florida.

Underlining (Italics)

Underlining is a line drawn under certain words. In books and laser-printed material, words that you underline when typing appear in italics. Italics are used for three purposes:

1. To indicate titles of books and newspapers.

> *Thriving on Chaos*   the San Francisco *Examiner*

2. To indicate words used as words or letters used as letters.

> That logo contains an attractive *M*.

> You used *there are* too many times in this paper.

Note: You may also use quotation marks to indicate words as words.

> You used "there are" too many times in this paper.

3. To emphasize a word.

> Make sure there are no empty spaces on the contract and that all the blanks have been filled in *before* you sign.

Abbreviations, Capitalization, and Numbers

Abbreviations

Use abbreviations only for long words or combinations of words that must be used more than once in a report. For example, if words such as *Fahrenheit* or phrases such as *pounds per square inch* must be used several times in a report, abbreviate them to save space. Several rules for abbreviating follow (*Chicago*).

1. If an abbreviation might confuse your reader, use it and the complete phrase the first time.

   This paper will discuss materials planning requirements (MPR).

2. Use all capital letters (no periods, no space between letters or symbols) for acronyms.

   NASA   NAFTA   COBOL   HUD   PAC

3. Capitalize just the first letter of abbreviations for titles and companies; the abbreviation follows with a period.

   Pres.   Co.

4. Form the plural of an abbreviation by adding just *s*.

   BOMs   VCRs   CRTs

5. Omit the period after abbreviations of units of measurement. Exception: use *in.* for *inch.*

6. Use periods with Latin abbreviations.

   e.g. (for example)   i.e. (that is)   etc. (and so forth)

7. Use abbreviations (and symbols) when necessary to save space on visuals, but define difficult ones in the legend, a footnote, or the text.

8. Do not capitalize abbreviations of measurements.

   10 lb   12 m   14 g   16 cm

9. Do not abbreviate units of measurement preceded by approximations.

   several pounds per square inch   15 psi

10. Do not abbreviate short words such as *acre* or *ton*. In tables, abbreviate units of length, area, volume, and capacity.

Capitalization

The conventional rules of capitalization apply to technical writing. The trend in industry is away from overcapitalization.

1. Capitalize a title that immediately precedes a name.

   Senior Project Manager Jones

But do not capitalize it if it is generic.

The senior project manager reviewed the report.

2. Capitalize proper nouns and adjectives.

Asia   American   French

3. Capitalize trade names, but not the product.

Apple computers   Cleanall window cleaner

4. Capitalize titles of courses and departments and the titles of majors that refer to a specific degree program.

The first statistics course I took was Statistics 1.

I majored in Plant Engineering and have applied for several plant engineering positions.

5. Do not capitalize after a colon.

The chair has four parts: legs, seat, arms, and back.

I recommend the XYZ lathe: it is the best machine for the price.

Numbers

The following rules cover most situations, but when in doubt whether to use a numeral or a word, remember that the trend in report writing is toward using numerals.

1. Spell out numbers below 10; use figures for 10 and above.

four cycles   1835 members

2. Spell out numbers that begin sentences.

Thirty employees received safety commendations.

3. If a series contains numbers above and below 10, use numerals for all of them.

The floor plan has 2 aisles and 14 workstations.

4. Use numerals for numbers that accompany units of measurement and time.

1 gram   0.452 minute

7 yards   6 kilometers

5. In compound-number adjectives, spell out the first one or the shorter one to avoid confusion.

   75 twelve-volt batteries

6. Use figures to record specific measurements.

   He took readings of 7.0, 7.1, and 7.3.

7. Combine figures and words for extremely large round numbers.

   2 million miles

8. For decimal fractions of less than 1, place a zero before the decimal point.

   0.613

9. Express plurals of figures by adding just *s*.

   21s   1990s

10. Place the last two letters of the ordinal after fractions used as nouns:

    $\frac{1}{10}$th of a second

    But not after fractions that modify nouns:

    $\frac{1}{10}$ horsepower

11. Spell out ordinals below 10.

    fourth part   eighth incident

12. For 10 and above, use the number and the last two letters of the ordinal.

    11th week   52nd contract

Works Cited

The Chicago Manual of Style: The Essential Guide for Writers, Editors, and Publishers.
15th ed. Chicago: University of Chicago Press, 2003.
U.S. Government Printing Office. *A Manual of Style.* 29th ed. New York:
Gramercy, 2001.

Documenting Sources

Documenting your sources means following a citation system to indicate whose ideas you are using. Three methods are commonly used: the American Psychological Association (APA) system, the Modern Language Association (MLA) system, and the numbered references system, shown here by the American Chemical Society (ACS) system. All three will be explained briefly. For more complete details, consult the *Publication Manual of the American Psychological Association* (5th ed.); the *MLA Handbook* (6th ed.); or *The ACS Style Guide* (2nd ed.).

How Internal Documentation Works

Each method has two parts: the internal citations and the bibliography, also called "References" (APA, ACS) or "Works Cited" (MLA). The internal citation works in roughly the same manner in all three methods. The author places certain important items of information in the text to tell the reader which entry in the bibliography is the source of the quotation or paraphrase. These items could be the author's last name, the date of publication, the title of an article, or the number of the item in the bibliography.

In the APA method, the basic items are the author's last name and the year of publication. In the ACS method, the basic item is the number of the item in the bibliography. In the MLA method, the basic item is the author's last name and sometimes the title of the work, often in shorthand form.

In each method, the number of the page on which the quotation or paraphrase appears goes in parentheses immediately following the cited material. Because the rest of the methods vary, the rest of this chapter explains each.

Here is an example of how each method would internally cite the following quotation from page 22 of the article "This Old Forest: The Home Depot Pitches in to Help Indonesia's Forests," by Katherine Sharpe and published in the magazine *Nature Conservancy* in Spring 2003.

> Most consumers have never heard of "certified" wood. Ron Jarvis, merchandising vice president for The Home Depot, knows this. Customers don't usually ask The Home Depot to carry wood products from forests that are sustainably managed and harvested—and independently verified as such.

APA Method

The APA method requires that you use just the author's last name and include the year of publication and a page number.

> According to Sharpe (2003), "Customers don't usually ask The Home Depot to carry wood products from forests that are sustainably managed and harvested—and independently verified as such" (p. 22).

To find all bibliographic information on the quotation, you would refer to "Sharpe" in the References section.

> Sharpe, K. (2003). This old forest: The Home Depot pitches in to help Indonesia's forests. *Nature Conservancy, 11,* 22.

MLA Method

The MLA method of citing the passage requires that you should include at least the author's last name with the page number.

> As the author noted, "Customers don't usually ask The Home Depot to carry wood products from forests that are sustainably managed and harvested—and independently verified as such" (Sharpe 22).

To find all the publication information for this quotation, you would refer to "Sharpe" in the Works Cited list.

> Sharpe, Karen. "This Old Forest: The Home Depot Pitches In to Help Indonesia's Forests." *Nature Conservancy* (Spring 2003): 22.

Numbered References Method

The numbered references method does not require you to use a last name, although you may. Every time you cite the source, whenever the citation occurs in your text, you place in the text the item's number in the bibliography. So if "Sharpe" were the second item in the bibliography, you use the number 2 to cite the source in the text. You also include a page number:

As the author noted, "Customers don't usually ask The Home Depot to carry wood products from forests that are sustainably managed and harvested—and independently verified as such" (2, p. 22).

And the Reference entry looks like this:

2. Sharpe, K. This Old Forest: The Home Depot Pitches In to Help Indonesia's Forests. *Nature Conservancy* Spring 2003, *11*(1) 22.

Note: The ACS Style Guide (like all other style guides for the numbered method) does not present a way to handle quotations. The ACS assumes that references in scientific literature are to ideas in essays and that quotations are never used. However, because the method is commonly used in academia, the quotation method of the APA is added to it here.

The "Extension" Problem

A common problem with internal documentation is indicating where the paraphrased material begins and ends. If you start a paragraph with a phrase like "According to Sharpe," you need to indicate which of the sentences that follow come from Sharpe. Or if you end a long paragraph with a parenthetical citation (Sharpe, 2003, pp. 22–23), you need to indicate which preceding sentences came from Sharpe. To alleviate confusion, place a marker at each end of the passage. Either use the name at the start and page numbers at the end or use a term like "one authority" at the start and the citation at the end.

According to Sharpe (2003), a "certified" wood product is one that comes from a "sustainable" forest. The maintenance and harvesting methods of such a forest must be independently verified. Marketing certified products is difficult because few consumers are aware enough of the products' existence to ask for them (22).

One authority explains that a "certified" wood product is one that comes from a "sustainable" forest. The maintenance and harvesting methods of such a forest must be independently verified. Marketing certified products is difficult because few consumers are aware enough of the products' existence to ask for them (Sharpe, 2003, p. 22).

The APA Method

APA Citations

Once you understand the basic theory of the method—to use names and page numbers to refer to the References—you need to be aware of the variations possible in placing the name in the text. Each time you cite a quotation or paraphrase, you give the page number preceded by *p.* or *pp.* Do not use *pg.* The following variations are all acceptable.

1. The author's name appears as part of the introduction of the quotation or paraphrase.

   As Sharpe (2003) noted, "Customers don't usually ask The Home Depot to carry wood products from forests that are sustainably managed and harvested—and independently verified as such" (p. 22).

2. The author is not named in the introduction to the quotation or paraphrase.

   It is noted that "customers don't usually ask The Home Depot to carry wood products from forests that are sustainably managed and harvested—and independently verified as such" (Sharpe, 2003, p. 22).

3. The author has several works listed in the References. If they have different dates, no special treatment is necessary; if an author has two works dated in the same year, differentiate them in the text and in the References with a lowercase letter after each date (2003a, 2003b).

   Sharpe (2003a) notes that "customers don't usually ask The Home Depot to carry wood products from forests that are sustainably managed and harvested—and independently verified as such" (p. 22).

4. Paraphrases are handled like quotations. Give the author's last name, the date, and the appropriate page numbers.

   Sharpe (2003) makes note that even though consumers don't ask for timber that is verified as sustainable, The Home Depot handles this type of wood (p. 22).

5. When citing block quotations, the period is placed *before* the page in parentheses. Do not place the quotation marks before and after a block quotation. Indent the left margin 5 spaces and double-space. Do not indent the right margin.

   According to Sharpe (2003),

   > Customers don't usually ask The Home Depot to carry wood products from forests that are sustainably managed and harvested—and independently verified as such. Despite the absence of significant consumer demand, The Home Depot recently threw its support behind a Nature Conservancy project that aims to create a supply of certified wood from Indonesia, where unlawfully harvested wood—including protected species and trees felled in national parks—account for two-thirds of the wood cut annually. "Part of our culture is doing the right thing," explains Jarvis. (p. 22)

6. If no author is given for the work, treat the title as the author and list the title first in the References.

> To learn the Internet, it is useful to know that "the two most important parts are search engines and Boolean/logical operators" ("Tips," 2006, p. 78).

> Tips for the Infohighway. (2006, July). *Cyberreal*, 78.

APA References

The references list (titled "References") contains the complete bibliographic information on each source you use. The list is arranged alphabetically by the last name of the author or the first important word of the title. Follow these guidelines.

- Present information for all the entries in this order: Author's name. Date. Title. Publication information.
- Double-space the entire list. Entries should have a hanging indent, with the second and subsequent lines indented.
- Use only the initials of the author's first and middle names. *Note:* Many local style sheets suggest using the full first name; if this is the style at your place, follow that style.
- Place the date in parentheses immediately after the name.
- Capitalize only the first word of the title and subtitle and proper nouns.
- The inclusion of *p.* and *pp.* depends on the type of source. In general, use *p.* and *pp.* when the volume number does not precede the page numbers (or for a newspaper article).
- Place the entries in alphabetical order

> Berkenkotter, C. (1991). Paradigm debates, turf wars, and the conduct of sociocognitive inquiry in composition. *College Composition and Communication, 42*(2), 151–169.

> Bostic, H. (2002). Reading and rethinking the subject in Luce Irigaray's recent work. *Paragraph: A Journal of Modern Critical Theory, 25,* 22–31.

> Cooper, M., Lynch, D., & George, D. (1997). Moments of argument: Agonistic inquiry and confrontational cooperation. *College Composition and Communication, 40,* 61–85

- If there are two or more works by one author, arrange them chronologically, earliest first.

> Berkenkotter, C. (1991).

> Berkenkotter, C. (1995).

Several common entries are shown below.

Book with One Author

Selfe, C. L. (1999). *Technology and literacy in the twenty-first century: The perils of not paying attention.* Carbondale: Southern Illinois University Press.

▶ Capitalize the first word after the colon.
▶ Use Zip Code abbreviations for states.

Book with Two Authors

George, D., & Trimbur, J. (2001). *Reading culture: Contexts for critical reading and writing* (4th ed.). New York: Longman.

Book with Editors

Hawisher, G., & Selfe, C. L. (Eds.). (1999). *Passions, pedagogies, and 21st century technologies.* Logan: Utah State University Press.

Essay in an Anthology

Sullivan, D. L., Martin, M. S., & Anderson, E. R. (2003). Moving from the periphery: Conceptions of ethos, reputation, and identity for the technical communicator. In T. Kynell-Hunt & G. Savage (Eds.), *Issues of power, status and legitimacy in technical communication: Evaluating the social and historical process of professionalization* (pp. 115–136). Amityville, NY: Baywood.

▶ Capitalize only the first word of the essay title and subtitle (and all proper nouns).
▶ Use *pp.* with inclusive page numbers.

Corporate or Institutional Author

American Telephone and Telegraph. (2005). *2004 annual report.* New York: Author.

▶ When the author is also the publisher, write *Author* for the publisher.
▶ In the text, the first citation reads this way (American Telephone and Telegraph [AT&T], 2005). Subsequent citations read (AT&T, 2005).
▶ This entry could also read

2004 annual report. (2005). New York: American Telephone and Telegraph.

▶ Cite this entry as (*2004 annual*).

Work Without Date or Publisher

Radke, J. (n.d.). *Writing for electronic sources.* Atlanta: Center for Electronic Communication.

▶ Use *n.p.* for no publisher or no place.

Brochure or Pamphlet

Teaching English in the 21st century [Brochure]. (n.d.). Houghton: Michigan Technological University.

▶ Treat brochures like books.
▶ Place any identification number after the title.
▶ Place the word *brochure* or *pamphlet* in brackets.
▶ This entry could also read

Michigan Technological University. (n.d.). *Teaching English in the 21st century* [Brochure]. Houghton: Author.

▶ In the text, reference this entry as (Michigan).

Later Edition of a Book

American Psychological Association. (2001). *Publication manual* (5th ed.). Washington, DC: Author.

Encyclopedia/Handbook

Phone recorder. (1991). In R. Graf (Ed.), *Encyclopedia of physical science and technology* (Vol. 3, pp. 616–617). Blue Ridge Summit, PA: Tab.

Posner, E. C. (1992). Communications, deep space. In *Encyclopedia of physical science and technology* (2nd ed., Vol. 3, pp. 691–711). San Diego: Academic Press.

▶ In the text, refer to the first entry and all works with no author this way (note the use of quotation marks): ("Phone," 1991).

Article in a Journal with Continuous Pagination

Sullivan, D. L., & Martin, M. S. (2001). Habit formation and story telling: A theory for guiding ethical action. *Technical Communication Quarterly, 10,* 251–272.

Article in a Journal Without Continuous Pagination

Fehler, B. (2003). Re-defining God: The rhetoric of reconciliation. *Rhetoric Society Quarterly, 33*(1), 105–128.

▶ Put the issue number in parentheses after the volume.
▶ You could also give the month or season, if that helps identify the work: (2001, Summer; 2003, January).

Article in a Monthly or Weekly Magazine

Simon, R. (2002, September). One year. *U.S. News and World Report*, 4–14.

▶ If the article has discontinuous pages, a comma indicates a break in sequence (4, 6–14).

Newspaper Article

Kerasotis, P. (2003, February 2). Florida's space coast grieves for NASA accidents in its own special way. *Green Bay Press Gazette*, p. A15.

▶ *Note:* If the article has multiple pages, use *pp.* (pp. A1, A15).

Personal Interview

Note: The Publication Manual of the American Psychological Association suggests that person and telephone interviews and letters should appear only in the text and not in the References. However, because these entries might be critical in research reports, a suggested form for their use in the References is given here.

1. In the text, reference personal communication material this way:

M. Anderson (telephone interview, December 7, 2005) suggests that . . .

2. In the References, enter it this way:

Anderson, M. (2005, December 7). [Personal interview].

▶ Arrange the date so the year is first.
▶ If the person's title is pertinent, place it in brackets.

Anderson, M. (2005, December 7). [Personal interview. CEO, Technology Innovations, San Diego, CA].

Telephone Interview

Anderson, M. (2005, December 7). [Telephone interview].

Personal Letter

Anderson, M. (2005, December 7). [Personal letter. CEO, Technology Innovations, San Diego, CA].

Professional or Personal Website—Homepage

Essential for citing webpages is that you give the date of the retrieval on which you viewed the site and the URL. Give as much other information as possible.

Anderson, M. (2006, February 17). *Techinnovations*. Retrieved March 14, 2006, from http://www.techinnovations.com

▶ Cite this version as (Anderson).

Explanation: Web owner; if available. (Date of the last update, if available). Title of article or document. Title of website. Date that you viewed the site and the site's URL. *Note:* If the owner and the date are not available, the above entry would look like this:

> *Techinnovations.* Retrieved March 14, 2006, from http://www.techinnovations.com

Cite this version as ("Techinnovations").

Professional or Personal Website—Internal Page

For an internal page of a website, give the URL of the document, not the homepage.

> Anderson, M. (2006, February 19). Single-sourcing and localization. *Techinnovations.* Retrieved March 16, 2006, from http://www.techinnovations.com./singlesrclocal.html

Explanation: Web owner, if available. Date of the last updating, if available—note that internal page updates and homepage updates can be different; use the date of the page whose information you use.) Title that appears on the document page. *Title that appears on that homepage.* Date that you saw the information and the URL.

E-Mail and Listservs

APA recommends that you cite e-mail and listserv postings in the same way as personal communication. In your text, such a citation would look like this:

> M. Anderson (personal communication, December 7, 2005) suggests that . . .

▶ Notice that the first name initial is presented first, unlike the order used in the reference list.
▶ APA recommends that personal communications not appear in the reference list, but if you are required to use them, follow this:

> Anderson, M. (2005, December 7). [E-mail]

Listserv Archives

Many listservs have archives where the original postings are stored more or less permanently, available to anyone who joins the listserv. If you use an archived version of a listserv message (and you should, because archived messages are more accessible), put the item in the reference list. Use this form:

> Anderson, M. (2005, December 7). A key to effective localization. Message posted to Global Answers electronic mailing list, archived at http://www.globalanswers.org/cgi-bin/enter7.12.05

▶ *Explanation:* Author: (Date of the original posting). Title from the subject line. Message posted to name of listserv, archived at the URL of the archive.

Article from an On-Line Service

Many libraries and companies use on-line services like EBSCOhost to find full-text articles. An entry in the References would look like this:

> Anderson, M. (2005, December 7). A key to effective localization. *Global Technical Communication, 25,* 23–28. Retrieved January 12, 2006, from http://search.epnet.com/comm-generic from EBSCOhost (Academic Search Elite)

▶ *Explanation:* Author. (Date of original publication). Article title. *Periodical title. Volume number of the periodical.* Date that you retrieved the article, the URL of the service, and name of the service (name of the database).

Note: Databases like EBSCOhost often present publisher information in a source line that does not give the complete page numbers of the article.

> Source: *Global Technical Communication,* 7 December 2005, Vol. 25, Issue 3, p. 23, 5p.

If the source line is the only information available, use this form for the page numbers: p. 23, 5p. In the text you will not be able to cite pages; just use the author's name and date.

Article Available from an On-Line Periodical

Treat an article from an on-line periodical like a hard copy article. Note that you must add the date of retrieval and the URL:

> Anderson, M. (2005, December). A key to effective localization. *e Global, 25.* Retrieved January 12, 2006, from http://eglobal.com/dec/articles/anderson.html

▶ *Explanation:* Author. (Date of original publication). Article title. *Periodical Title. Volume number of the periodical.* Date you retrieved the article and the URL of the periodical.

The MLA Method

. . . .

The following section describes variations in MLA citation and explains entries in the MLA Works Cited section.

MLA Citations

Once you understand the basic theory of the method—to use names and page numbers to refer to the Works Cited—you need to be aware of the possible variations of placing the name in the text. In this method, unlike APA, each time you refer to a quotation or paraphrase, you give the page number only; do not use *p.* or *pg.*

1. The author's name appears as part of the introduction to the quotation or paraphrase.

   Sharpe notes, "Customers don't usually ask The Home Depot to carry wood products from forests that are sustainably managed and harvested—and independently verified as such" (22).

2. Author is not named in introduction to quotation.

   What seems quite evident is that "customers don't usually ask The Home Depot to carry wood products from forests that are sustainably managed and harvested—and independently verified as such" (Sharpe 22).

3. Author has several sources in the Works Cited.

   Sharpe points out that "customers don't usually ask The Home Depot to carry wood products from forests that are sustainably managed and harvested—and independently verified as such" (*Nature* 22).

4. Paraphrases are usually handled like quotations. Give the author's last name and the appropriate page numbers.

   Sharpe notes that even though consumers don't ask for timber that is verified as sustainable, The Home Depot handles this type of wood (22).

5. In block quotations, place the period before the page parentheses. Do not place quotation marks before and after the block quotation. Indent the left margin 10 spaces and double-space. Do not indent the right margin.

   According to Sharpe (2003),
   > Customers don't usually ask The Home Depot to carry wood products from forests that are sustainably managed and harvested—and independently verified as such. Despite the absence of significant consumer demand, The Home Depot recently threw its support behind a Nature Conservancy project that aims to create a supply of certified wood from Indonesia, where unlawfully harvested wood—including protected species and trees felled in national parks—account for two-thirds of the wood cut annually. "Part of our culture is doing the right thing," explains Jarvis. (22)

6. If no author is given for the work, treat the author as the title because the title is listed first in the Works Cited list.

   To learn the Internet, it is useful to know that "the two most important parts are search engines and Boolean/logical operators" ("Tips" 78).

   "Tips for the Infohighway." *Cyberreal* July 2004: 78.

7. If the title of the book is very long, you may shorten the title when you discuss it in the text. For instance, Hawisher and Selfe's book is titled *Passions, Pedagogies, and 21st Century Technology.* In the text, you may simply refer to the book as *Passions.*

MLA Works Cited List

The Works Cited list contains the complete bibliographic information on each source you use. The list is arranged alphabetically by the last name of the author or, if no author is named, by the first important word of the title.

Follow these guidelines:

▶ Present information for all the entries in this order: Author's name. Title. Publication information (including date).
▶ Capitalize the first letter of every important word in the title.
▶ Enclose article titles in quotation marks.
▶ Double-space an entry if it has two or more lines.
▶ Indent the second and succeeding lines 5 spaces.
▶ If the author appears in the Works Cited list two or more times, type three hyphens and a period instead of repeating the name for the second and succeeding entries. Alphabetize the entries by the first word of the title.

Several common entries appear below. For more detailed instructions, use the *MLA Handbook,* 6th ed., by Joseph Gibaldi (New York: MLA, 2003).

Book with One Author

Selfe, Cynthia L. *Technology and Literacy in the Twenty-First Century: The Perils of Not Paying Attention.* Carbondale: Southern Illinois University Press, 1999.

Winsor, Dorothy A. *Writing Like an Engineer: A Rhetorical Education.* Mahwah: Lawrence Erlbaum, 1996.

▶ Only the name of the publishing company needs to appear: You may drop "Co." or "Inc."

Book with Two Authors

George, Diana, and John Trimbur. *Reading Culture: Contexts for Critical Reading and Writing.* 4th ed. New York: Longman, 2001.

A long title may be shortened in the text, in this case to *Reading Culture.*

Book with Editors

Hawisher, Gail, and Cynthia L. Selfe, eds. *Passions, Pedagogies, and 21st Century Technologies.* Logan: Utah State University Press, 1999.

Mirel, Barbara, and Rachel Spilka, eds. *Reshaping Technical Communication: New Directions and Challenges for the 21st Century.* Mahwah: Lawrence Erlbaum, 2001.

Essay in an Anthology

Sullivan, Dale L., Michael S. Martin, and Ember Anderson. "Moving from the Periphery: Conceptions of Ethos, Reputation, and Identity for the Technical Communicator." *Issues of Power, Status and Legitimacy in Technical Communication: Evaluating the Social and Historical Process of Professionalization.* Ed. Teresa Kynell-Hunt and Gerald Savage. Amityville, NY: Baywood, 2003. 115–36.

▶ In the text, both the article title and the book title may be shortened; for example the article title could be "Conceptions" and the book title could be "Issues."

Corporate or Institutional Author

American Telephone and Telegraph. *2004 Annual Report.* New York: Author, 2005.

▶ When the author is also the publisher; write *Author* for the publisher.
▶ In the text, the first citation reads this way (American Telephone and Telegraph [AT&T]). Subsequent citations read (AT&T).
▶ This entry could also read

2004 Annual Report. New York: American Telephone and Telegraph, 2005.

▶ Cite this entry as (*2004 Annual*).

Work Without Date or Publisher

Radke, Jean. *Writing for Electronic Sources.* Atlanta: Center for Electronic Communication, n.d.

▶ Use *n.p.* for no publisher or no place.
▶ If neither publisher nor place is given "N.p.: n.p., 2004."

Brochure or Pamphlet

Teaching English in the 21st Century. Houghton: Michigan Technological University, n.d.

▶ If the pamphlet has an identification number, place it after the title.

Later Edition of a Book

American Psychological Association. *Publication Manual*. 5th ed. Washington, DC: APA, 2001.

Encyclopedia/Handbook

"Phone Recorder." *Encyclopedia of Physical Science and Technology*. Ed. Rudolph F. Graf. Vol. 3. Blue Ridge Summit: Tab, 1991. 616–17.

Posner, Edward C. "Communications, Deep Space." *Encyclopedia of Physical Science and Technology*. 2nd ed. Vol. 3. San Diego: Academic, 1992.

▶ No page numbers appear in Posner because entries in the book are arranged alphabetically.

Article in a Journal with Continuous Pagination

Sullivan, Dale L., and Michael S. Martin. "Habit Formation and Story Telling: A Theory for Guiding Ethical Action." *Technical Communication Quarterly* 10 (2001): 251–72.

Article in a Journal Without Continuous Pagination

Fehler, Brian. "Re-defining God: The Rhetoric of Reconciliation." *Rhetoric Society Quarterly* 33.1 (2003): 105–28.

Article in a Monthly or Weekly Magazine

Simon, Roger. "One Year." *U.S. News and World Report* Sept. 2001: 4+.

▶ If the article has discontinuous pages, give the first page only, followed by a plus sign: 4+.

Newspaper Article

Kerasotis, Peter. "Florida's Space Coast Grieves for NASA Accidents in Its Own Special Way." *Green Bay Press Gazette* 2 Feb. 2003: A15.

▶ Identify the edition, section, and page number: A reader should be able to find the article on the page.
▶ Omit the definite article (*the*) in the title of the newspaper in the text of the article: If the newspaper is a city newspaper and the city is not given in the title, supply it in brackets after the title (e.g. *Globe and Mail* [Toronto]).

Personal Interview

▶ In the text, interviews are cited like any other source: (Schmidt).
▶ In the Works Cited list, enter it this way:

Anderson, Marlon. Personal interview. 7 Dec. 2005.

▶ If the person's title or workplace are important, add them after the name:

Anderson, Marlon, CEO, Technology Innovations. San Diego: 7 Dec. 2005.

▶ Use this rule for telephone interviews and letters, too.

Telephone Interview

Anderson, Marlon. Telephone interview. 7 Dec. 2005. [Add title and workplace if necessary.]

Personal Letter

Anderson, Marlon. Letter to author. 7 Dec. 2005. [Add title and workplace if necessary.]

Professional or Personal Website—Homepage

Essential for citing webpages is that you give the date of the retrieval on which you viewed the site and the URL. Give as much other information as possible.

Anderson, Marlon. *Techinnovations.* 17 Feb. 2006. 14 March 2006 <http://www.techinnovations.com>.

Explanation: Web owner, if available. *Title of the Website.* Date of last update. Date of retrieval and <URL of the site>.

Note: If the owner and the date are not available, the above entry would look like this:

Techinnovations. 14 March 2006 <http://www.techinnovations.com>.

Cite this site as ("Techinnovations").

Professional or Personal Website—Internal Page

For an internal page of a website, give the URL of the document, not the homepage.

Anderson, Marlon. "Single-sourcing and Localization." 19 February 2006. *Techinnovations* 16 March 2006 <http://www.techinnovations.com/ta.html>

Explanation: Author, if available. "Title of Internal Page." Date of last updating, if available. *Title of the Entire Website* (from the homepage) date of retrieval and <URL of the internal page, if possible>.

E-Mail

Treat e-mail like personal communication. Use this form in the Works Cited section:

> Anderson, Marlon. "Ways to Localize Translations." E-mail to author. 7 Dec. 2005.

Explanation: Author. "Title (taken from the subject line)." Description of the message, including recipient. Date of the message.

In text the citation would read:

> (Anderson)

Listservs

Although listservs are basically collections of e-mails, the entry for a listserv posting requires more data. In the Works Cited section, use this form.

> Anderson, Marlon. "A Key to Effective Localization." On-line posting. 7 Dec. 2005. Globalanswers listserv. 12 Jan. 2006 <globalwork-l@global answers.org>.

Note: If you can use an archived version of the document (and you should do so, if you can), give the URL of the archive, e.g., http://globalanswers.org/ cgi-bin/enter.

Explanation: Author. "Title" (use the subject line). The phrase "On-line posting." Date of the posting. Name of the listserv. Date of the retrieval. <The on-line address of the listserv's website, or if that address is not available, the email address of the list's moderators>.

Article Available from an On-Line Source

Many libraries and companies use on-line services like EBSCOhost to find full-text articles. In your text, cite the full-text articles by using the author's last name. Usually you cannot present a page number, because the full-text articles are seldom paginated; just skip the page information if it is not available. An entry in the Works Cited section would look like this:

> Anderson, Marlon. "A Key to Effective Localization." *Global Technical Communication.* 7 Dec. 2005: 23–28. *Academic Search Elite.* EBSCOhost. Michigan Technological University. 12 Jan. 2006 <http://search .epnet.com/comm-generic> Keyword used: localization

Explanation: Author. "Title of Article." *Title of the Hard-Copy Periodical* date of original publication: page numbers, if available. *Title of the database.* Title of service. Library you used to reach the on-line service. Date you retrieved the article <URL of the on-line service>. Keyword you used to find the article (optional).

Note: Databases like EBSCOhost often present the publishing information in a source line that does not give the complete page numbers of the article.

> Source: *Global Technical Communication,* 7 Dec. 2005, Vol. 25, Issue 12, p. 23, 5p.

If the source line is the only information available, use this form for the page numbers: p. 23, 5p. In the text you will not be able to cite pages; just use the author's name.

Article Available from an On-Line Periodical

Treat an article from an on-line periodical like a hard copy article. Note that you must add a date of retrieval and the URL.

Anderson, Marlon. "A Key to Effective Localization." *eGlobal* Dec. 2005. Jan. 12, 2006 <http://eGlobal.com/dec/articles/anderson.html>.

Explanation: Author. "Title of Article." *Title of On-Line Periodical* date of original publication. Date of retrieval and <URL of the article>.

Numbered References

The numbered method uses an arabic numeral, rather than a name or date, as the internal citation. The numeral refers to an entry in the bibliography. *Use APA form for the bibliographic entries.* The bibliography may be organized one of two ways:

- Alphabetically
- In order of their appearance in the text, without regard to alphabetization (ACS suggests this method)

Numbered references are another method of citation. Many periodicals have adopted this method because it is cheaper to print one number than many names and dates. The difficulty with the method is that if a new source is inserted into the list, all the items in the list and all the references in the text need to be renumbered. This issue has become less of a concern with the advanced abilities of word processing programs.

The following sample shows the same paragraph and bibliography arranged in two different ways. Note that the author's name may or may not appear in the text.

Alphabetically

The inclusion of phthlates in toys has caused a major controversy. According to researchers, phthlates cause kidney and liver damage in rats (2). As a result of pressure brought by Greenpeace (1), the European Union outlawed phthlates in toys, especially teething toys, like teeth rings (2). As a result of the action, two alternate plasticizers, adipate and epoxidized soy bean (EOS), will be used more. EOS seems very promising because it has FDA approval (3). Many authorities, however, feel that the ban could politicize science (4). Another authority says that the concern is ungrounded because many earlier toxicological conferences concluded that the threats from phthlates to humans are minuscule (1).

1. Fanu, J. (1999, November 22). Behind the great plastic duck panic. *New Statesman, 128,* 11.
2. Melton, M. (1999, December 20). Lingering troubles in toyland. *U.S. News and World Report, 127,* 71.
3. Moore, S. (1999, December 1). Phthlate ban could boost demand for alternatives. *Chemical Week, 161,* 17
4. Scott, A. (1999, October 27). EU warns on Sevesco directive. *Chemical Week 161,* 24.

By Position of the First Reference in the Text

The inclusion of phthlates in toys has caused a major controversy. According to researchers, phthlates cause kidney and liver damage in rats (1). As a result of pressure brought by Greenpeace (2), the European Union outlawed phthlates in toys, especially teething toys, like teeth rings (1). As a result of the action, two alternate plasticizers, adipate and epoxidized soy bean (EOS), will be used more. EOS seems very promising because it has FDA approval (3). Many authorities, however, feel that the ban could politicize science (4). Another authority says that the concern is ungrounded because many earlier toxicological conferences concluded that the threats from phthlates to humans are minuscule (2).

1. Melton, M. (1999, December 20). Lingering troubles in toyland. *U.S. News and World Report, 127,* 71.
2. Fanu, J. (1999, November 22). Behind the great plastic duck panic. *New Statesman, 128,* 11.
3. Moore, S. (1999, December 1). Phthlate ban could boost demand for alternatives. *Chemical Week, 161,* 17.
4. Scott, A. (1999, October 27). EU warns on Sevesco directive. *Chemical Week 161,* 24.

Examples

The following examples present three sample papers, one in each of the three formats. The first two (APA and Numbered) are excerpted from much longer papers. The MLA document briefly shows the use of the MLA format.

Example B.1

Excerpt in APA
Format

MECHANICAL PROPERTIES

The mechanical properties of a film or coating describe how they will perform in the distribution environment. A thorough evaluation of edible films by a packaging engineer will include a look at their mechanical properties and a comparison of these attributes against other packaging materials. This section will describe two important mechanical properties: tensile strength and elongation.

Tensile Strength

Tensile strength can be described as the amount of force required to break a material. Knowing the package's tensile strength can help the packaging engineer decide if the material will remain intact as it flows through the packaging machinery. It will also help the engineer predict whether the material will break as it is stretched around a product. Table 4 summarizes the tensile strengths of various edible films. Banjeree and Chen (1995) found that whey protein films withstood 5.94 MPa of pressure before breaking. However, the addition of lipids to the whey film lowered the tensile strength to 3.15 MPa (p. 1681).

Table 4
Tensile Strength and Elongation

Film Material	Tensile Strength (MPa)	Elongation Thickness (%)	Source
Proteins			
Whey	5.94	22.74	Banerjee (1995)
Whey/Lipid	3.15	10.78	Banerjee (1995)
Milk	8.6	22.1	Maynes (1994)
Zein	38.3	—	Yamada (1995)
Rice	31.1	2.9	Shih (1996)
Soybean	7.2	0.75	Stuchell (1994)
Polysaccharides			
Cellulose	66.33	25.6	Park (1993)
Synthetics			
LDPE	13.1–27.6	100–965	Park (1993)
PVDC	48.4–138	20–40	Maynes (1994)

The tensile strengths of cellulose and grain-based edible films have also been measured and compared to synthetic plastics. Park, Weller, and Vergano (1993) found that cellulose films exhibited a tensile strength value of 66.33 MPa. In the same study, LDPE required from 13.1 to 27.6 MPa to break up (p. 1362). Yamada, Takahashi, and Noguchi (1995) discovered that zein protein films exhibit similar or higher tensile strengths than polyvinylide

(continued)

chloride (PVDC) films. Shih (1996) found that rice protein films resisted breaking until 31.1 MPa of force was applied. Lastly, Brandenburg, Weller, and Testin (1993) and Stuchell and Krochta (1994) found that soy protein films took 7.2 MPa of force to break.

The thickness of the film can have a bearing on its tensile strength. Park et al. (1993) found that the tensile strength of cellulose did not improve as the thickness increased. However, no direct comparison of thicknesses between cellulose films and LDPE films was made. Therefore, it is difficult to make a true comparison. Yamada et al. (1995) discovered that in order to achieve tensile strengths similar to PVDC, zein protein films 7 times thicker than the PVDC had to be used. Most researchers do not list the thickness of the product when testing for tensile strength. This lack of completeness in reporting their results will cause some confusion on the part of packaging professionals.

Elongation

Elongation refers to the amount that a material will stretch before it breaks. Table 4 lists the percent of elongation of various edible films. Park et al. (1993) found that the cellulose elongation percentages varied widely among different molecular weights of films. Chen (1995) along with Maynes and Krochta (1994) found that milk protein films had significantly lower percentages of elongations than traditional plastic films. They found that the milk proteins elongated anywhere from 1 to 75 percent of their original length, whereas LPDE elongated to 5 times its original length before breaking. It is widely believed that the structure of proteins and the way that they crystallize negatively affects the film's elongation properties (McHugh & Krochta, 1994).

Polysaccharides have better elongation characteristics than proteins. Park et al. (1993) discovered that cellulose films can elongate from 10% to 200% of their original length. They compared these figures to LDPE, which was found to elongate 1 to 10 times its original length before breaking. As with tensile strength, a certain amount of elongation is dependent on the thickness of the material. The amount of elongation should be reported on the basis of the thickness of the sample tested.

References

Banerjee, R., & Chen, H. (1995). Functional properties of edible films using whey protein concentrate. *Journal of Dairy Science, 78,* 1673–1683.

Brandenburg, A. H., Weller, C. L., & Testin, R. F. (1993). Edible films and coatings from soy protein. *Journal of Food Science, 58,* 1086–1089.

Chen, H. (1995). Functional properties and applications of edible films made of milk proteins. *Journal of Dairy Science, 78,* 2563–2583.

Maynes, J., & Krochta, J. (1994). Properties of edible films from total milk protein. *Journal of Food Science, 59,* 909–911.

McHugh, T. and Krochta, J. (1994) Milk-protein-based edible films and coatings. *Food Technology, 48,* 97–103.

Park, H. J., Weller, C. L., & Vergano, P. J. (1993). Permeability and mechanical properties of cellulose-based edible films. *Journal of Food Science, 58,* 1361–1364, 1370.

Shih, F. (1996). Edible films from rice protein concentrate and pullulan. *Cereal Chemistry, 73,* 406–409.

Stuchell, Y., & Krochta, J. (1994). Enzymatic treatments and thermal effects on edible soy protein films. *Journal of Food Science, 59,* 1332–1337.

Yamada, K., Takahashi, H., & Noguchi, A. (1995). Improved water resistance in zein films and composites for biodegradable food packaging. *International Journal of Food Science and Technology, 30,* 599–608.

Example B.2

Excerpt with Numbered References

BRIEF HISTORY OF HTML

A Need for HTML

In 1989 Tim Berners-Lee, a CERN (Conseil Européen pour la Recherche Nucléaire) (1) employee, wrote a proposal for the creation of a system that would easily allow scientists to locate and browse one another's research documents, as well as post their own (2). CERN had employees all over the world and having such a system would have benefited the center's research greatly (3). This project was later dubbed "The World-Wide Web" or "WWW" for short. This should not be confused with the Internet, which is simply a large network that supports many things, of which the WWW is one.

The WWW proposal set forth certain requirements that the system was to meet. Embedded in these requirements and throughout the proposal was the call for a formatting language for the hypertext documents that would be shared on the WWW, as well as a transfer protocol for transmitting them over a network (3). The formatting language was named HTML, and the network protocol was called the HyperText Transfer Protocol, or HTTP for short (4). The proposal also called for a method of locating documents within the new system and was implemented via the Uniform Resource Locator, or URL (5). Of course, by having all these documents shared, one would need a way to locate a document. Therefore, a program was written to accomplish this and was referred to as the "search engine."

The Creation of HTML

By 1990 the WWW project was well underway. The first known code written by Berners-Lee to process a hypertext file was dated September 25, 1990, and

(continued)

Example B.2

(continued)

the first known HTML document on the Web was dated November 13, 1990 (6). By December of 1990 a simple text browser, as well as some simple hypertext documents, were available for viewing within the CERN community (7).

With initial work on both a hypertext language and browser behind him, Berners-Lee realized that no current hypertext system was adequate to his needs and officially proposed that a new language be developed to support the WWW system (4). The new language was to be called HTML and was to be a subset extracted from the broader SGML (4). In the summer of 1991 the specification of HTML and the first simple browser were put on the Internet for public download and use, and CERN launched the World Wide Web (3). Once the specification for HTML was released, other organizations were able to create websites and write their own browsers such as Cello, Viola, and MidasWWW (8). The number of documents available on the WWW quickly increased as more and more organizations were able to compose and post documents on any subject they wished.

REFERENCES

1. Wilson, B. (1997). Glossary of terms. Retrieved March 20, 2003, from http://jeffcovey.net/web/html/reference/html/misc/glossary.htm
2. Koly, W. Quick HTML history. Retrieved March 20, 2003, from http://www.highlatitude.com/comdex99/sld005.htm
3. Wilson, B. (1997). HTML overview. Retrieved March 20, 2003, from http://jeffcovey.net/web/html/reference/html/history/html.htm
4. The history of HTML, p. 2. Retrieved March 20, 2003, from http://howdyyall.com/HTML/HISTORY/HTMLhist2.htm
5. The history of HTML, p. 3. Retrieved March 20, 2003, from http://howdyyall.com/HTML/HISTORY/HTMLhist3.htm
6. Palmer, S. B. The early history of HTML. Retrieved March 20, 2003, from http://infomesh.net/html/history/early/
7. The history of HTML, p. 1. Retrieved March 20, 2003, from http://howdyyall.com/HTML/HISTORY/HTMLhist.htm
8. Some early ideas for HTML. (January 1, 2003). Retrieved March 20, 2003, from http://www.w3.org/MarkUp/historical

Example B.3

Excerpt in MLA Format

HTML 2.0 AND BEYOND

By late 1993, HTML had become a collage of standard tags and a wide range of unique tags created for individual browsers. In April 1994, Dan Connolly, a professional in on-line documentation and on-line formatting systems, proposed bringing HTML back to its roots in SGML and wrote a draft of what would become HTML 2.0. This draft was revised and rewritten, which caused it to lose much of its SGML tone, by Karen Muldrow in July 1994 and shortly thereafter was presented to the IETF (Internet Engineering Task Force) for approval (Wilson, "Glossary").

Unfortunately, the goal of the draft became muddled, and its main accomplishment was to support all of the browser tags in existence at the time, rather than a shift towards SGML. In September 1995, a finalized official version of HTML 2.0 was released, and most of its tags were supported by the majority of browsers (Wilson, "HTML"; Koly).

In 1994, while HTML 2.0 was being developed by a myriad of groups and individuals, Tim Berners-Lee and others formed what is now known as the W3C or the World Wide Web Consortium. The W3C was involved in the finalization of HTML 2.0 and is to date the primary organization involved in all work on the HTML standard. It was created in order to keep HTML standardization on the right track, but it took some time before they became effective (Shannon).

While under the W3C, HTML continued to evolve from 2.0 to 3.0. In fact, by the time version 2.0 was approved, a draft for 3.0 had already been written! HTML 3.0 was released as a draft in September 1995. It was a very unpopular version because it made huge changes in what tags should be used and what they should be used for. Version 3.2, which implemented fewer changes, quickly appeared in May 1996 and was much more widely accepted (Koly).

Most of the major changes from version 3.0 were reserved for 4.0, except for the creation of CSS (Cascading Style Sheets). After this point in time all HTML modifications were to be modular in nature, allowing the changes to take place slowly. At this point in time, Netscape was the dominant browser. Its implementation of numerous proprietary tags that were not in the specification caused a major setback for the standardization of HTML (Shannon).

HTML 4.0 was finalized by the W3C in 1997. Shortly thereafter, it was revised again into version 4.01. This version is the current official standard, but work is ongoing at the W3C to further improve HTML. These further improvements incorporate the idea of storing formatting information away from the HTML itself (Shannon).

Works Cited

Koly, William. "Quick HTML History." 1999. 20 Mar. 2003 <http://www.highlatitude.com/comdex99/sld005.htm>.

Shannon, Ross. "HTML Source: The History of HTML." 1999. 20 Mar. 2003. <http://www.yourhtmlsource.com/comdex99/sld005.htm>.

Wilson, Brian. "HTML 2.0." 1997. 20 Mar. 2003 <http://jeffcovey.net/web/html/reference/html/history/html20.htm>.

———. "Glossary." 1997. 20 Mar. 2003 <http://jeffcovey.net/web/html/reference/html/misc/glossary.htm>.

Exercises

1. Edit the following sentences to place an APA citation correctly and/or to place a MLA citation correctly.

   a. On page 29 of his 2002 article on the *ethos* of technical communicators, Mr. George Carlson notes that "Technical communicators often find it difficult to establish their *ethos* in the corporate environment."

   b. In 2003, Dr. Ellen Keenan said on page 46 that computer software has been both a bane and blessing for workers because of the updating of software versions and the lack of corresponding hardware updates.

   c. Dr. Tim Rongstad noted (p. 21, 2001) that "Generally speaking, it is significant that research demonstrates that good writing skills are learned as early as preschool."

   d. "The importance of oral communication when connected to written and visual communication should not be underestimated," according to Dr. Pamela Poole, a noted communication scholar, on p.112 in 2001.

2. Pick a paragraph from one of the three examples in this appendix and rewrite it in one of the other two citation styles (e.g., change APA to numbered or MLA).

3. Turn these sets of data into the APA References list and an MLA Works Cited list.

   Brett Peruzzi/Building bridges between marketing and technical publications teams/Retrieved 11 Aug. 2003/EServer TCLibrary/2001/TECHWR-L/ <http://tc.eserver.org/14499.html>.

   Dànielle DeVoss, Julia Jasken, Dawn Hayden/January 2002/pages69–94/ the journal is *Journal of Business and Technical Communication*/ Volume 16(1)/the title is Teaching intra and intercultural communication: A critique and suggested method.

4. Use the following two excerpts and the information presented in Example 9.1 (p. 223) to create a brief research report on pixels. Your audience is people who are just beginning to use color in documents.

PIXELS AND COLOR DEPTH

From Patrick Lynch and Sarah Horton, *Web Style Guide: Basic Design Principles for Creating Web Sites* 2nd ed. New Haven: Yale University Press, 2001. 154–156.

To control the color of each pixel on the screen, the operating system must dedicate a small amount of memory to each pixel. In aggregate, this memory dedicated to the display screen is often referred to as "video RAM" or "VRAM" (Video Random Access Memory). In the simplest form of black-and-white computer displays, a single bit of memory is assigned to each pixel. Because each memory bit is either positive or negative (0 or 1), a 1-bit display system can manage only two colors (black or white) for each pixel on the screen.

If more bits of memory are dedicated to each pixel in the display, more colors can be managed. When 8 bits of memory are dedicated to each pixel, each pixel could be one of 256 colors. (256 = 2 to the eight power; in other words, the maximum number of unique combinations of zeros and ones you can make with 8 bits.) This kind of computer display is called "8-bit" or "256-color" display, and is common on older laptop computers and desktop machines. Although the exact colors that an 8-bit screen can display are not fixed, there can never be more than 256 unique colors on the screen at once.

If still more memory is dedicated to each pixel, nearly photographic color is achievable on the computer screen. "True-color" or "24-bit" color displays can show millions of unique colors simultaneously on the computer screen. True color images are composed by dedicating 24 bits of memory to each pixel; 8 each for red, green, and blue components (8 + 8 + 8 = 24).

The amount of VRAM dedicated to each screen pixel in the display is commonly referred to as the "color depth" of the monitor. Most Macintosh and Windows computers sold in recent years can easily display color depths in thousands (16-bit) or millions (24-bit) of simultaneous colors. To check in your computer system for the range of color depth available to you, use the "Display" control panel (Windows) or the "Monitors" control panel (Macintosh).

From: Optimizing web pages WebNotes.com (2001–2004). Retrieved August 12, 2003, from http://www.webdevelopersnotes.com/design/24.php3/

Resolution for web graphics should be not more than 72 dpi, which is the resolution of most monitors. There are many high end workstations that support higher resolutions (96 dpi) but I guess the users of such high end computers would not use these machines to surf the Net often :-) Use a graphics program to change the resolution of the image if necessary.

Color depth determines the number of colors present in the image. It's an important factor when optimizing GIF images. The higher the number of colors, the larger the image size. Remember, there is always a trade-off between the image quality and size.

[I]t's best to check the size of a GIF image saved under different color depths. As the color depth decreases so does the quality.

JPEG employs a loosy algorithm for compressing images. This means that when you save a high quality image as a JPEG, some information is lost forever. Standard graphics programs allow you to change the JPEG compressions levels.

> There are some programs such as the **JPEG Optimizer** (www.xat.com) that allow you to apply different compression levels to different parts of the same image.

5. Rewrite the tensile strength section of Example 1 in order to emphasize that one film type is best.

Writing Assignment

Find three articles on a similar topic and write a memo in which you give the gist of all three to your supervisor. Do not just summarize them each in turn; blend them so they support a main point that you want to call to the supervisor's attention. Use any of the three methods of documenting.

Works Cited

Dodd, Janet S., ed. *The ACS Style Guide: A Manual for Authors and Editors*. 2nd ed. Washington, DC: American Chemical Society, 1997.

Gibaldi, Joseph. *MLA Handbook for Writers of Research Papers*. 6th ed. New York: MLA, 2003.

Publication Manual of the American Psychological Association. 5th ed. Washington, DC: APA, 2001.

Index